THE NATURE OF GEOLOGY

contemporary readings

THE NATURE OF GEOLOGY

contemporary readings

Brainerd Mears, Jr.

University of Wyoming

VNR VAN NOSTRAND REINHOLD COMPANY
New York / Cincinnati / Toronto / London / Melbourne

To "Uncle" Walter Bucher and
"Doc" Elwyn Perry.
Two outstanding human beings and
each in his own way an outstanding
geologist.

Van Nostrand Reinhold Company Regional Offices:
Cincinnati   New York   Chicago   Millbrae   Dallas

Van Nostrand Reinhold Company Foreign Offices:
London   Toronto   Melbourne

Manufactured in the United States of America

Published by Van Nostrand Reinhold Company
450 West 33rd Street, New York, N.Y. 10001

Published simultaneously in Canada by
Van Nostrand Reinhold Ltd.

10   9   8   7   6   5   4   3   2   1

PREFACE

This collection of readings is specifically for beginning students of geology, and it is intended to demonstrate that geology is a dynamic, living science. Introductory textbooks excel in explaining the forces that shape the landscape, and some of them succeed in breathing life into the fossil record; but very few move beyond a description of phenomna to relate geology to the interests and experiences of students. These selections serve as a complement to any standard elementary text by presenting accounts of human endeavor, and not results only.

Some of the selections were included for their vivid imagery and style. Although geology is clearly not a humanity (because it is directed outward toward a better understanding of the physical world), Rachel Carson demonstrates that science need not rob nature of its beauty. To think that the coast of Maine or the Grand Canyon of the Colorado River is more awe-inspiring to the layman than to a worker trained in the mental images of geology is nonsense. Outstanding scientists (as well as the best humanists) are characterized by excellent imaginations, coupled necessarily with disciplined thinking. Geology is certainly not a social science. However, geologists are involved increasingly with such pressing social problems as environmental pollution, as the timely essay by David Evans indicates.

Many selections were chosen to show geology as a human occupation, a way of life. That it involves adventure, and also plain physical work, is well documented from classical to recent times: as in Pliny the Younger's description of the 79 A.D. eruption of Vesuvius, John Wesley Powell's account of his boat trip through a Uinta Mountain Canyon a century ago, and Leonard Engel's report of work on a modern oceanographic research vessel. The objective reporting in textbooks allows no space to review the personalities involved in the evolution of techniques and concepts, as Ruth Moore does for the history of radiometric dating.

The facts and theories of geology (or any science) are the work of real people—often groping, sometimes wrong, occasionally inspired. That geologists frequently have healthy scientific disagreements, a normal aspect of the scientific testing of theories, can be seen in such articles as Wilson's and Beloussov's recent exchange of opposing views on the new theory of sea-floor spreading. That rivalry and ambition can sometimes create nasty situations is fortunately rare, but not unheard of, as the Cope-Marsh feud, discussed by Edwin Colbert, indicates.

Geologists, by and large, have not been a particularly philosophical lot. They have mainly busied themselves with the immediate goal of describing and interpreting the earth, rather than reflectively contemplating the logic and methods underlying their work. Yet there are thought-provoking, and not too technical, articles about scientific method in geology, such as G. K. Gilbert's "classic" description of his approach to a problem of the abandoned shorelines above the Great Salt Lake. Such writings, if more accessible to beginning students and interested laymen, would have long ago dispelled the misconception that geology is some-how an easy and dogmatic science.

Many other excellent writings could have been used in this collection. Those selected are just a few, which I have enjoyed, and which are not likely to duplicate the material in elementary texts. Together they constitute a straightforward theme—the presentation of geology in a more human light.

CONTENTS

Section III Geology and Mankind

THE NATURE OF GEOLOGY

contemporary readings

Photograph by Herb Pownall, University of Wyomir

1

IMAGES OF THE EARTH

1

Prologue

HANS CLOOS

The *"Prologue" from the autobiography of Hans Cloos presents thought-provoking views of man and geology. One of the world's outstanding structural geologists, Cloos was born and trained in Germany, travelled widely, headed several geology departments, and edited the leading German geologic journal. He had a beautiful literary style and a rare sense of the harmony of geologic relationships.*

> ... *it was during my enchanted days of travel that the idea came to me which, through the years, has come into my thoughts again and again and always happily—the idea that geology is the music of the earth.*

Earth; beautiful, round, colorful planet. You carry us safely through the emptiness and deadness of space! Graciously you cover the black abyss with air and water. You turn us towards the sun, that we may be warm and content, that we may wander, with open eyes, through your meadows, and look upon your splendor. And then you turn us away from the too fiercely burning sun, that we may rest in the coolness of the night from life's heat and the struggle of the day.

The present is the era of man. Our struggle for survival dominates the scene; we increase or diminish other forms of life to nourish our own. A thousand other ages have preceded us, a thousand more will follow. This patient earth has offered its growing life thousands upon thousands of times to the warmth of the sun, that it might thrive and be transformed, and eventually vanish to provide space, light, and sustenance for new and different kinds of life. Restlessly the earth has changed, like the ever-moving sea. Lands and mountains rose out of the sea, only to be returned again to the ocean.

But the present is the era of man. Today knowledge reigns supreme. For the first time since its beginning our planet, earth, sees and understands itself. For a billion years the earth rolled on, quite blind and mute. It has used up all this enormous period of time in forming, out of plants and animals, through millions of unfinished experiments, the organ through which it will recognize itself. For a billion years the patient earth amassed documents and inscribed them with signs and pictures which lay unnoticed and unused. Today, at last, they are waking up, because man has come to rouse them. Stones have begun to speak, because an ear is there to hear them. Layers become history and, released from the enchanted sleep of eternity, life's motley, never-ending dance rises out of the black depths of the past into the light of the present.

Today is the day of man! We wander, walking upright, with our gaze directed into the distance, over the lands of the earth. At river banks and at mountain peaks we stop before rocky walls, and we peer through creviced windows into the depths. For the first time, we decipher the earth's diary that has been left us as a legacy. We read with trained senses and interpret with the tool of disciplined thinking. We translate the earth's language into our own, and enrich the already bright and colorful surface of the present with the knowledge of the inexhaustible abundance of the past.

The well-marked path to knowledge is open to anyone willing to make the effort to follow it, though no one will ever quite reach its end.

Yet there is another, inner way, a way that is not accessible to everyone. It leads from the unconscious within ourselves to the imponderable and invisible in the earthly environment. It is this way which binds the artist to the world. He who walks this trail sees the beauty of the earth, and hears its music.

Why does man find beauty in a landscape? Is it not because he is a part of nature, inwardly subject to nature's laws, because he has an unconscious insight into the internal order of the earth, into the rhythm of its repetitions, the harmony of its lines and surfaces and the balanced interplay of its parts? And does not our delight in the contemplation of nature grow out of the harmony between the music of our own soul and the music of the earth?

If we dislike the way mountains are depicted in old paintings it is because this harmony is lacking. If the artificial lake in the darkness of the forest is less appealing than a natural lake in an abandoned glacial valley or volcanic crater, what alienates us is the artificial disturbance of an equilibrium that has ripened through thousands of years. However, if we see a clearly graded mountain unmarred by mechanical intrusion, if we look over vast and rolling plains, or let our eyes wander over a wide panorama of open country with silver streams winding, undammed, through hills and woods, then we rest in the immense and complete order of the earth.

And then it may well happen that what we have been seeing as beautiful repose changes into motions; lines seem to rise and fall, surfaces wave and swing. Folding, which long ago buckled horizontal strata into huge subterranean waves of stone (folds) is revived. The rounded crests of folds rise and fall rhythmically. Giant smoking volcanoes stand in a row like the pipes of a cosmic organ through which the earth's mighty breath blows its roaring music. Granite domes in the deserts, the broadly arched shields of the Rocky Mountains become the chimes waiting only for the earthquake's stroke to awaken the mighty symphony that drowns out all human tones and resounds throughout the world. The strangely rhythmic alternation of layers in prehistoric oceans— thick and thin, light and dark, hard and soft, succeeding each other many thousands of times—becomes a restless tremolo passage in the epic music of geological history.

But man, even the thinking, exploring man who has made it his life's work and duty to listen to nature's voice, can scarcely hear "music" every moment of the day. It is human custom, and geology affirms the practice, to explore the natural order by sober, patient observation and by logical deduction, and to describe what has been found so that anyone can readily understand

and enjoy it. And so it is in this book. For just as the two-hour concert by the interpreting virtuoso is preceded by weeks of practice and years of training, so are the musical consummations in geology the precious result of decades of laborious work which is itself based on centuries of earlier work by others.

* * *

"The present is the age of man." What does this geological "present" mean? When did the era of man begin and how long will it last?

Ever since the beginning of life on the earth, man, too, has lived. But he is entitled to call himself a human being only since the time he first walked upright, first succeeded in making fire and in fashioning tools. And so the age of man has already lasted some hundreds of thousands of years. During this time man has engaged in an exchange with the earth in which he has taken nearly everything and given little in return.

At the beginning each man's relationship with the earth was probably of a rude and simple sort. He knew where to get the raw materials for his weapons and tools, where to find drinking water, in which kind of rock to find caves for habitation; but the earth itself seemed to be teeming with dangerous demons.

Later, there must have been experts, skilled workmen and workshops, and every hundredth man knew what the signs were that indicated the presence of flint or clay, iron or copper in a particular place. During this still remote period the fear of demons might already have been transformed into the worship of Gods. An earth-goddess may at that time have been the spiritual link between man and his planet.

Today there is one geologist to about every half-million people. Only two or three among them may have a faint notion that human beings are anything but leaves on the tree of life or, even in their highest forms, anything more than bloom and fruit. Only a few accept, on behalf of the many, the responsibility for preserving, reading, and translating the scriptures of the earth. As a rule, only man's *conscious* relationship with his planet is considered to fall within the province of a science that has become specialized and virtually cryptic. The *unconscious* relationship has been left to the domain of superstition, of the divining rod, and the fairy tale.

Finally, we come to our "today," when a few hundred or thousand years are only minutes in the sweeping geological sense. The contemporary geologist is aware of a great and responsible task—to look back, from the brief today into the long "yesterday" and the immeasurable "day before yesterday," in order that he may tell the many what he, the solitary and privileged one, sees, and reads into, and learns from the geologic past.

The experienced observer does more than merely report and recite. He guides the eager student to an understanding of the earth. He may chart the scientist's steep, barren road of sober observation and strict deduction, or the artist's gentle road of contemplation and empathy. And, finally, he may point out his own unique way, the path of the initiated, which leads him from the laboratories and libraries to the meadows and flower gardens of the living earth.

* * *

Other sciences long ago found the way to their roots. The fruits of zoology and botany have become familiar to most of us. These sciences have taught us to know and love the animals and plants that are our companions on this planet, and even to understand them a little. Through the sciences of physics and chemistry laws have been discovered which have given us an enormous power over our environment. Astronomy, on the other hand, shows how small we are as measured against the immensities of space and time.

The study of the planet on which we live is different from other sciences. The earth is large and old enough to teach us modesty; and yet it is small enough to be comprehended and to be learned from, as our understanding of it increases. The earth teaches us the detail and the whole in immeasurable abundance. Two fundamental principles stand out above all others:

As parts of the earth we depend on its inorganic substance and on the eternal change which it undergoes. And as children of the earth we are subordinate, dependent particles in the unceasing stream of all life.

To talk to people about the earth is in some respects easier, in others more difficult, today, than it would have been hundreds or thousands of years ago when men wore skins and used stone axes. It is more difficult because the modern world-picture has become infinitely more complicated and profound than it was then. And it is easier since man himself, who has to comprehend the information, has grown mentally. Man's technological development in the last few minutes of our geological present has increased the length of our stride and the range of our vision a thousandfold, and has thus reduced the disparity between the tempo of the planet and its inhabitants. An entire mountain range can be seen at a glance, a whole continent is viewed in a few hours. Traveling by car we feel the waves of the stony crust below us. By plane we can reach heights from which we see the earth in a manner once possible only with the aid of maps— which are but dry substitutes for actual travel.

Today the earth gives man home and nourishment, and so it seems worthwhile to present him with a picture of his planet, and to recommend to man a more intimate mental and physical acquaintance with his earth.

Writing about the earthquake of 1755, which destroyed the city of Lisbon, Albert Schweitzer remarks that this event "raised the question among the multitude whether the world really was ruled by a wise and good Creator. Voltaire, Kant, and many other thinkers of that period availed themselves of the event to formulate their ideas, in part admitting their perplexity, in part seeking new outlets for their optimism based on the scientific control of nature." The Lisbon earthquake, like all others before and after it, was a normal occurrence in the geological development of our planet. The time is coming when man will be so well-versed in the earth's habits that he will be able to anticipate earthquakes and prepare for them. When this happens, a unity between man and earth will have been achieved, a unity without which a consistent concept of the world is impossible.

Since the Lisbon earthquake, different and far greater catastrophes have assailed mankind. Some have been introduced by man himself and are more difficult than any earthquake to reconcile with the all-too-simple idea of a beneficent Creator. But precisely because of this it has also become quite clear for the first time that Albert Schweitzer's question is not only pertinent

Albert Schweitzer: "Verfall und Wiederaufbau der Kultur" (P. Haupt, Bern, 1923); and "Kultur und Ethik" (P. Haupt, Bern, 1924).

to the man-nature relationship, but that it goes far beyond the realm of sensory experience.

How closely akin are music, the purest and most ethereal of the arts, and mathematics, most sober of the sciences, how ever unlike their forms may be. He who hears the music of the earth will find that pleasure in its melodies is more than a light and gladsome enjoyment. He will find, indeed, that the experience furnishes another and deeper understanding of the language in which the world and its Creator speak to us.

2

Fragments From The Edge Of The Sea

RACHEL L. CARSON

Much of the earth's later history has been deciphered from fossils and sedimentary rocks laid down in seas of the last 600 million years. Thus the present-day shore and near-shore environments are of particular interest to geologists who seek to interpret the ceaseless flood and ebb of ancient seas across the continental blocks, and the adaptations of life to conditions along the margins of sea and land. Here in Paleozoic time, the first marine plants and animals tentatively crossed the strand line and became ancestors to the life that evolved and conquered the land. These fragments from Rachel Carson's book, The Edge of the Sea, *vividly describe a variety of coasts and relate them to their geologic history and to an important episode in the evolution of life.*

from the Preface

Like the sea itself, the shore fascinates us who return to it, the place of our dim ancestral beginnings. In the recurrent rhythms of tides and surf and in the varied life of the tide lines there is obvious attraction of movement and change and beauty. There is also, I am convinced, a deeper fascination born of inner meaning and significance.

When we go down to the low-tide line, we enter a world that is as old as the earth itself—the primeval meeting place of the elements of earth and water, a place of compromise and conflict and eternal change. For us as living creatures it has special meaning as an area in or near which some entity that could be distinguished as Life first drifted in shallow waters—reproducing, evolving, yielding that endlessly varied stream of living things that has surged through time and space to occupy the earth.

To understand the shore, it is not enough to catalogue its life. Understanding comes only when, standing on a beach, we can sense the long rhythms of earth and sea that sculptured its land forms and produced the rock and sand of which it is composed; when we can sense with the eye and ear of the mind the surge of life beating always at its shores—blindly, inexorably pressing for a foothold. To understand the life of the shore, it is not enough to pick up an empty shell and say "This is a murex," or "That is an angel wing." True understanding demands intuitive comprehension of the whole life of the creature that once inhabited this empty shell: how it survived amid surf and storms, what were its enemies, how it found food and reproduced its kind, what were its relations to the particular sea world in which it lived.

The seashores of the world may be divided into three basic types: the rugged shores of rock, the sand beaches, and the coral reefs and all their associated features. Each has its typical community of plants and animals. The Atlantic coast of the United States is one of the few in the world that provide clear examples of each of these types. I have chosen it as the setting for my pictures of shore life, although—such is the universality of the sea world—the broad outlines of the pictures might apply on many shores of the earth.

from Chapter III, "The Rocky Shores"

Up and down the coast the line of the forest is drawn sharp and clean on the edge of a seascape of surf and sky and rocks. The softness of sea fog blurs the contours of the rocks; gray water and gray mists merge offshore in a dim and vaporous world that might be a world of creation, stirring with new life.

The sense of newness is more than illusion born of the early morning light and the fog, for this is in very fact a young coast. It was only yesterday in the life of the earth that the sea came in as the coast subsided, filling the valleys and rising about the slopes of the hills, creating these rugged shores where rocks rise out of the sea and evergreen forests come down to the coastal rocks. Once this shore was like the ancient land to the south, where the nature of the coast has changed little during the millions of years since the sea and the wind and the rain created its sands and shaped them into dune and beach and offshore bar and shoal. The northern coast, too, had its flat coastal plain bordered by wide beaches of sand. Behind these lay a landscape of rocky hills alternating with valleys that had been worn by streams and deepened and sculptured by glaciers. The hills were formed of gneiss and other crystalline rocks resistant to erosion; the lowlands had been created in beds of weaker rocks like sandstones, shale, and marl.

Then the scene changed. From a point somewhere in the vicinity of Long Island the flexible crust of the earth tilted downward under the burden of a vast glacier. The regions we know as eastern Maine and Nova Scotia were pressed down into the earth, some areas being carried as much as 1200 feet beneath the sea. All of the northern coastal plain was drowned. Some of its more elevated parts are now offshore shoals, the fishing banks off the New England and Canadian coasts— Georges, Browns, Quereau, the Grand Bank. None of it remains above the sea except here and there a high and isolated hill, like the present island of Monhegan, which in ancient times must have stood above the coastal plain as a bold monadnock.

Where the mountainous ridges and the valleys lay at an angle to the coast, the sea ran far up between the hills and occupied the valleys. This was the origin of the deeply indented and exceedingly irregular coast that is characteristic of much of Maine. The long narrow estuaries of the Kennebec, the Sheepscot, the Damariscotta and many other rivers run inland a score of miles. These salt-water rivers, now arms of the sea, are the drowned valleys in which grass and trees grew in a geologic yesterday. The rocky, forested ridges between them probably looked much as they do today. Offshore, chains of islands jut out obliquely into the sea, one beyond another—half-submerged ridges of the former land mass.

But where the shore line is parallel to the massive ridges of rock the coast line is smoother, with few indentations. The rains of earlier centuries cut only short valleys into the flanks of the granite hills, and so when the sea rose there were created only a few short, broad bays instead of long winding ones. Such a coast occurs typically in southern Nova Scotia, and also may be seen in the Cape Ann region of Massachusetts, where the belts of resistant rock curve eastward along the coast. On such a coast, islands, where they occur, lie parallel to the shore line instead of putting boldly out to sea.

As geologic events are reckoned, all this happened rather rapidly and suddenly, with no time for gradual adjustment of the landscape; also it happened quite recently, the present relation of land and sea being achieved perhaps no more than ten thousand years ago. In the chronology of Earth, a few thousand years are as nothing, and in so brief a time the waves have prevailed little against the hard rocks that the great ice sheet scraped clean of loose rock and ancient soil, and so have scarcely marked out the deep notches that in time they will cut in the cliffs.

For the most part, the ruggedness of this coast is the ruggedness of the hills themselves. There are none of the wave-cut stacks and arches that distinguish older coasts or coasts of softer rock. In a few, exceptional places the work of the waves may be seen. The south shore of Mount Desert Island is exposed to heavy pounding by surf; there the waves have cut out Anemone Cave and are working at Thunder Hole to batter though the roof of the small cave into which the surf roars at high tide.

In places the sea washes the foot of a steep cliff produced by the shearing effect of earth pressure along fault lines. Cliffs on Mount Desert—Schooner Head, Great Head, and Otter—tower a hundred feet or more above the sea. Such imposing structures might be taken for wave-cut cliffs if one did not know the geologic history of the region.

On the coasts of Cape Breton Island and New Brunswick the situation is very different and examples of advanced marine erosion occur on every hand. Here the sea is in contact with weak rock lowlands formed in the Carboniferous period. These shores have little resistance to the erosive power of the waves, and the soft sandstone and conglomerate rocks are being cut back at an annual rate averaging five or six inches, or in some places several feet. Marine stacks, caves, chimneys, and archways are common features of these shores.

Here and there on the predominantly rocky coast of northern New England there are small beaches of sand, pebbles, or cobblestones. These have a varied origin. Some came from glacial debris that covered the rocky surface when the land tilted and the sea came in. Boulders and pebbles often are carried in from deeper water offshore by seaweeds that have gripped them firmly with their "holdfasts." Storm waves then dislodge weed and stone and cast them on the shore. Even without the aid of weeds, waves carry in a considerable volume of sand, gravel, shell fragments, and even boulders. These occasional sandy or pebbly beaches are almost always in protected, incurving shores or dead-end coves, where the waves can deposit debris but from which they cannot easily remove it.

When, on those coastal rocks between the serrate line of spruces and the surf, the morning mists conceal the lighthouses and fishing boats and all other reminders of man, they also blur the sense of time and one might easily imagine that the sea came in only yesterday to create this particular line of coast. Yet the creatures that inhabit the intertidal rocks have had time to establish themselves here, replacing the fauna of the beaches of sand and mud that probably bordered the older coast. Out of the same sea that rose over the northern coast of New England, drowning the coastal plain and coming to rest against the hard uplands, the larvae of the rock dwellers came—the blindly searching larvae that drift in the ocean currents ready to colonize whatever suitable land may lie in their path or to die, if no such landfall is their lot.

Although no one recorded the first colonist or traced the succession of living forms, we may make a fairly confident guess as to the pioneers of the occupation of these rocks, and the forms that followed them. The invading sea must have brought the larvae and young of many kinds of shore animals, but only those able to find food could survive on the new shore. And in the beginning the only available food was the plankton that came in renewed clouds with every tide that washed the coastal rocks. The first permanent inhabitants must have been such plankton-strainers as the barnacles and mussels, who require little but a firm place to which they may attach themselves. Around and among the white cones of the barnacles and the dark shells of the mussels it is probable that the spores of algae settled, so that a living green film began to spread over the upper rocks. Then the grazers could come—the little herds of snails that laboriously scrape the rocks with their sharp tongues, licking off the nearly invisible covering of tiny plant cells. Only after the establishment of the plankton-strainers and the grazers could the carnivores settle and survive. The predatory dog whelks, the starfish, and many of the crabs and worms must, then, have been comparative late-comers to this rocky shore. But all of them are there now, living out their lives in the horizontal zones created by the tides, or in the little pockets or communities of life established by the need to take shelter from surf, or to find food, or to hide from enemies.

The pattern of life spread before me when I emerge from that forest path is one characteristic of exposed shores. From the edge of the spruce forests down to the dark groves of the kelps, the life of the land grades into the life of the sea, perhaps with less abruptness than one would expect, for by various little interlacing ties the ancient unity of the two is made clear.

from Chapter IV, "The Rim of Sand"

On the sands of the sea's edge, especially where they are broad and bordered by unbroken lines of wind-built dunes, there is a sense of antiquity that is missing from the young rock coast of New England. It is in part a sense of the unhurried deliberation of earth processes that move with infinite leisure, with all eternity at their disposal. For unlike that sudden coming in of the sea to flood the valleys and surge against the mountain crests the drowned lands of New England, the sea and the land lie here in a relation established gradually, over millions of years.

During those long ages of geologic time, the sea has ebbed and flowed over the great Atlantic coastal plain. It has crept toward the distant Appalachians, paused for a time, then slowly receded, sometimes far into its basin; and on each such advance it has rained down its sediments and left the fossils of its creatures over that vast and level plain. And so the particular place of its stand today is of little moment in the history of the earth or in the nature of the beach—a hundred feet higher, or a hundred feet lower, the seas would still rise and fall unhurried over shining flats of sand, as they do today.

And the materials of the beach are themselves steeped in antiquity. Sand is a substance that is beautiful, mysterious, and infinitely variable; each grain on a beach is the result of processes that go back into the shadowy beginnings of life, or of the earth itself.

The bulk of seashore sand is derived from the weathering and decay of rocks, transported from their place of origin to the sea by the rains and the rivers. In the unhurried processes of erosion, in the freighting seaward, in the interruptions and resumptions of that journey, the minerals have suffered various fates—some have been dropped, some have worn out and vanished. In the mountains the slow decay and disintegration of the rocks proceed, and the stream of sediments grows—suddenly and dramatically by rockslides—slowly, inexorably, by the wearing of rock by water. All begin their passage toward the sea. Some disappear through the solvent action of water or by grinding attrition in the rapids of a river's bed. Some are dropped on the riverbank by flood waters, there to lie for a hundred, a thousand years, to become locked in the sediments of the plain and wait another million years or so, during which, perhaps, the sea comes in and then returns to its basin. Then at last they are released by the persistent work of erosion's tools—wind, rain, and frost— to resume the journey to the sea. Once brought to salt water, a fresh rearranging, sorting, and transport begin. Light minerals, like flakes of mica, are carried away almost at once; heavy ones like the black sands of ilmenite and rutile are picked up by the violence of storm waves and thrown on the upper beach.

No individual sand grain remains long in any one place. The smaller it is, the more it is subject to long transport—the larger grains by water, the smaller by wind. An average grain of sand is only two and one half times the weight of an equal volume of water, but more than two thousand times as heavy as air, so only the smaller grains are available for transport by wind. But despite the constant working over of the sands by wind and water, a beach shows little visible change from day to day, for as one grain is carried away, another is usually brought to take its place.

The greater part of most beach sand consists of quartz, the most abundant of all minerals, found in almost every type of rock. But many other minerals occur among its crystal grains, and one small sample of sand might contain fragments of a dozen or more. Through the sorting action of wind, water, and gravity, fragments of darker, heavier minerals may form patches overlying the pale quartz. So there may be a curious purple shading over the sand, shifting with the wind, piling up in little ridges of deeper color like the ripple marks of waves—a concentration of almost pure garnet. Or there may be patches of dark green— sands formed of glauconite, a product of the sea's chemistry and the interaction of the living and the non-living. Glauconite is a form of iron silicate that contains potassium; it has occurred in the deposits of all geologic ages. According to one theory, it is forming now in warm shallow areas of the sea's floor, where the shells of minute creatures called foraminifera are accumulating and disintegrating on muddy sea bottoms. On many Hawaiian beaches, the somber darkness of the earth's interior is reflected in sand grains of olivine derived from black basaltic lavas. And drifts of the "black sands" of rutile and ilmenite and other heavy minerals darken the beaches of Georgia's St. Simons and Sapelo Islands, clearly separated from the lighter quartz.

In some parts of the world the sands represent the remains of plants that in life had lime-hardened tissues, or fragments of the calcareous shells of sea creatures. Here and there on the coast of Scotland, for example, are beaches composed of glistening white "nullipore sands"—the shattered and sea-ground remains of coralline algae growing on the bottom offshore. On the coast of Galway in Ireland the dunes are built of sands composed of tiny perforated globes of calcium carbonate—the shells of foraminifera that once floated in the sea. The animals were mortal but the shells they built have endured. They drifted to the floor of the sea and became compacted into sediment. Later the sediments were uplifted to form cliffs, which were eroded and returned once more to the sea. The shells of foraminifera appear also in the sands of southern Florida and the Keys, along with coral debris and the shells of mollusks, shattered, ground, and polished by the waves.

From Eastport to Key West, the sands of the American Atlantic coast, by their changing nature, reveal a varied origin. Toward the northern part of the coast mineral sands predominate, for the waves are still sorting and rearranging and carrying from place to place the fragments of rock that the glaciers brought down from the north, thousands of years ago. Every grain of sand on a New England beach has a long and eventful history. Before it was sand, it was rock—splintered by the chisels of the frost, crushed under advancing glaciers and carried forward with the ice in its slow advance, then ground and polished in the mill of the surf. And long ages before the advance of the ice, some of the rock had come up into the light of the sun from the black interior of the earth by ways unseen and for the most part unknown, made fluid by subterranean fires and rising along deep pipes and fissures. Now in this particular moment of its history, it belongs to the sea's edge—swept up and down the beaches with the tides or drifted alongshore with the currents, continuously sifted and sorted, packed down, washed out, or set adrift again, as always and endlessly the waves work over the sands.

On Long Island, where much glacial material has accumulated, the sands contain quantities of pink and red garnet and black tourmaline, along with many grains of magnetite. In New Jersey, where the coastal plain deposits of the south first appear, there is less magnetic material and less garnet. Smoky quartz predominates at Barnegat, glauconite at Monmouth Beach, and heavy minerals at Cape May. Here and there beryl occurs where molten magma has brought up deeply buried material of the ancient earth to crystallize near the surface.

North of Virginia, less than half of one per cent of the sands are of calcium carbonate; southward, about 5 per cent. In North Carolina the abundance of calcareous or shell sand suddenly increases, although quartz sand still forms the bulk of the beach materials. Between Cape Hatteras and Lookout as much as 10 per cent of the beach sand is calcareous. And in North Carolina also there are odd local accumulations of special materials such as silicified wood—the same substance that is contained in the famous "singing sands" of the Island of Eigg in the Hebrides.

The mineral sands of Florida are not of local origin but have been derived from the weathering of rocks in the Piedmont and Appalachian highlands of Georgia and South Carolina. The fragments are carried to the sea on southward-moving streams and rivers. Beaches of the northern part of Florida's Gulf Coast are almost pure quartz, composed of crystal grains that have descended from the mountains to sea level, accumulating there in plains of snowlike whiteness. About Venice there is a special sparkle and glitter over the sands, where crystals of the mineral zircon are dusted over its surface like diamonds; and here and there is a sprinkling of the blue, glasslike grains of cyanite. On the east coast of Florida, quartz sands predominate for much of the long coast line (it is the hard-packing quartz grains that compose the famous beaches of Daytona) but toward the south, the crystal sands are mingled more and more with fragments of shells. Near Miami the beach sands are less than half quartz; about Cape Sable and in the Keys the sand is almost entirely derived from coral and shell and the remains of foraminifera. And all along the east coast of Florida, the beaches receive small contributions of volcanic matter, as bits of floating pumice that have drifted for thousands of miles in ocean currents are stranded on the shore to become sand.

Infinitely small though it is, something of its history may be revealed in the shape and texture of a grain of sand. Wind-transported sands tend to be better rounded than water-borne; furthermore, their surface shows a frosted effect from the abrasion of other grains carried in the blast of air. The same effect is seen on panes of glass near the sea, or on old bottles in the beach flotsam. Ancient sand grains, by their surface etchings, may give a clue to the climate of past ages. In European deposits of Pleistocene sand, the grains have frosted surfaces etched by the great winds blowing off the glaciers of the Ice Age.

We think of rock as a symbol of durability, yet even the hardest rock shatters and wears away when attacked by rain, frost or surf. But a grain of sand is almost indestructible. It is the ultimate product of the work of the waves—the minute, hard core of mineral that remains after years of grinding and polishing. The tiny grains of wet sand lie with little space between them, each holding a film of water about itself by capillary attraction. Because of this cushioning liquid film, there is little further wearing by attrition. Even the blows of heavy surf cannot cause one sand grain to rub against another.

from Chapter V, "The Coral Coast"

I doubt that anyone can travel the length of the Florida Keys without having communicated to his mind a sense of the uniqueness of this land of sky and water and scattered mangrove-covered islands. The atmosphere of the Keys is strongly and peculiarly their own. It may be that here, more than in most places, remembrance of the past and intimations of the future are linked with present reality. In bare and jaggedly corroded rock, sculptured with the patterns of the corals, there is the desolation of a dead past. In the multicolored sea gardens seen from a boat as one drifts above them, there is a tropical lushness and mystery, a throbbing sense of the pressure of life; in coral reef and mangrove swamp there are the dimly seen foreshadowings of the future.

This world of the Keys has no counterpart elsewhere in the United States, and indeed few coasts of the earth are like it. Offshore, living coral reefs fringe the island chain, while some of the Keys themselves are the dead remnants of an old reef whose builders lived and flourished in a warm sea perhaps a thousand years ago. This is a coast not formed of lifeless rock or sand, but created by the activities of living things which, though having bodies formed of protoplasm even as our own, are able to turn the substance of the sea into rock.

The living coral coasts of the world are confined to waters in which the temperature seldom falls below 70° F. (and never for prolonged periods), for the massive structures of the reefs can be built only where the coral animals are bathed by waters warm enough to favor the secretion of their calcareous skeletons. Reefs and all the associated structures of a coral coast are therefore restricted to the area bounded by the Tropics of Cancer and Capricorn. Moreover, they occur only on the eastern shores of continents, where currents of tropical water are carried toward the poles in a pattern determined by the earth's rotation and the direction of the winds. Western shores are inhospitable to corals because they are the site of upwellings of deep, cold water, with cold coastwise currents running toward the equator.

In North America, therefore, California and the Pacific coast of Mexico lack corals, while the West Indian region supports them in profusion. So do the coast of Brazil in South America, the tropical east African coast, and the northeastern shores of Australia, where the Great Barrier Reef creates a living wall for more than a thousand miles.

Within the United States the only coral coast is that of the Florida Keys. For nearly 200 miles these islands reach south-westward into tropical waters. They begin a little south of Miami where Sands, Elliott, and Old Rhodes Keys mark the entrance to Biscayne Bay; then other islands continue to the southwest, skirting the tip of the Florida mainland, from which they are separated by Florida Bay, and finally swinging out from the land to form a slender dividing line between the Gulf of Mexico and the Straits of Florida, through which the Gulf Stream pours its indigo flood.

To seaward of the Keys there is a shallow area three to seven miles wide where the sea bottom forms a gently sloping platform under depths generally less than five fathoms. An irregular channel (Hawk Channel) with depths to ten fathoms traverses these shallows and is navigable by small boats. A wall of living coral reefs forms the seaward boundary of the reef platform, standing on the edge of the deeper sea.

The Keys are divided into two groups that have a dual nature and origin. The eastern islands, swinging in their smooth arc 110 miles from Sands to Loggerhead Key, are the exposed remnants of a Pleistocene coral reef. Its builders lived and flourished in a warm sea just before the last of the glacial periods, but today the corals, or all that remains of them, are dry land. These eastern Keys are long, narrow islands covered with low trees and shrubs, bordered with coral limestone where they are exposed to the open sea, passing into the shallow waters of Florida Bay through a maze of mangrove swamps on the sheltered side. The western group, known as the Pine Islands, are a different kind of land, formed of limestone rock that had its origin on the bottom of a shallow interglacial sea, and is now raised only slightly above the surface of the water. But in all the Keys, whether built by the coral animals or formed of solidifying sea drift, the shaping hand is the hand of the sea.

In its being and its meaning, this coast represents not merely an uneasy equilibrium of land and water masses; it is eloquent of a continuing change now actually in progress, a change being brought about by the life processes of living things. Perhaps the sense of this comes most clearly to one standing on a bridge between the Keys, looking out over miles of water, dotted with mangrove-covered islands to the horizon. This may seem a dreamy land, steeped in its past. But under the bridge a green mangrove seedling floats, long and slender, one end already beginning to reach down through the water, ready to grasp and to root firmly in any muddy shoal that may lie across its path. Over the years the mangroves bridge the water gaps between the islands; they extend the mainland; they create new islands. And the currents that stream under the bridge, carrying the mangrove seedling, are one with the currents that carry plankton to the coral animals building the offshore reef, creating a wall of rocklike solidity, a wall that one day may be added to the mainland. So this coast is built.

To understand the living present, and the promise of the future, it is necessary to remember the past. During the Pleistocene, the earth experienced at least four glacial stages, when severe climates prevailed and immense sheets of ice crept southward. During each of these stages, large volumes of the earth's water were frozen into ice, and sea level dropped all over the world. The glacial intervals were separated by milder interglacial stages when, with water from melting glaciers returning to the sea, the level of the world ocean rose again. Since the most recent Ice Age, known as the Wisconsin, the general trend of the earth's climate has been toward a gradual, though not uniform warming up. The interglacial stage preceding the Wisconsin glaciation is known as the Sangamon, and with it the history of the Florida Keys is intimately linked.

The corals that now form the substance of the eastern Keys built their reef during that Sangamon interglacial period, probably only a few tens of thousands of years ago. Then the sea stood perhaps 100 feet higher than it does today, and covered all of the southern part of the Florida plateau. In the warm sea off the sloping southeastern edge of that plateau the corals began to grow, in water somewhat more than 100 feet deep. Later the sea level dropped about 30 feet (this was in the early stages of a new glaciation, when water drawn from the sea was falling as snow in the far north); then another 30 feet. In this shallower water the corals flourished even more luxuriantly and the reef grew upward, its structure mounting close to the sea surface. But the dropping sea level that at first favored the growth of the reef was to be its destruction, for as the ice increased in the north in the Wisconsin glacial stage, the ocean level fell so low that the reef was exposed and all its living coral animals were killed. Once again in its history the reef was submerged for a brief period, but this could not bring back the life that had created it. Later it emerged again and has remained above water, except for the lower portions, which now form the passes between the Keys. Where the old reef lies exposed, it is deeply corroded and dissected by the dissolving action of rain and the beating of salt spray; in many places the old coral heads are revealed, so distinctly that the species are identifiable.

While the reef was a living thing, being built up in that Sangamon sea, the sediments that have more recently become the limestone of the western group of Keys were accumulating on the landward side of the reef. Then the nearest land lay 150 miles to the north, for all the southern end of the present Florida peninsula was submerged. The remains of many sea creatures, the solution of limestone rocks, and chemical reactions in the sea water contributed to the soft ooze that covered the shallow bottoms. With the changing sea levels that followed, this ooze became compacted and solidified into a white, fine-textured limestone, containing many small spherules of calcium carbonate resembling the roe of fish; because of this characteristic it is sometimes known as "oolitic limestone," or "Miami oolite." This is the rock immediately underlying the southern part of the Florida mainland. It forms the bed of Florida Bay under the layer of recent sediments, and then rises above the surface in the Pine Islands, or western Keys, from Big Pine Key to Key West. On the mainland, the cities of Palm Beach, Fort Lauderdale, and Miami stand on a ridge of this limestone formed when currents swept past an old shore line of the peninsula, molding the soft oozes into a curving bar. The Miami oolite is exposed on the floor of the Everglades as rock of strangely uneven surface, here rising in sharp peaks, there dropping away in solution holes. Builders of the Tamiami Trail and of the highway from Miami to Key Largo dredged up this limestone along the rights of way and with it built the foundations on which these highways are laid.

Knowing this past, we can see in the present a repetition of the pattern, a recurrence of earth processes of an earlier day. Now, as then, living reefs are building up offshore; sediments are accumulating in shallow waters; and the level of the sea, almost imperceptibly but certainly, is changing.

from Chapter I, The Marginal World

The edge of the sea is a strange and beautiful place. All through the long history of Earth it has been an area of unrest where waves have broken heavily against the land, where the tides have pressed forward over the continents, receded, and then returned. For no two successive days is the shore line precisely the same. Not only do the tides advance and retreat in their eternal rhythms, but the level of the sea itself is never at rest. It rises or falls as the glaciers melt or grow, as the floor of the deep ocean basins shifts under its increasing load of sediments, or as the earth's crust along the continental margins warps up or down in adjustment to strain and tension. Today a little more land may belong to the sea, tomorrow a little less. Always the edge of the sea remains an elusive and indefinable boundary.

The shore has a dual nature, changing with the swing of the tides, belonging now to the land, now to the sea. On the ebb tide it knows the harsh extremes of the land world, being exposed to heat and cold, to wind, to rain and drying sun. On the flood tide it is a water world, returning briefly to the relative stability of the open sea.

Only the most hardy and adaptable can survive in a region so mutable, yet the area between the tide lines is crowded with plants and animals. In this difficult world of the shore, life displays its enormous toughness and vitality by occupying almost every conceivable niche. Visibly, it carpets the intertidal rocks; or half hidden, it descends into fissures and crevices, or hides under boulders, or lurks in the wet gloom of sea caves. Invisibly, where the casual observer would say there is no life, it lies deep in the sand, in burrows and tubes and passageways. It tunnels into solid rock and bores into peat and clay. It encrusts weeds or drifting spars or the hard, chitinous shell of a lobster. It exists minutely, as the film of bacteria that spreads over a rock surface or a wharf piling; as spheres of protozoa, small as pinpricks, sparkling at the surface of the sea; and as Lilliputian beings swimming through dark pools that lie between the grains of sand.

The shore is an ancient world, for as long as there has been an earth and sea there has been this place of the meeting of land and water. Yet it is a world that keeps alive the sense of continuing creation and of the relentless drive of life. Each time that I enter it, I gain some new awareness of its beauty and its deeper meanings, sensing that intricate fabric of life by which one creature is linked with another, and each with its surroundings.

In my thoughts of the shore, one place stands apart for its revelation of exquisite beauty. It is a pool hidden within a cave that one can visit only rarely and briefly when the lowest of the year's low tides fall below it, and perhaps from that very fact it acquires some of its special beauty. Choosing such a tide, I hoped for a glimpse of the pool. The ebb was to fall early in the morning. I knew that if the wind held from the northwest and no interfering swell ran in from a distant storm the level of the sea should drop below the entrance to the pool. There had been sudden ominous showers in the night, with rain like handfuls of gravel flung on the roof. When I looked out into the early morning the sky was full of a gray dawn light but the sun had not yet risen. Water and air were pallid. Across the bay the moon was a luminous disc in the western sky, suspended above the dim line of distant shore—the full August moon, drawing the tide to the low, low levels of the threshold of the alien sea world. As I watched, a gull flew by, above the spruces. Its breast was rosy with the light of the unrisen sun. The day was, after all, to be fair.

Later, as I stood above the tide near the entrance to the pool, the promise of that rosy light was sustained. From the base of the steep wall of rock on which I stood, a moss-covered ledge jutted seaward into deep water. In the surge at the rim of the ledge the dark fronds of oarweeds swayed, smooth and gleaming as leather. The projecting ledge was the path to the small hidden cave and its pool. Occasionally a swell, stronger than the rest, rolled smoothly over the rim and broke in foam against the cliff. But the intervals between such swells were long enough to admit me to the ledge and long enough for a glimpse of that fairy pool, so seldom and so briefly exposed.

And so I knelt on the wet carpet of sea moss and looked back into the dark cavern that held the pool in a shallow basin. The floor of the cave was only a few inches below the roof, and a mirror had been created in which all that grew on the ceiling was reflected in the still water below.

Under water that was clear as glass the pool was carpeted with green sponge. Gray patches of sea squirts glistened on the ceiling and colonies of soft coral were a pale apricot color. In the moment when I looked into the cave a little elfin starfish hung down, suspended by the merest thread, perhaps by only a single tube foot. It reached down to touch its own reflection, so perfectly delineated that there might have been, not one starfish, but two. The beauty of the reflected images and of the limpid pool itself was the poignant beauty of things that are ephemeral, existing only until the sea should return to fill the little cave.

Whenever I go down into this magical zone of the low water of the spring tides, I look for the most delicately beautiful of all the shore's inhabitants—flowers that are not plant but animal, blooming on the threshold of the deeper sea. In that fairy cave I was not disappointed. Hanging from its roof were the pendent flowers of the hydroid Tubularia, pale pink, fringed and delicate as the wind flower. Here were creatures so exquisitely fashioned that they seemed unreal, their beauty too fragile to exist in a world of crushing force. Yet every detail was functionally useful, every stalk and hydranth and petal-like tentacle fashioned for dealing with the realities of existence. I knew that they were merely waiting, in that moment of the tide's ebbing, for the return of the sea. Then in the rush of water, in the surge of surf and the pressure of the incoming tide, the delicate flower heads would stir with life. They would sway on their slender stalks, and their long tentacles would sweep the returning water, finding in it all that they needed for life.

And so in that enchanted place on the threshold of the sea the realities that possessed my mind were far from those of the land world I had left an hour before. In a different way the same sense of remoteness and of a world apart came to me in a twilight hour on a great beach on the coast of Georgia. I had come down after sunset and walked far out over sands that lay wet and gleaming, to the very edge of the retreating sea. Looking back across that immense flat, crossed by winding, water-filled gullies and here and there holding shallow pools left by the tide, I was filled with awareness that this intertidal area, although abandoned briefly and rhythmically by the sea, is always reclaimed by the rising tide. There at the edge of low water the beach with its reminders of the land seemed far away. The only sounds were those of the wind and the sea and the birds. There was one sound of wind moving over water, and another of water sliding over the sand and tumbling down the faces of its own wave forms. The flats were astir with birds, and the voice of the willet rang insistently. One of them stood at the edge of the water and gave its loud, urgent cry; an answer came from far up the beach and the two birds flew to join each other.

The flats took on a mysterious quality as dusk approached and the last evening light was reflected from the scattered pools and creeks. Then birds became only dark shadows, with no color discernible. Sanderlings scurried across the beach like little ghosts, and here and there the darker forms of the willets stood out. Often I could come very close to them before they would start up in alarm— the sanderlings running, the willets flying up, crying. Black skimmers flew along the ocean's edge silhouetted against the dull, metallic gleam, or they went flitting above the sand like large, dimly seen moths. Sometimes they "skimmed" the winding creeks of tidal water, where little spreading surface ripples marked the presence of small fish.

The shore at night is a different world, in which the very darkness that hides the distractions of daylight brings into sharper focus the elemental realities. Once, exploring the night beach, I surprised a small ghost crab in the searching beam of my torch. He was lying in a pit he had dug just above the surf, as though watching the sea and waiting. The blackness of the night possessed water, air, and beach. It was the darkness of an older world, before Man. There was no sound but the all-enveloping, primeval sounds of wind blowing over water and sand, and of waves crashing on the beach. There was no other visible life—just one small crab near the sea. I have seen hundreds of ghost crabs in other settings, but suddenly I was filled with the odd sensation that for the first time I knew the creature in its own world—that I understood, as never before, the essence of its being. In that moment time was suspended; the world to which I belonged did not exist and I might have been an onlooker from outer space. The little crab alone with the sea became a symbol that stood for life itself—for the delicate, destructible, yet incredibly vital force that somehow holds its place amid the harsh realities of the inorganic world.

The sense of creation comes with memories of a southern coast, where the sea and the mangroves, working together, are building a wilderness of thousands of small islands off the south-western coast of Florida, separated from each other by a tortuous pattern of bays, lagoons, and narrow waterways. I remember a winter day when the sky was blue and drenched with sunlight; though there was no wind one was conscious of flowing air like cold clear crystal. I had landed on the surf-washed tip of one of those islands, and then worked my way around to the sheltered bay side. There I found the tide far out, exposing the broad mud flat of a cove bordered by the mangroves with their twisted branches, their glossy leaves, and their long prop roots reaching down, grasping and holding the mud, building the land out a little more, then again a little more.

The mud flats were strewn with the shells of that small, exquisitely colored mollusk, the rose tellin, looking like scattered petals of pink roses. There must have been a colony nearby, living buried just under the surface of the mud. At first the only creature visible was a small heron in gray and rusty plumage—a reddish egret that waded across the flat with the stealthy, hesitant movements of its kind. But other land creatures had been there, for a line of fresh tracks wound in and out among the mangrove roots, marking the path of a raccoon feeding on the oysters that gripped the supporting roots with projections from their shells. Soon I found the tracks of a shore bird, probably a sanderling, and followed them a little; then they turned toward the water and were lost, for the tide had erased them and made them as though they had never been.

Looking out over the cove I felt a strong sense of the interchangeability of land and sea in this marginal world of the shore, and of the links between the life of the two. There was also an awareness of the past and of the continuing flow of time, obliterating much that had gone before, as the sea had that morning washed away the tracks of the bird.

The sequence and meaning of the drift of time were quietly summarized in the existence of hundreds of small snails—the mangrove periwinkles—browsing on the branches and roots of the trees. Once their ancestors had been sea dwellers, bound to the salt waters by every tie of their life processes. Little by little over the thousands and millions of years the ties had been broken, the snails had adjusted themselves to life out of water, and now today they were living many feet above the tide to which they only occasionally returned. And perhaps, who could say how many ages hence, there would be in their descendants not even this gesture of remembrance for the sea.

The spiral shells of other snails—those quite minute—left winding tracks on the mud as they moved about in search of food. They were horn shells, and when I saw them I had a nostalgic moment when I wished I might see what Audubon saw, a century and more ago. For such little horn shells were the food of the flamingo, once so numerous on this coast, and when I half closed my eyes I could almost imagine a flock of these magnificent flame birds feeding in that cove, filling it with their color. It was a mere yesterday in the life of the earth that they were there; in nature, time and space are relative matters, perhaps most truly perceived subjectively in occasional flashes of insight, sparked by such a magical hour and place.

There is a common thread that links these scenes and memories—the spectacle of life in all its varied manifestations as it has appeared, evolved, and sometimes died out. Underlying the beauty of the spectacle there is meaning and significance. It is the elusiveness of that meaning that haunts us, that sends us again and again into the natural world where the key to the riddle is hidden. It sends us back to the edge of the sea, where the drama of life played its first scene on earth and perhaps even its prelude; where the forces of evolution are at work today, as they have been since the appearance of what we know as life; and where the spectacle of living creatures faced by the cosmic realities of their world is crystal clear.

3

Silesia

HANS CLOOS

Far beneath our familiar surroundings, where sedimentary rocks can be seen in the making, lies the geologic environment where high temperatures and pressures create plutonic igneous and metamorphic rocks. Although extensively covered by the surface "veneer" of sediments and visible only in shield areas, mountain roots, and deep canyons, the plutonic rocks form the bulk of the earth's crust. Since they can be seen only after prolonged erosion, millions of years after they were created, these rocks are difficult to interpret. Yet, as Hans Cloos shows, they contain their own magnificent story for those able to read it.

Serpentine

FOR ERICH BEDERKE

Gray hills—stony, sad-looking, barren—rise out of golden fields and dense green forests. It is as if a curse lay on them. On one of these hills, a hundred years ago, stood the gallows of the little town which runs from the crest to the plain below. Does the wicked spirit of the hanged keep the spot bare of flowers and greenness? No, the gallows was erected here because the land was good for nothing else, and because it could be seen from afar.

The hill consists of serpentine, a rare, greenish-white, tough, unstratified, curious rock derived by chemical addition of water from peridotite, a basic (that is, poor in silicic acid) igneous rock.

Peridotite itself is composed of the minerals olivine and diopside, a variety of augite. The weathering of serpentine is a beautiful example of the way constituent materials are differentiated and form ore deposits.

Throughout geological history plants have been unable to adjust themselves to the lean and alien nourishment this particular serpentine offers. All around the hill there have always been other kinds of rock — granite or slate, limestone or sandstone — more readily available and more appealing to plants. Therefore to this day scarcely any growing thing likes to strike root and live on the Galgenberg (Gallows Mountain).

The reason serpentine is so rare and peculiar is simple and understandable enough: it belongs properly deep below the surface. It is material of the third crustal shell, reckoning downward, and it is really a miracle that in some places it ever reaches the surface through the middle basaltic and the uppermost granitic layers. It is heavier than most other rocks. Its specific weight would be still greater had it not, when coming up as a hot melt out of closely compressed depths into the roomier regions of the upper crust, absorbed water and swelled, so to speak. Here and there "shallower" serpentines have grown in the outermost crust itself. But the rock in this Silesian hill is of deep origin. This is shown in the way it has modestly fitted itself to its strange surroundings. The long, narrow hills follow one another like drops oozing from a long and narrow fissure. The line along which they are strung like pearls is actually a deep fracture which so sharply separates two parts of the earth's crust that it must reach down into very great depths. It is one of those fissures which have cut through the whole crust from the fluid under-layer to the brittle surface. At one time a long, winding strip of crust heaved up and sank down more quickly and violently than adjacent areas. It was more wildly shaken and suffered deeper and more serious wounds, which finally healed into a rude, swollen welt. Serpentine is found along the center line of just such ranges. A similar line can be traced through the Alps to Greece, and on through Asia Minor to the South Asiatic highland. In the Urals it separates Europe from Asia.

This is, broadly speaking, what geology knows about the nature, location, and origin of the rare rock, serpentine.

The mining engineer, however, adds more significant information: of the useful materials found in the earth, no less than six, among them three of the most important, are found either exclusively or predominantly in serpentine, or in its water-free mother rock known as peridotite. These are: the precious diamond and chrysoprase; the metals platinum and chromium; and the technologically important magnesite and asbestos. The platinum and chromium are both furnished by the peridotite. They were precipitated during the molten stage if the melt contained a surplus of these elements. Platinum is found in grains, chromium ore in lumps, in the cooled residue. Where this surplus comes from, and why there should be platinum in the Urals and chromium in the Balkans and Asia Minor, is a mystery that may never be explained. Certain it is that the earth's interior was not mixed uniformly. It would be astonishing if that had been the case. In any event, man profits thereby.

Here, however, nothing valuable has been precipitated by the rising melt; in other words, nothing better has developed than a uniformly black, heavy peridotite. Where this peridotite took up water, it became a dirty green or brown or gray mottled "serpentine." Here the geological development should have come to an end. However, we meet with a completely unexpected phenomenon, something that astonishes us in a practical, a theoretical, and an aesthetic sense.

Just as white light is broken up by a prism into a spectrum of colors, the plain gray serpentine disintegrates into a colorful series of beautiful and novel minerals and crystals as soon as it comes in contact with the uppermost—and, to it, unfamiliar—forces of the earth's surface. They bloom in the gloomy rock like a bouquet of bright flowers and it takes the combined efforts of mining engineer and geologist to track down the lovely yield.

Not far from Breslau, near the little old Silesian town of Frankenstein, are these hills of serpentine, covered by short grass and spruces. Small quarries have been opened, for the tough rock resists weathering and serves fairly well as road metal. If we break off a fresh piece it shows nothing but a dirty mottled gray or bluish-green surface. We could analyze it chemically, and after several weeks of laborious and expensive effort would find that most of it is silica, like all volcanic rocks. The proportion of silica is relatively small, corresponding to the very deep origin of this "most basic" of all rocks. It usually also contains some iron, very little alumina, and a small but ever-present trace of nickel. The second most important part of the mixture would be magnesia. Whereas in granites, diorites, and basalts, magnesia is present only in traces, here it constitutes a third. The common chemical title for our rock is hydrous magnesium silicate.

But we need neither a chemical nor a miscroscopic analysis. We can see all these things, and more, in their delightful natural splendor thanks to the miners, who have cut into the interior of the mountain.

There are some large open pits with tracks, ramps, and tailing dumps; small shafts with old-fashioned windlasses; and all kinds of cuts and openings. On the walls we see how the mother rock of serpentine, under the influence of circulating water and solutions, has broken up into its constituent parts, and how these have grown and put out shoots. Each substance shows itself in its characteristic color and appearance. It seems as if all this beauty has been awaiting human discovery, so that it may be freed from subterranean darkness and finally be carried out to its proper place in the blinding sunlight.

Where the originally scant nickel content of the serpentine has been concentrated and exposed to the addition of water and oxygen, the formerly invisible element blooms as a deep, juicy green, like grass after a spring rain. Now it can be recognized everywhere, as it forms a green "nodular ore" but of thousands of tiny grains contrasting against the red background. It enters veins of limestone or magnesite and colors them, growing and branching out like green weeds.

Today the colorful picture is preserved only in remnants. It is almost two hundred years since the mines were first opened when the Prussian king, Frederick the Great, wanted the precious Silesian stone, chrysoprase, for the palace of Sans Souci. During World War I whole mountains of serpentine were moved to extract the nickel at any cost. Most abundant today is magnesite, which grows into white tree-like forms in the rude jungle of decomposed serpentine. The never quite abandoned mining of this rare carbonate, used in the manufacture of fireproof slabs and bricks, gives assurance that this impressive mineralization will remain visible for a long time to come.

But even if some time in the future all this glory should have disappeared, the geological archives of the last fifty years would still contain the memory of this most exquisite example of one of nature's curiosities: beautifully colored forms growing not only in the sunlight, but also in the earth's dark womb, where there are no eyes to see them, and where such brilliant lavishness seems to be utterly wasted. Perhaps it is not just an accident that the richest wreaths of color unfold at the spot where the very deepest and least surface-like materials meet the waters and solutions from above, at places where the substances of deep origin have but recently risen and tension between their chemical potentials is greatest.

When the mountains are leveled either by erosion or by human hands and when the flowery cellar is opened, the bright colors become traitors. The miner recognizes nickel by the green of its bloom, copper by another shade of green and a lapis lazuli blue, iron by its rusty-red color. The ores are then doomed, because along with the ores of the surface the less spectacular ones at greater depths are also removed. Millennia may pass before more of this gaily decorative treasure is ready, millions of years before the deeper ore deposits themselves may be replenished in step with the sluggish changes of this ancient earth.

Granite
For the noblest of all the human senses is sight.

—ALBRECHT DURER

The mountains of the Riesengebirge in Silesia rise in two steps from the friendly hill country into a world of lonely highland plateaus and white summer clouds. The valley is a patchwork of fields and meadows, mirrored ponds, and light-colored trees. Half-way up is the realm of deep, dark forests. On top are meadows like green lakes, with low islands of dark Scotch firs. Beyond the meadows the mountains again drop steeply into blue depths.

Big rocks are strewn here and there. In the bottomland they look like huge toys. Bushes grow out of them, children play hide-and-seek in their clefts. On the higher wooded slopes they are more rare. One stumbles across them in the fir thickets and only the tallest show above the trees. Some of them rise on the high ridge, landmarks visible from afar. At close range, they look like towers. They defy the storm and in winter are draped with snow and ice.

These cliffs look as if they had been built up like masonry. To the inhabitants the bizarre contours suggest animal heads and human shapes. Their names are found on signposts. But to the geologist they do not represent the caprice of giants. For him their forms express the structure of the ground. They are the facial expression of the mountain and show its character.

The rocks of the mountain are also exposed in the gorges and brooks which descend from the heights to the valleys. They tower high and steep above the cirques (circular depressions) worn into the crest by the ice of the Glacial Period. In these pockets, even in July, remnants of the winter's snow still linger. The rocks are also exposed in the cuts made for the winding roads and paths, down which brown-skinned wood-cutters bring their harvest, and up which climb pallid city-dwellers, carrying their cares and sorrows.

A trail runs the length of the crest. In six hours one can walk all the way from one end to the other. Crossing the ridge transversely takes almost as long. The distance is shorter, but between the old German city at the northern foot of the range and the valleys to the south, there is a rise of 3,000 feet. There are some small towns on the mountain, countless farms and wood-cutters' huts, hotels, and summer cottages. It is a large, bright, special kind of world. And everything in it is built of one kind of rock. This rock is granite.

This granite is not especially hard, but it is beautiful. Its bright color derives from the pink shining domino-sized plates of pure, well-developed crystals of potash feldspar scattered among the fine-grained "dough" of gray quartz and little leaves of black mica. The feldspar is lacking only on the ridge; there the rock is harder, and that is why the old upland plateau is preserved at the great height to which it was lifted long ago.

The pink granite is not hard enough for paving blocks, but because of its beauty it has been used in a bridge which spans a distant stream. It has also been used in the pedestals and steps of some well-intentioned monuments. But the most beautiful structure made of this lovely granite is the mountain range itself. The whole substructure is pink granite, down to great and undetermined depths.

For more than a hundred years geologists have found this rock pleasing and studied its crystals, finding many peculiarities and special kinds containing feldspar and black mica in different proportions from those of the main body. A very light kind is called aplite; a very coarse one, pegmatite. These types penetrate the main body like roots or branches. Layers very rich in mica lie in the main mass like dark strata. At an early date geologists also found a dark, almost black rock and some of a brownish-red color, intruding as foreign bodies into the main granite. Geologists call them dikes, meaning fissures that have been filled by later intrusions of rock.

Studies of the rock here led to a simple formulation applicable to all granitic bodies. The granite was once hot, full of gas, and molten; and in this condition it forced its way up and out of unknown depths into its present place. It heated and toasted the surrounding rock, and was cooled in doing so. Later, it was exposed by the erosion of its cover. How the hot mass had made a place for itself in the cold crust, whether by melting its way in or by brutal mechanical force, or whether the granite was produced by in-place transformation of rock already there—these and similar important questions petrography was unable to answer. For geologists, who were less interested in the chemical composition than in the movements of parts of the crust, little had been achieved. The surrounding country had spoiled the geologists. There they found mountains replete with signs which showed, everywhere, the intensity, direction, and magnitude of motion. They had become accustomed to studying structures and history in the manner of architects for whom each capital and column is an eloquent witness.

Insofar as it had failed to yield similar signs of motion, the granite area was a geological wasteland, a large, blank, uninvestigated and undescribed spot on the map.

Or are there actually veiled signs of motion which might be deciphered and interpreted?

We take a heavy and a light hammer to break off pieces of rock and to expose small characteristic features, and also a good compass and a clinometer to measure them, and a stack of topographic maps on which to plot our measurements. And then one bright morning in July we ride to the blue mountains. Before November wraps them in snow and darkness, we shall know more than we do today!

The first weeks are spent in orientation and a general survey. We use tourist trails and see the same things they see, yet in a different perspective. No, that is not really it: we see altogether different things. The tourist looks at a group of rocks, and thinks he sees the outline of a horse's head, or what seem to be dwarfs, or an oven. Incidental outlines mean nothing, however, to the geologist. He goes beyond them, supplies missing parts, and connects the many outcrops seen in hours of wandering along the trails. The tourist is delighted with the cover of vines and moss, and the colorful oxidation on the rock surface. To the geologist they are a nuisance. With the hammer he tries to get below the surface, so that he may free the ancient, eternal rock from the temporary cover that masks it. He kneels, puts his hand lens to the rock and his eye to the lens, measures a direction with his clinometer, takes notes, sketches, moves on a few steps, and repeats the process. The idle onlooker shakes his head, smiles at this queer business and the rock-collecting, and keeps on climbing to the nearest inn. We let him go, and follow the geologist's laborious path.

His task is essentially simple. He meditates: in the adjacent sedimentary rocks there are surfaces and lines whose arrangement or disarrangement, formation or distortion, show movements. Therefore he must look in the granite for directional patterns, however confused they may appear at first. The geologist finds swarms of feldspar dominoes that are parallel to each other, mica leaves laid together in neat, even layers, or packets; also foreign bodies, fragments of the rock which surrounds the granite, which lie frequently one behind the other in a line or in a plane. Then there are schlieren; that is, streaks which differ in composition and color or grain size from the main rock, but which merge imperceptibly into it. When struck with hammer or chisel the rock cleaves more easily in some directions than in others, as wood does, and at times shows a fibrous texture. The rock seems to have grown, like a tree. All these signs are obviously traces of the flow of the melt before it finally hardened. Now the groundwork has been laid.

However, the geologist has to keep on searching, observing, and measuring, and must record on the maps as many structures as possible, plot them, and make notes on form, direction and angles of inclination. In that way a network of structural symbols will be laid over the whole mountain range. This entails alertness as well as physical endurance, and takes months to complete.

Hard work finally brings results: a general pattern appears. The structures not only show a local order, but a constant relationship from cliff to cliff, and from one valley to the other. Indeed, as mapping progresses, the structural picture becomes simpler rather than more complicated. Ultimately thousands of structure symbols are on the map and show two large, simple domes, each made of many shells. They are contiguous and include the whole mountain range.

A surprising result! I had counted on evidence of many small, local magmatic movements, as shown in vortexes or whorls, with upward or downward streams such as are well known in hardened lava. But this is quite different and incomparably more impressive. It is the picture of one gigantic upwelling of magma, driven upward by enormous forces, comparable to a vast bubble rising through mud or dough. The viscous melt did not flow upward or downward in channels or streams; no, it rose along a large, broad front, stretched out and expanded sideways and yielded to a force from the depths which pushed and drove it.

The material of such a widely spanning arch must have had an extraordinary viscosity and effective strength. Rapid cooling then stiffened the dough to a crust, the melt into rock.

But the hot depth still continued to exert pressure, and slowly the plutonic phase reached a second stage. Another group of structures was being formed.

Almost all rocks on this earth are jointed or fractured. Were this not so, rivers and quarry men would have to work ten times as hard to cut into them and carry the fragments away. Some rocks, like basalt and porphyry, have wide joints because they shrank fast when they cooled. Everyone knows their beautiful geometrical columns. In other rocks, joints appeared later when they were pressed or bent during deformation of the earth's crust. Granite combines these two effects. While cooling and shrinking it was also subjected to the pressures and tensions of its moving frame.

So then we had to measure joints, and not merely hundreds of them, but thousands, a boring and simply mechanical job. Signs of relationships and connection began to appear. After some weeks a rough pattern evolved, which henceforth remained unchanged except for the filling in of some detail. This joint pattern of the second plutonic phase, like the flow-structure of the first, was the result of one slow and uniform uplift. In the center of the dome, the joints and the granite pillars or plates are exactly vertical. Toward the two sides they are inclined inward, at first steeply, then flattening out toward the periphery, and continuing into the wall rock.

For the second time simple facts gain importance through mapping and addition. The joints are arranged as a great fan which penetrates the whole mountain. It stands erect and closed in the center segment, but spreads out at both ends.

This radial, spoke-like jointing supports an arch which has been torn apart. We cannot actually see this arch, but can reconstruct it theoretically according to the laws of physics. For the second time an imaginary rainbow spans the whole range. This time, however, it is not the result of an upwelling flow, but of the rising of the cover that had hardened into brittle armor, and that carried a large segment of the surrounding rock with it as it rose.

In its primary, flowing, or plastic state the tough melt (magma) welled up vertically, pushing the crustal cover of older rock into a dome. The traces of this movement were found in schlieren and parallel mineral veins or mica-rich layers. In the later and more brittle condition the dome broke up, forming joints, fissures, and fractures, that are partly filled by dikes.

Now the geologists' and the tourists' ways of looking at rocks begin to converge. The fact that a cliff assumed the outlines of a horse's head, or of a dwarf, depended, of course, on very limited local circumstances, and was entirely accidental. Nonetheless, the general tendency to take on fantastic shapes is definitely geological, in a three-fold sense:

The pink granite, with its changing colors and shapes, was born in the deep melt crucible; its internal structure dates from the cooling and hardening period; water from the sky, however, throughout millennia of erosion, has modeled the external forms.

We let our imagination wander amid the ghostly company of these natural walls and towers. We do not see animals, gnomes, or giants' toys. No, we hear out of the past the ceaseless cracking, groaning, and snapping as fine joints appeared and spread through the still warm mass. Blocks broke off cliffs and fragments crashed into yawning fissures. Then movement ceased. As weathering progressed, the forest invaded the mountain. Mosses and fir trees sent roots into the opening joints, and where rainwater lingered on flat surfaces, the rock was scooped out into basins by slow disintegration.

But the final act of the drama is yet to come. Summer has moved on into September. Having refreshed ourselves at the clear spring below the round white mountain chapel, we step out in the mild sunshine onto the flat top of a cliff that rises above the forest. This crag is made of dense, smooth, purple porphyry, and not of granite. It is one of the towers in a natural, upright wall running straight from the ridge across hills and through valleys into the lowland. In places the wall is deeply fragmented. The porphyry rises vertically out of one of the cracks in the granite into which it once intruded from below, this re-enforcing, as with ribs, the more friable granite. For a good hour's walk farther to the west, the boundary of the granite runs parallel with the porphyry dike. Still farther to the west there are two or three similar dikes. These walls are not vertical, but inclined toward the center of the granite dome. The nearer we approach the granite boundary, the gentler is the angle of inclination of the dikes. They form another fan structure like the vastly more numerous joints of the granite. They are larger, over twenty feet thick, many miles long, and occur hundreds of yards apart.

By the time the granite dome was split open, it must already have been very thick and strongly arched. We can calculate the curvature of the arch from the opening of the fan structure and the thickness of the individual dikes. During the progressive up-doming, the fissures split open more widely and filled with melt. This later melt gradually became more basic in composition, and finally ended up as a dark basaltic rock. The black bands are beautiful to see as they run through the almost white granite. With their formation the granitic drama has come to a close. Slowly the epilogue unfolds.

This lovely September morning on the Kraebersteine in the eastern Riesengebirge, for the first time the whole enormous structure of the range stands out light and clear and understandable before us. We see two images simultaneously in one broad space. One is the actual landscape, through which we climbed two months ago, one of the most beautiful spots on earth. Towering beyond is the bald ridge, terminating at the left in the cone of the Schneekoppe. The sky is deep blue, the earth is green, but with autumn's first shimmer of red. At the same time we see another, imaginary landscape, a phantom structure of pink granite, weaving through and arching over the actual one. It is a colossal structure of the granitic pluton, one of the great examples of the extrusions of the earth's hot interior into the cold top layer of our living space. Dangerous was the unfolding of its tremendous powers; frightful the way it rose irresistibly, carrying the whole vast load of its cover of layered rock before it and expanding it to the breaking point. At any moment there might have been an eruption; tremendous masses of melt might have spewed up, burning and burying the ground beneath.

However, nothing of the sort happened. The cold of the earth's surface and of the atmosphere brought the upward movement to a halt. It also prompted the growth of the first crystals and schlieren, dictating their orientation in the flow of the melt and hardening the melt into rock. During this process the granite burst apart and fractured. Had this not happened, separation of the melts into granite and porphyry dikes could not be interpreted. Everything we see is the result of the subjugation of the internal forces of the earth to external conditions. The super-position of the present landscape has put its seal on the slow and tenacious conquest of the fearsome subterranean world by the world of sunlight, clear water, and cool air.[1]

[1] A scientific description of these phenomena is found in Hans Cloos's **Einfuhrung in die tektonische Behandlung magmatischer Erscheinungen,** vol. 1, "Das Riesengebirge in Schlesien" (Borntrager, Berlin, 1925). See Figs. 17 and 18.

4

Still In The Making

JOSEPH WOOD KRUTCH

*A rare geologic showplace, Grand Canyon is a mile-deep monu-
ment to slow but persistent erosion by the Colorado River where
it slices through the great Kaibab uplift. Many chapters of earth
history can be read in the magnificent receding cliffs and slopes
on colorful Paleozoic sedimentary rocks, and in the steep, somber
inner gorge cut deep into Precambrian igneous and metamorphic
rocks. Joseph Krutch is not a professional geologist, but this may
be an advantage for us. A distinguished English teacher, drama
critic, and a reflective naturalist, Joseph Krutch gives an excellent
description and poses some stimulating questions.*

When did all this happen? When did the land begin to rise and the
river to cut downward? How long have at least the beginnings of
the Canyon been there?

No one knows exactly. Estimates have varied all the way from
one to several million years. But on any scale except the human
the discrepancy is not very great, and today the one-million-year
estimate is generally regarded as untenable. A million or seven
million years is a short time as geology goes and not so very long
ago. In any case, the Canyon is what has been described as "a
youthful geomorphic feature." In general valleys and canyons are
likely to be the youngest grand features of any landscape except
those for which recent volcanic action is responsible.

Take even the discredited million-year estimate, and that
would mean that the beginnings of the Canyon go back to about

the time of the earliest half-man. Take the seven million, and it would already have been imposing by the time the first recognizable human existed anywhere. But in either case, the presence or absence of the nearly human would be the only really major difference so far as life on earth is concerned. Seven million years ago mammals were already becoming dominant, and flowering plants were already more successful than the fernlike and horsetail-like vegetation of earlier times. Seven million years ago there may have already been pony-sized horses on the American plains, though they were later to disappear and the horse was re-introduced by Europeans. A million years ago one of the great ice ages was about to begin and the woolly mammoth to flourish.

At whichever time the cutting may have begun, it was long after the last of the giant reptiles had perished and after his hardened footprints had already been buried beneath hundreds of feet of the shale and sandstone which had to be washed away before the footprints were again exposed. They can still be seen not many miles away. And as the river sawed slowly through the rising strata, its deepening walls exposed again to sight older and older formations going back more and more millions of years until, finally, they add up to more than a billion. Nowhere on the earth's surface (except possibly on the Canadian Shield) are to be seen rocks older than those which form the sides of the Canyon's inner gorge.

What has all this to do with the beauty of the Canyon or with the peace and quiet and solitude to be found on its rim? Some would answer "nothing," and for them that answer is perhaps correct. But it is not the only one. Spaciousness has a great deal to do with the sense of peace and quiet and solitude, and spaciousness can be temporal as well as dimensional. Seated at any point on the rim, I look up and down as well as east and west, and the vista is one of the most extensive ever vouchsafed to man. But I am also at a point in time as well as in space. The one vista is as grandiose as the other. I am small and alone in the middle of these great distances, vertical as well as horizontal. But the gulf of time over which I am poised is inconceivably more vast and much more dizzying to peer into.

There is also another, less intangible reason why even as a spectacle the Canyon is absorbing almost in direct proportion to one's understanding of its structure. The fantastic—and at first sight it seems *merely* fantastic—is only momentarily arresting. Nothing that is without rhyme or reason can hold the attention for long. Hence the Canyon takes a firmer and firmer grip as the logic behind the seem-

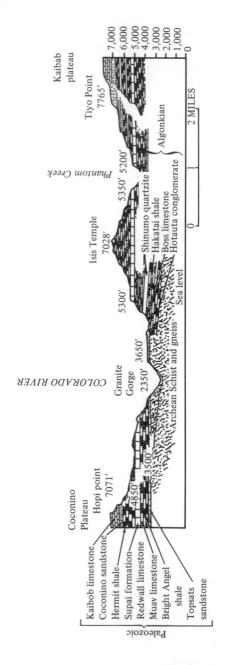

ingly illogical begins to reveal itself. Why, one begins to ask, these varying colors, these oddly sculptured pinnacles, these walls, slopes and terraces sometimes sloping steeply, sometimes lying almost flat in broad plateaus and sometimes dropping vertically down?

The first few hundred feet below the rim are nearly vertical; next comes a gentle slope broken by towers and turrets; then, after another vertical cliff, a broad, almost level plateau; and finally, the sheer drop into the inner gorge, at the bottom of which the river races and foams over rapids and shallows. The schematized outline is shown in the cross section.

And the most obvious questions are simply these: Why is it so much wider at the top than at the bottom? Why is there in some places an inner gorge into which the river fits snugly and which is not very much wider at the top than at the bottom? Surely the river was never ten miles wide and surely it has not gone on shrinking progressively as it cut deeper and deeper.

Of course not. All valleys widen as time goes on. In fact, that is why most rivers make valleys, not canyons. But if most rivers make valleys, then why did the Colorado make a canyon, and why has it obviously widened so much more at the top than toward the bottom?

To the first of these questions, "Why a canyon and not a valley?" the answer is that special conditions other than the slowly rising land existed here. River courses usually become valleys because, as the courses cut downward, water, frost, and the other forces of erosion break down the sides so rapidly that the valley becomes wide faster than it grows deep and, other things being equal, the less the cutting force of the stream, the wider the valley will be in proportion to its depth.

To make imposing canyons you need a considerable river carrying a large amount of abrasive material. But that river must flow through an arid country where the breaking down and widening of the sides will take place more slowly in proportion to the downward cut than it does in regions of normal rainfall. Because the rainfall in the West is so much less than in most other parts of the United States, the West is a country of canyons, great and small, as the East is a country of valleys. And because all the conditions for canyon-making were realized more extravagantly by the Colorado River and the region through which it ran than anywhere else on the globe, its canyon is the most triumphant example of what a river can do. There was the rising land; the swift, sand-filled river; and the arid country.

But why the deep, steep-sided V in the inner gorge in the Bright Angel area and the sudden opening out at its rim, so that the gorge makes almost a canyon within a canyon? Of course, the inner gorge was more recently cut, but not much more recently than the broad plateau which seems to separate it from the higher, wider portions might suggest. The answer, as we shall see when we get down there, is that the gorge is cut in a different, much harder, much more resistant rock. The plateau and all the successive strata above it are sedimentary rocks laid down as sand, mud and lime either under water or on the surface of some ancient desert. But the rocks of the inner gorge are black, terribly hard, and so ancient that their earliest history is largely a matter of conjecture. Molten rock forced up through cracks from below has made great vertical seams of granite, but they are young by comparison with the older mass which goes back to that most ancient of all times, called the Archean. Undoubtedly, parts of it, too, were sediment in some past, a billion years ago, but is has been so heated and compressed and torn that its aboriginal character has been lost and it is very hard. The river must have cut very slowly through it, and weather has been able to affect very little its nearly vertical walls. Had all the rocks of the Canyon been equally hard, it might be now much less deep but much more narrow than it is.

Every cliff and terrace and pinnacle above the gorge is the result of differential weathering—the flats where some homogeneous stone has worn evenly away, the cliffs where something more resistant has stood boldly up, the pinnacles and mesas often the result of some cap of hard stone which has protected softer layers underneath, though they have been washed away everywhere else, thus leaving a pillar or a table still standing because protected from above. Or, if you prefer more technical language, a geologist will tell you that the plateaus are "geomorphically homogeneous" while "the stepped topography is due to the fortuitous alternation of beds having widely different resistance to erosion." He will add that drainage systems flowing down the sides have increased the variety with side canyons and amphitheaters; also that the varying colors which give to the exposed walls a sort of tuttifrutti appearance result from the equally fortuitous circumstance that there was "an alternation of light colored beds and dark colored beds, with the striking red beds in intermediate position."

When one has been talking in terms of millions of years, it is difficult to realize that the Canyon is not finished, that it is not made but in the making. Since the river still runs swift and still carries a tremendous load of cutting material it must still be slowly wearing away the very hard Archean rock over which it runs—though unless the land is still rising (as it may be), it will cut more and more slowly as the gradient diminishes. On the other hand, the Canyon is getting wider much more rapidly that it is getting deeper, and the widening, unlike the deepening, is visually evident. Many of the large rock masses which form the promontories upon which visitors walk out for the best views into the depths are obviously separating from the rim by large cracks. The timid sometimes refuse for that reason to trust themselves to the rocks, though many will no doubt still be there hundreds of years hence. Other, smaller, boulders are more precariously attached, and looking over the rim one may see where still others broke loose, fell hundreds of feet down, rolled a hundred more along some slope, and finally came temporarily to rest—temporarily because it is inevitable that the new support will someday be washed away from under them.

Even more striking are the still fresh scars on the sides of the vertical cliffs where great slabs, many tons of weight, have sloughed off, crashed into fragments where they struck, and then poured a stream of debris across terraces and down new cliffs. Inevitably, Bright Angel Lodge, if it stands long enough, will someday tumble into the chasm which now opens perhaps a hundred feet from its edge; the lodge on the north rim will fall even sooner, because it is closer to the disintegrating rim.

The cause of the widening, operative wherever any canyon or valley exists, is running or falling water, wind, the expanding roots of plants which get a foothold in crevasses and, where the weather is cold enough, the expansive force of freezing water which, like the roots, widens the cracks in which it gathers. The geologist, Edwin D. McKee, who has closely studied the process at Grand Canyon, thinks that the running and falling water is here very much the most important factor—especially the water which falls in the cloudbursts of summer and that which runs over the edge during the spring melting of the snow. At both such times sheets of water pour over the rim, remorselessly nibbling away limestone and sandstone, and shale, most of which were laid down under the same element that will reduce them ultimately to sand again.

How fast is this happening? No one, so far as I know, has attempted to measure or even to estimate the rate at which the average distance between the two rims is increasing. But at least it is not so very slow as earth changes are measured. Major rockfalls, not to mention the minor ones, are relatively frequent. Some years ago, Emory Kolb, who had spent ten years on the rim, could report seven major falls within sight of his house, besides many others which could be heard. One, on the north rim just opposite Grand Canyon Village and plainly visible from there, was also audible across the ten airline miles between Village and opposite rim. During a thunderstorm in December, 1932, a great promontory just west of the Kolb studio dropped as a single mass and came to rest on the Supai formation more than a thousand feet below the rim. On a recent visit I noticed a huge fresh scar where a tremendous block which must have weighed many tons had broken off in 1954 from near the top of the Coconino sandstone (i.e., some five hundred feet below the rim). The major part of it lies now in fragments about three hundred feet lower, while a diminishing cascade of smaller and smaller fragments stream several hundred feet still farther down.

If the earth's crust in this region lies quiet for a time long enough, Grand Canyon will become a wide valley and finally, if time is still longer, a flat plain, all the successive layers of stone washed away, leaving the river to meander again as it once did—like an over-sized brook across an oversized meadow.

Improbable? Already hundreds of feet of rock, laid down before the plateau was lifted to its present height, have disappeared completely, though they are still to be found on higher ground not far away. Time and time again in the earth's history "geomorphic features" more massive than the Grand Canyon have been leveled and obliterated only to have equally imposing features built again, either because the surface was forced up or because lava streams welled up from the earth's bowels.

As a matter of fact, this sort of thing has happened twice right where the Canyon now lies. To see the evidence of that we need to descend the mile below the rim and meet some of the oldest rocks face to face. But that journey had best be postponed for a while.

Meanwhile, a good way to dispel the human-aid delusion and the false sense of scale upon which it is founded is to probe the depth gingerly by strolling a mile and a half down the least used of the two principal trails which lead from the south rim to the bottom—remembering as you go that though you may stroll down the bottom, you cannot stroll back up again. Such a mile and a half's probe will carry you only about eight hundred feet in vertical distance, but that is far enough to make the wall at one's back tower as high as all but the tallest skyscrapers, while the depths have not been brought detectably nearer. The broad Tonto Plateau, almost three thousand feet below the rim, looks just as far away as it ever did, and the bottom of the gorge is still usually invisible.

But though you seem to have made no progress toward plumbing the depths, you have already passed through quite a variety of changing scenes. For one thing, it is noticeably warmer, since to go eight hundred feet down is equivalent to going nearly five hundred miles southward, and the vegetation is noticeably different. So, too, are the color and the texture of the rocks and the shapes into which they have been sculptured. At first one zigzagged down two almost vertical walls, of grayish-white limestone separated by weak red sandstone. At about five hundred feet is sandstone of a slightly different color and so obviously different a texture that the line which marks the division between the two is clearly visible. This third layer is not quite so thick as the first two combined—something more than three hundred instead of something more than five hundred feet—and at just about the end of the mile-and-a-half walk it also suddenly ends where it rests upon a wall of red stone more startlingly different from either of the three previous walls than they are from one another.

The tops of the first mesas and buttes into which certain other rocks have been formed are mostly still below, and for some distance yet the steep wall down which one has been climbing is a wall and no more—sometimes almost perfectly vertical and everywhere cleanly rather than fantastically cut. Your walk will end near the top of this fourth layer, and on the exposed surface of a slab just by the side of the trail you will see very plainly impressed the neat double row of small footprints left by some four-footed creature when the rock was sand.

Any geologist will tell you that when you stood at the rim you were standing upon limestone laid down at the bottom of a shallow sea during the Permian period or, according to recently revised estimates, about 200,000,000 years ago; that your descent of some eight hundred feet had carried you back only a few million years farther without taking you beyond the limits of the Permian; and that these prints were made by some beast—perhaps reptilian, perhaps amphibian—who passed by there before the dinosaurs had got a start. They are real footprints in the sands of time; by comparison, the most enduring metaphorical one left by the most ambitious and most successful of men are written in water.

Suppose there had been no geologist conveniently at our elbow. What could a reasonably observant person see for himself, what sort of notions could he form of the meaning of what he saw; what tentative theories advance of the how and the why?

This is the sort of question I have frequently asked of myself when faced with some natural phenomena I could glibly explain on the basis of what other men had learned. And the answer I am compelled to give is usually sufficiently humiliating to myself. I would not know much about the world I live in had I been compelled to depend upon my own observations. The only consolation is that most men would have to admit that, and even the best men have seldom advanced knowledge more than a short step.

Nevertheless, we may, I think, pay ourselves the compliment of assuming that our short walk would have taught us something. The grandness of the scale, which is obvious as soon as you get into the Canyon instead of merely looking at it, would have disposed of the notion that man had had anything to do with the formation, and I like to think that the theory of a great crack in the cooling surface of a molten earth would also have been disposed of by the obvious fact that the rock wall is not homogeneous and that the different kinds of stone are obviously in layers, one on top of the other.

In most places on earth exposed rock strata are exposed only to a shallow depth and are also so tilted, broken or twisted that their character does not thrust itself upon the attention. But at Grand Canyon they have lain so quietly while they were sliced through like a layer cake that it seems almost as though they had been provided for the special purpose of demonstrating to an unobservant mankind that one great class of rocks are not the result of the cooling of a molten mass but that such rocks were, as the geologist says, "laid down," often one on top of the other. As the distinguished geologist, Charles Schuchert, once wrote on the Canyon, "Such a geological insight into the structure of the earth's outer shell is nowhere else to be had."

Just possibly I might have noticed for myself that embedded in the topmost layer are fragments which upon close examination are obviously bits of sea shell and of coral, and if I had noticed that, I would probably have wondered, as men wondered for so many years after fossils had become quite well known, how the devil they got there. But I doubt very seriously that I would have noticed a difference between the texture of the upper two and the third layers which are otherwise so similar—a difference obvious enough when pointed out and consisting in the fact that whereas the topmost formations are composed of thin layers separated by almost perfectly horizontal lines, similar lines in the layer below run in sloping curves which seem to outline subsequently buried humps—so that it is, as the geologists say, "cross-bedded." And even if I had noticed that, I am pretty sure the explanation, again convincing enough once it has been suggested, would not have occurred to me. The top two layers are solidified lime which fell quietly to the bottom of quiet seas and lay there undisturbed; the next layer is of wind-blown sand, the outline of whose dunes, later covered by more sand, is so clear that from their outline—steeper on one side than on the other— one can see even from which direction the prevailing winds once blew over that desert.

Would I have done any better, or done even so well, with the question raised earlier, not how the rocks got there, but how they had been slashed through? Looking through binoculars at the opposite wall, ten miles away, I could hardly have missed the fact that rock layers in the two walls match. Each conspicuous layer corresponds to a similar layer on the opposite side. The conclusion that they continue one another, or rather that they would if the great gash had not been cut, is irresistible. Would I have assumed some violent catastrophe which cut or broke it in a relatively brief time? Or would the fact, evident enough in the great masses of rock cracked off from the sides and waiting to fall, or the other great masses lying where they have fallen, have suggested that since the whole face of the region is certainly still slowly changing, then the grand features themselves might have resulted from slow change? If I had indeed concluded something of the sort, then would I have gone on to suppose that the river, glimpsed from some point on the rim, had done the job?

Give me the benefit of a large doubt and suppose that I would have. The largest difficulty would still remain. I would still have been faced with the problem of getting the river "up there" where, as a matter of fact, it never had been. Would I have solved that problem, too? I very much doubt it. But it is some consolation to know that much simpler problems in geology did not suggest their obvious answers until a few generations ago, and that details concerning the Canyon are still being discussed as new facts come to light.

5

On Foot And In The Saddle

EDWIN D. MC KEE

Edwin D. McKee writes about the intriguing Grand Canyon region with a purpose different from that of Krutch in the previous selection. No one is better qualified to discuss the Grand Canyon country than McKee, now with the U.S. Geological Survey and previously a naturalist at Grand Canyon National Park, Chairman of the Geology Department at the University of Arizona, and Director of Geologic Research at the Museum of Northern Arizona. From his extensive field work and countless trips into the canyon, he gives a remarkable picture of its natural history as well as practical tips for amateur and professional geologists who would explore the region.

The canyons of semiarid regions are hostile, desolate places. More than any other, the Grand Canyon is hostile and desolate. Such was the opinion of James O. Pattie, who, with his father, was the first American to see this canyon, for he spoke of its walls as "those horrid mountains that forever onward go." Similar thoughts were expressed by Lieutenant J.C. Ives while making a notable exploration of the region in 1857 and 1858, three decades after the Pattie trip, for he wrote in his diary about this "profitless locality" and made the prophecy that "the Colorado River, along her lonely and majestic way, shall be forever unvisited and undisturbed." Even

Clarence Dutton, eminent geologist and masterful describer of scenery, tacitly admits that according to usual standards of beauty the Grand Canyon has for the newcomer forms that seem grotesque, magnitudes that appear coarse, and color esteemed unrefined, immodest, and glaring.

In spite of the overwhelming aspect of the Grand Canyon and the difficulties of climbing except on the approved trails, there surges up in one the same strong urge to climb and to explore that great mountains invoke in all real lovers of the out-of-doors. True, it differs from mountain climbing in many respects and is, in a physical sense, the reverse. But it holds as reward that same satisfying feeling of conquest. Proof is found in the degree of pride attained by the person who has descended the depths and returned, whether on muleback or by foot. He has accomplished something that cannot be achieved except by one who is willing to undergo certain physical discomforts and to exert himself beyond the norm of everyday life.

It is obvious that other factors enter into consideration as reasons for impelling the "out-of-doors" person to descend this canyon. High in importance among these is the desire to experience personally the great changes that are encountered in climate, in vegetation, in animal life, and in scenery. The contrasts are amazing, and simply to be told about them by some individual who went before is no substitute for experiencing them oneself. It is essentially the difference between reading about Europe in a geography book and visiting Europe.

Nor are contrasts of the Canyon to be fully appreciated merely by being alternately on the Canyon rim and at the bottom. If this were true, one might ride serenely down in a tram car and avoid all the bother and annoying experiences, not to mention occasional real hardships, experienced in hiking or riding the trails. But de luxe methods and time-saving devices are not substitutes. It is the little things of nature and of scenery all along the trail that go to make up the experience that is lasting. Actually tram cars have been built and used in connection with construction projects in lower portions of Grand Canyon, but, fortunately, the trams were later removed by government order, for they were not intended to be and never could be substitutes for the experiences of trail trips into the depths.

Problems of the Trail

Canyon climbing, at its best, entails difficulties. Among these is the fact that the uphill portion, the most strenuous part, comes last instead of first as in mountain climbing. Frequently, therefore, the hiker has to start the ascent when already weary. This, of course, applies primarily to those attempting a round trip to the river and back in a single day. Such individuals are also handicapped by reaching the high altitudes, with accompanying rare atmosphere, near the trip's end when fatigue is already in evidence.

A second and equally significant factor in causing Grand Canyon climbing to be difficult, at least in the summer season, is the heat. This affects the mule-rider as well as the hiker. Quite normally, in the Canyon bottom, the thermometer reaches a daily maximum well above one hundred degrees Fahrenheit in the shade during most of late June, July, and August. Although a dry heat and, therefore, not oppressive as it would be under moisture conditions, nevertheless it taxes one's strength and causes much trouble for those not properly equipped and outfitted. The factor of thirst—for there is little or no water along most trails—is also an important consideration, and, even where it is plentiful, water must be used sparingly and with discretion. Sunburn may likewise cause much discomfort, for at high altitudes the sun burns deeply, and in the Canyon bottom it has added potency through reflection from the rock surfaces.

The conditions encountered in traveling a desert canyon like Grand Canyon, whether on foot or by horse, and in camping in it, are so different from those along mountain trails of better-watered regions that the beginner should consider them with care when equipping. I have known of experienced explorers who suffered greatly because of poor calculations in this connection. One biologist, fresh from a month of climbing in the High Sierra, took with him on a hard canyon trip the type of food used in the mountains, and suffered in consequence. Many articles were too dry for him to eat after strenuous climbing in the heat. There was another case of a famous arctic explorer who carried several cans of peanut butter with him on a week's trip in Grand Canyon only to find the sight of this once cherished luxury becoming more and more distasteful after each day of hiking in the hot sun.

Camping in the semidesert canyons of the Southwest has both advantages and disadvantages over camping in most mountain areas. Among the definite assets is an almost complete absence of mosquitoes and of fog or dew. Furthermore, the rainfall is of such a character that it is possible to sleep out under the stars for long periods without using any tent or other shelter, and, even during the time of summer cloudbursts, which come with violence nearly every afternoon, the evaporation is such that much of the ground surface dries off within a few hours.

On the debit side of canyon camping are several factors which one must expect and get used to or else one will derive little pleasure from such trips. These include the dust, which may become very troublesome if the wind blows, the heat of summer, and the flies, which may be bad in local areas. One must, of course, learn to watch out for the cactus, yucca, and other plants that stick, as well as for a few animals such as the scorpion and the rattlesnake. Most persons learn quickly through a few experiences. Fortunately, there are no animals whose poison is apt to be really dangerous. Old-timers in the Arizona deserts like to advise, "Everything that moves can bite or sting; everything that doesn't move sticks or pricks."

Most persons visit the canyons of the Southwest during the summer season. The reason is that most people take their vacations at that time, partly because schools are out, partly because it is the "normal" vacation period, but it does not follow that this is the best season to explore the canyons. On the contrary, the early summer, especially June, is usually the hottest, driest time of the entire year, and late summer is the time of daily cloudbursts. In the writer's opinion, the autumn is the ideal time for planning trips in this area. In the winter it is likely to be cold and stormy, even in the low altitudes of the canyon bottoms, and at higher levels there is almost always snow. The late spring months may be very pleasant in the canyon, but the weather is normally rather uncertain and subject to fluctuation. In October and November, on the other hand, it is usually cool, dry, and delightfully pleasant, and I say this despite the fact that on a recent ten-day pack trip made in October my party encountered rain nearly every day.

How best to equip for trail trips and for camping in the Canyon Country is a subject that may have an unlimited number of answers. The many ramifications of the problem include various situations dependent on the length of trip, the method of travel, the time of year, and the type of trail. Then, too, the human element always enters in the matter of difference of opinion. Despite all of the variable factors it is, nevertheless, possible to make certain broad generalizations and some detailed suggestions that are applicable under most conditions.

Tips for Camp and Trail

All sorts of people wear all sorts of shoes. Some suffer discomfort, some don't. The shoe is perhaps the most important single item of clothing to be considered for canyon trips, yet even among experienced hikers there is far from general agreement as to what is most suitable. Personally, I like shoes that are ankle high. Boots add far too much extra weight, confine the legs unnecessarily, and above all are too hot for comfort in summer weather. Advantages usually ascribed to boots, such as protection from snakes or from dense brush and assistance in wading shallow streams, may be largely discounted in this region because of the general lack of such features. On the other hand, low shoes and sneakers are not desirable except as camp slippers, both because they fail to support the ankles and because they allow easy access of dust, dirt, and small stones. Hobnails are good for travel on the trails but poor for rock work and, considering the usual dry character of the region, are perhaps less desirable than good composition soles.

Insofar as the remainder of one's wardrobe is concerned, the principal governing factor is the season. During much of the year, a broad-brimmed hat is not only desirable but a virtual necessity for sun protection. In this connection one may be reminded of the slogan, "When in Rome, do as the Romans do." The local cowboys and Indians demonstrate a wide variety of styles in hats, all suitable to the conditions, although it might be added that a pith helmet of the type usually ascribed to African big-game hunters, or a Mexican sombrero, will serve equally well.

In the rainy season a raincoat or slicker should be carried by those traveling on horse or mule. For hikers, however, the extra weight and the additional heat when the coat is worn make this impractical. Most persons will agree that it is better to take one's chances when hiking. Usually a rock shelter in the canyon can be found for the duration of a typical short but violent cloudburst. On the other hand, if one does get caught in an August rain, it is apt to feel cool and refreshing, and the rate of evaporation is such that clothes dry off very quickly afterward.

For traveling and camping in late fall and winter, it is well to remember that the weather may be quite cold even in the depths of Grand Canyon—and this applies especially to the nights. Furthermore, not only is the *actual* length of the nights considerable at this season but the *apparent* length is still greater because of the few hours each day during which sunlight can penetrate the lower portions of narrow canyons. With this in mind it may not be amiss to suggest gloves and, for sleeping purposes, an extra pair of wool stockings and wool underwear. One experienced camper I know always carries a small chemical heating pad, which, upon addition of a few spoonfuls of water, generates enough heat to warm feet all night.

The problem of sunburn should never be considered lightly at the high altitudes found in the Canyon Country. In addition to the broad-brimmed hat already recommended, one does well to carry some protective grease and a tube of lipstick (this applies to men as well as women, and colorless varieties of lipstick are on the market especially for the former). Dark glasses are helpful for many people and may save considerable eyestrain.

As for general equipment for a canyon trip, selection depends largely on the amount that can readily be carried by the method of travel used. If one is hiking and must depend entirely on one's own back, naturally the story is very different from when pack animals are used. A few items which have proved of great value on canyon trips but which, for the most part, can be taken only when horses or mules are employed for packing are a short-handled shovel, a canvas water bag, a collapsible canvas bucket, a carbide "miner's" lamp, an air mattress, and a light-weight but sturdy grill. Also tweezers, for pulling out stickers and thorns, and adhesive tape for blisters, often prove very valuable.

Certain of the items listed above may seem superfluous to anyone uninitiated in ways of desert camping, but they have all proved their worth under various conditions. The short-handled shovel, for instance, is not merely a convenience used in setting up camp but may be a necessity in repairing or building trail on a trip off the main path. Also, it is sometimes needed in digging out a water hole in a dry stream bed or wash. The canvas water bag is desirable because it allows some evaporation and therefore keeps water much cooler than does the ordinary canteen. Air mattresses usually improve one's spirits by allowing a good sleep and are especially useful in some parts of any canyon where the rockiness of the terrain makes it difficult to find or dig a really smooth surface to lie on. One geologist I worked with liked his mattress especially because it gave him an extra five minutes each morning to lie on it while waiting for deflation to be complete! A light grill is recommended as a convenience because in many localities wood is scarce, and the usual type of rod for holding pots is difficult to rig.

Water is the big problem of the desert, and the Arizona-Utah canyons are in a general desert region. It is always advisable to carry some water, whether the trip be long or short. It is also desirable to learn the location and character of springs and water holes before starting, wherever this is possible. Many springs and streams in the region contain salt or alkali in sufficient quantity to make the water unpalatable. As the local people say, "The water is wet only on one side," or "The water is so thick with alkali that you must cut it with a knife." Such waters must, of course, be carefully avoided even if one is very thirsty. I once had the unpleasant experience of being very dry and without drinking water while actually following for many hours and frequently wading a shallow stream of this kind in a narrow Utah canyon. At the bottom of Grand Canyon the Colorado River, muddy though it is, furnishes good water for all uses. The sediment which is ever present in this stream is carried principally in suspension and will readily settle when the water is quiet. Good camp practice, therefore, is to fill all available receptacles as soon as camp is started, and within an hour or slightly more clear water will be available for use.

In planning food for the dry Canyon Country, especially during warm weather, two important needs should be kept in mind. One of these is the furnishing of sufficient liquid material and the other is the replacement in the human system of salt lost through perspiration. For the former purpose canned tomatoes, canned fruit, and juices are very good and should constitute an important part of any larder. For supplying salt to the body, bouillon cubes are especially effective and at the same time form a very easily prepared and pleasant drink. Some hikers like the commercial concentrated salt tablets that are made for desert use. While climbing, in order to keep away thirst and avoid dryness, it is often desirable to suck lemons or to chew on strips of dried meat known as "jerky," which is sold in some stores of the region. Application of a few of these little "tricks of the trail" may go a long way toward making the difference between a pleasant and a disagreeable trip.

The Elements Have Their Say

The elements have always constituted a favorite subject of conversation, partly because of their influence on man's enterprises, partly also because of their unpredictable whims. They are constantly changing. Sometimes they play pranks, other times they are even and steady. Always they serve as outer garments that tend to change the very nature and complexion of the features that they clothe. Even as a woman takes on new appearance with change from sports attire to evening gown and from dinner dress to lounging frock, so the scenery varies with each change in the weather.

In the Canyon Country of the Southwest the elements often find expression in extremes. Violent cloudbursts, powerful wind storms, snow, and shimmering heat transform the scenery with suddenness and completeness. It is little wonder that human attempts to reproduce the canyons—especially Grand Canyon—on canvas usually seem inadequate and the results inaccurate. Even the best pictures can hope only to show a single mood, the view of a specific place at a specific time, whereas the scene is constantly changing and will never twice be the same. By skillful use of words, a few great writers have given some suggestion of the constant change and the extremes of scenic grandeur brought about by the elements, but these must be rather vague and generalized, for they cannot attempt to cover all of the variety to be found. Little wonder is it that persons who come back time and time again always find the scenery of the Canyon Country different and intriguing.

One of the really great spectacles to be witnessed in the arid Southwest is a canyon storm. Such storms occur almost daily during late July and August. The day begins clear and cloudless; gradually billowy clouds float in on a constant southwest wind; by noon many parts of the blue sky are hidden, and soon after, here and there, yet seldom for long in one place, the rain pours down with violence, the thunder roars, and the lightning flashes. The evening brings dispersal of the clouds, the stars come forth, and the landscape appears fresh and clean — and then the entire procedure is duplicated in form if not in detail the following day.

From a canyon trail the experience of a summer cloudburst is both awesome and wonderful. The suddenness of arrival, the concentrated power of the rain, and the floods that follow are marvelous to behold. A storm may rage for twenty minutes, then lift as rapidly as it arrived. Every side canyon in the area and every depression near by will contain a stream of liquid mud, where shortly before was only bare rock or dry alluvium. In this land of sparse vegetation a single torrential shower often does more destruction than a season's rainfall on the densely covered slopes of a humid region. And should the sun's rays appear at this time, hundreds of streamlets heading toward the canyon bottoms laden with upland mud will appear silvery.

In most canyons of the Southwest numerous beautiful, though short-lived, waterfalls develop over the sides of cliffs following every heavy downpour. Few of these flow for more than two or three hours following a torrential thunderstorm, but many have a considerable drop. In Zion Canyon, where exceptionally high cliffs have been formed in rocks of uniform hardness, waterfalls making a clear leap of over seven hundred feet develop, and "sliding falls" of two thousand feet form in several places. At Grand Canyon the vertical Redwall cliff, which averages about five hundred feet, and the Coconino cliff, of three hundred feet, are responsible for many large waterfalls during every hard storm. As many as fifty such temporary falls — active at the same time — have been reported from within sight of the floor of Zion Canyon, and similar records of large numbers come from nearly all of the high-walled canyons of the region. In addition to size and number, color is an interesting feature of these canyon waterfalls. Few of them have the foamy white color commonly associated with swift currents. They are heavily charged with sand, mud, and other debris, and so appear red, brown, yellow, and various other colors.

While on the trails or camping in the canyons there is not, for the most part, despite the rather terrifying character of these storms, any great danger from cloudbursts, provided normal precautions are taken. The imprudent person, however, may find them very serious, for they represent the unleashing of tremendous power.

A never-to-be-forgotten experience of the writer occurred partway down the Grand Canyon where he climbed under a large overhanging cliff for protection from rain only to find to his dismay that a great waterfall soon developed in front of that particular spot. The water landed with a terrific splash but a few feet away, bringing with it mud, then pebbles, and finally boulders, and even though these missiles bounced harmlessly outward, their nearness gave a most unpleasant feeling. The real dangers thus lie both in the rush of swollen streams and in the sliding and rolling of rock debris. At times of storm the hiker should stay out of canyon bottoms and stream courses and avoid possible areas of moving rocks among the cliffs.

The effects of cloudbursts are experienced not alone within the canyons. Every year people report having been witness to spectacular "walls of water" moving rapidly down and filling dry stream beds or arroyos, blocking or destroying roadways, and even wrecking automobiles that are unfortunate enough to be in the way. But it is in the canyons, where gradients are mostly greater, and where enclosing walls are more confining, that such actions reach a peak. In August of 1938 a great flood that came down Bright Angel Canyon in the bottom of Grand Canyon was determined to be twenty feet high and rolled boulders up to six feet in diameter along the Canyon floor for half a mile. Such storms are relatively scarce, but nonetheless impressive.

An interesting feature of some desert storms is that in their wake certain areas that are normally dry and desert-like may become alive with toads. These creatures apparently aestivate for long periods in the dry ground and then miraculously come forth with the rain. Little wonder that people speak of "raining frogs"!

Rockfalls and landslides resulting from work of the elements are spectacular features that are occasionally witnessed by travelers of the trail. These slides come not only at times of torrential downpour but also frequently when the snow is melting. Most of them occur where weathering has undermined a cliff to such an extent that a portion of its face suddenly breaks loose and falls forward with the pull of gravity. One sunny February day following a night of snowfall, the writer heard the roars of sixteen sizable slides in the jagged cliffs of Iceberg Canyon near the Arizona-Nevada line. Again one may go for months without noting any. In the Grand Canyon, because of the large range of vision, rockfalls are witnessed not uncommonly, and there the roar of a large one may be heard as far as ten miles away, while the dust is seen rising above the Canyon rim like the ejecta of a volcano.

The winter season is one in which remarkable and rather weird scenic effects are produced in the Canyon Country as a result of snow and misty clouds. As nearly the entire region is formed of essentially horizontal or flat-lying rock strata, little covered with vegetation or debris and weathered into alternations of cliff and slope, snow accentuates the banded character. Slopes are covered with white; the cliffs, which naturally remain bare, are mostly red; the result is a series of red and white bands. Although this bizarre color pattern is developed more or less throughout the region when snows are heavy, it is especially striking as seen in the walls of Grand Canyon.

Winter clouds are also remarkable for the effect that they have on canyon scenery. Often they are misty and travel low. They may appear as little wisps or floating islands; again they may accumulate until they fill an entire canyon, blotting out all features beyond the foreground. Many a visitor to Grand Canyon has been immeasurably disappointed upon arriving on the rim to find the view completely obliterated, but when after a few hours, as is normally the case, clouds begin to move apart, giving vistas of gorgeous scenery and brilliant colors ten or twelve miles away, he obtains one of the greatest thrills imaginable. Corresponding experiences are often obtained along the trails at this season and add much to the richness of one's memories.

Clear atmosphere extends the range of vision over vast distances in the Canyon Region. Much of the area is characterized by intense quiet and solitude during large parts of the year, yet one of the elements is always on the move. It is the wind. Streams of air are constantly twisting and swirling in and about the canyons. Above the Plateau a remarkably regular southwest wind is present at all seasons, but down in the gorges air currents are different. Warm drafts move upward; cascades of cool air go down, constantly fluctuating with time and place. Aviators passing over the rim of Grand Canyon at one thousand feet sometimes find their planes steadily elevated, and clouds may be seen to rise after a storm. Cool air settles toward the Canyon floor until lightened by warmth and expansion, then it rises. Camping along the Colorado River one often is impressed by the regularity and power of the air currents that develop each day at about sunrise and again at sunset, frequently blowing sand along the shore. On the trails, and also on the rims, one frequently encounters "pockets" or drafts of definite warm or cold air.

The "dust devil" or whirlwind is a manifestation of the eccentricities of air currents commonly displayed in the Canyon Country wherever quantities of loose sand or dust have accumulated. These "sand geysers" are caused by wind swirling around and upward with a spiral motion into a funnel-shaped column; they are moving atmospheric vortexes which lift sediment high in the air as spouts or fountains. The Navajo Indians of northern Arizona believe that an evil spirit is within each dust devil and will take pains to avoid or appease these spirits whenever possible. White men will avoid them, too, if they realize the sting that comes with the swirling sand!

Natural History of the Trailside

There are in this world many who would "improve" on nature at every turn. I once knew a rancher who, while among the wonders and beauty of Kanab Canyon, could talk of little but how that region might be blocked off into a magnificent pasture for Brahma steers. I have met engineers who practically dreamed of the great "contribution" they could make to civilization by building a road into Grand Canyon. I have seen the forester who scorned the desert because good forests would not grow there. An amazingly widespread notion is entertained by civilized man that it is his duty to remodel and rearrange all things in nature and to tame all that which is wild.

One of the real pleasures of traveling through most of the canyons in the Southwest is that so little evidence of man's endeavors is found within them. Much of their charm is due to the fact that rocks and plants and animals are still where nature placed them. Each form of life seems perfectly adapted to its environment—the result of ages of trial-and-error experiment in the game of life. On every side are demonstrated ingenious stratagems, contrived in the never-ending struggle for existence. There are reptiles that match the rocks, plants that survive drought because of small leaf surface or large water-storage space, mammals developed to walk easily on loose sand, others that can eke out a livelihood from plants despite spines and thorns. These and many other forms of life meet the special requirements of their local communities and therefore add interest and appeal to the trailside.

With knowledge comes appreciation. A visitor to the Grand Canyon once told me that he found it easier and more entertaining as he climbed the trail to count his steps in terms of geologic formations rather than of miles. He realized that he was literally passing through the ages as he progressed and that his trip took him through strata representing many millions of years in the earth's history. Such thoughts doubtless would have been even more intriguing had he been aware of the many ancient plants and animals buried in the rocks about him and of the stories they had to tell concerning climates and landscapes of the past.

In similar manner the botanist or ecologist obtains a certain feeling of satisfaction as he recognizes his approximate elevation through the appearance of certain plants or groups of plants along the way. He knows, for instance, that the catclaw or Acacia greggii seldom grows above four thousand feet in this latitude, that the lower level of the ponderosa or yellow pine is at about seven thousand feet, and that each of a host of other species will furnish him with information. Likewise, the bird enthusiast, the mammalogist, the student of reptiles will all find satisfaction in the knowledge that enables them to understand and appreciate various features along the trail. But one does not need to be a specialist to derive such pleasures. The amateur natural-history student who observes with care and enthusiasm will find an abundance of drama being enacted all about him on any of the canyon trails.

Glance about that portion of trail located at high elevation—say eight thousand or nine thousand feet—as found at Bryce Canyon or in the Kaibab on the north side of Grand Canyon. Here a person familiar with trails in the mountains of New England or in the lake region of Minnesota will recognize many familiar friends among the plants and animals, but with them will be some new and strange associates. Here one may enjoy the game of trying to recognize the many types of evergreens by their shapes or of testing cross-sections of needles to see whether they are round as in the pine, square as in the spruce, or flat as in the fir. One also may become interested in the groves of quaking aspen, whose white bark suggests from a distance the birch and whose leaves quiver with the slightest breeze, thus justifying the species name of *tremuloides*. If it is early summer these aspens may be causing a local "snow storm" by filling the air and covering the ground with their white, fluffy, cotton-like seeds. If it is autumn, the brilliant golden-yellow leaves may be bringing joy to some color photographer or artist.

In among these trees of higher altitudes are birds and mammals that also bring memories of the northland or of mountain peaks in other regions. The bright red or yellow crossbill, whose name is derived from its peculiar scissor-like beak, adapted to cutting seeds from the spruce cone, is apt to fly past. The noisy Clark's nutcrackers of the jay clan may call attention to their presence by raucous sounds. Red-breasted nuthatches, three-toed woodpeckers, solitaires, and many feathered associates of cool climates bring to mind the fact that these are trails due to be covered deep with snow during a long winter ahead. Here and there a tree trunk chewed by porcupine or the scattered remains of a chipmunk's feast of pine cones are reminders that four-legged forest dwellers also are about, and, if one is lucky, a deer or a spruce squirrel may be seen.

Canyon trails at lower levels in this region traverse what are commonly known as the "pygmy forests." These forests are composed of the small nut pines or pinons and the junipers, together with various shrubs such as the cliff rose, serviceberry, and wild currant. Trees grow well apart, and in most places there is little underbrush. Readily apparent is the interesting fact that in this region one goes down, instead of up, to approach timber line, for the lower limit of the pygmy forest is the beginning of the open desert. Of course, there is an upper timber line, too, but at this latitude it is attained only on the highest mountain peaks—above eleven thousand feet.

A majority of the canyons in the Southwest are at least partly within the belt of pinons and junipers. Many of these canyons have a pygmy forest along their rims and upper slopes but extend downward into the desert zone. The trails passing through such forests contain many features of natural history that are unique to the region and for that reason of especial interest to the visitor from afar. It is here that one finds large flocks of garrulous pinon jays, wearing Prussian blue but no crest, and making a terrific noise as they come picnicking through the woods. Here also one normally encounters groups of plump, short-tailed pygmy nuthatches, hanging upside down or in other acrobatic poses, talking continuously in friendly fashion. The more reserved Rocky Mountain nuthatches, the cheery mountain chickadees, and others that call to mind similar forms throughout the United States are also present among the trees.

Where the canyon walls become rocky and rugged, other types of birds appear abundantly. The canyon wren is especially conspicuous because of its weird song that starts high and comes tripping down the chromatic scale with a great display of exuberance. Frequently, even when the wren itself is too far away to be clearly seen with the unaided eye, its rich, vibrant song carries back and forth among the rocks. Another equally typical sound of the deep, narrow canyons is the coarse croak of the raven. This call echoes between canyon walls where the large, black birds, looking like overgrown crows, fly about singly or in pairs from cliff to cliff. Then there is the startling sound, like a bullet passing near, made by white-throated swifts as they circle and wheel and dart about with terrific speed among the cliffs. It seems almost unbelievable that any living creature can go so fast and yet keep such perfect control as is demonstrated by these long curving wings that sometimes move, first one and then the other, like a man swimming the crawl. It is a thrilling experience of the Canyon Country to watch and hear these birds zip past one's head and "power dive" into the depths below.

Back among the pinons and junipers again, along the ground, many other creatures of strange habit but of more silent ways come to the attention of the observant hiker. On relatively level stretches, especially along canyon rims, a horny lizard with circular body plan and flattened profile, commonly known as the horned toad, makes his abode. The color scheme of his body blends well with the ground about. Disturb him, and he will puff up as though suddenly inflated and may even emit a hissing sound. He would look as ferocious as any reptile of the ancient Mesozoic were he enlarged a few dozen times, yet for the most part he is a mild-mannered, gentle creature.

Not far away the peculiar home of another catcher-of-insects may attract attention. It is a conical-shaped hole in the sand about an inch deep and half again as wide at the top. Actually it is more than a house—it is a trap used by the wily and cunning flesh-eating larva of an innocent adult insect. Down at the bottom of the cone, hidden under loose sand, stays this insect larva—clumsy, covered with spiny hairs, and having mousetrap jaws with which it grabs any hapless ant or similar creature that falls into the pit. Many a fascinating half-hour may be spent in watching the movements of this creature commonly known as the ant lion or doodlebug.

Everywhere along the trail are other fascinating features typical of the pygmy forest area. Usually in spring and summer there are brilliant flowers here and there, varying in type and color with the season. There may be clumps of trumpet-shaped scarlet buglers, and, if there are, hummingbirds are likely to be near by, "treading" the air and whistling shrilly. There may be patches of bluebonnets, the state flower of Texas, or deep blue larkspur, orange mallow, and red or yellow paintbrush scattered about. Later in the season these flowers are largely replaced by others such as the snakeweed and rabbitbrush, and the blues and reds give place to yellows. As in all regions of semiarid climate, the display of trailside flowers varies tremendously with the local rainfall, but under optimum conditions it is almost unbelievably brilliant and beautiful.

So far nothing has been said concerning the main varieties of shrubs that are characteristic of the pinon-juniper belt. There is the cliff rose, which is very common and very beautiful when covered with yellow-centered, cream-colored blossoms, and which, despite a bitter taste that gives it the name of quinine bush, is a favorite food of the deer. It is an evergreen with waxy leaves and, like many other members of the rose family, develops lovely little plumes for transporting its seeds when the wind blows. Even more striking are the larger seed-plumes of a close relative called the Apache plume. Then there is the mountain mahogany of the same family—interesting because the Navajo Indians formerly made a lovely red dye from its roots. Representing the apple family is the serviceberry, very common along many trails and, when in bloom, a mass of white. Its berries are edible. Most beautiful and fragrant of all the wayside shrubs in the area, however, is the syringa or mock orange.

Descending any canyon trail toward the lower limit of the evergreens, one encounters many changes in the plant and animal life. Lizards are found in greater abundance, especially the small-scaled Utas, while some of the larger types, such as the brilliant orange or green collared lizards, make an appearance. The birds are semidesert types, including the rock wren, the black-throated desert sparrow, and the sage thrasher. Mammals are seen less frequently here than in the wooded areas above because most forms are nocturnal, but abundant evidence of their presence usually can be found. Under rock ledges or beneath large cactus plants huge piles of sticks, branches, and various kinds of rubbish testify to the industry of the so-called "pack rat," while on sandy surfaces the tracks of silky-haired, jumping pocket mice or of trim little white-footed mice are common. The mammal most likely to be seen and almost sure to be heard any time any day is the rock squirrel, which rushes about among the crags whistling lustily.

In the zone of dwarfed pinons and junipers and of increased desert conditions, various types of plants peculiarly adapted to the dry environment appear in great abundance. Among these is the century plant or mescal, which forms veritable "forests" in places. This interesting plant grows for many years with only a clump of sharp, saw-toothed leaves above the ground, and then some spring will send up a slender flower-bearing stalk, twelve or fifteen feet high, in a matter of days. After the flowers come the seeds, and then the entire plant dies, leaving only a brown hollow spear to wave and rattle like a ghost for months afterward. The tender shoots of these mescals have long been considered a delicacy by various Indians of the region, and circular rock pits in which the plants have been roasted are visible near many trails.

The yucca—swordlike but friendly—is another conspicuous plant common near the "lower timber line." Its leaves form queer-looking clusters about the base of a stalk as do those of the century plant, but they are narrower and without teeth on the sides, while the stalk normally extends upward but a few feet. Its larger blossoms are creamy-white and at once suggest relationship to the lily family. Of all the many plants in this semidesert region, the yucca has been one of the most useful to aboriginal man. Its pods form a food, its leaves are used for basketry, its fibers for thread, and its roots for soap.

If variety is the spice of life, these canyon trails must have an abundance of spice. A number of them, especially those in Zion and Grand Canyon, extend their lower reaches well down into a true desert environment, and there the features of natural history along the way make a remarkable contrast with those seen at higher elevation. Down where spring comes early and plants grow far apart, the vegetation is of a typical desert type. The little spotted skunk and raccoon-like ring-tailed cat leave tracks in the sand and often examine camp sites at night. Now and then a large scorpion with brown body and yellow legs is found beneath a rock, or a centipede crawls by—reason enough why the camper should shake out his shoe before putting it on in the morning. But by and large there is little to worry about and nothing to fear among the queer denizens of this lower realm, and there is much of beauty and of interest to be found. Large bluish pipe-vine swallowtail butterflies, friendly but noisy ash-throated flycatchers, harmless king snakes with black and white stripes, and a host of other creatures soon convince the observant hiker that this desert is far from desolate and lifeless.

Wherever water comes to the surface, or even near it, in this land of drought, an oasis is formed. Like most oases these are regular paradises for all forms of life, and their beauty is enhanced many times by contrast with surrounding dry areas. The largest of the oases in the canyons of the Southwest are those formed along permanent stream courses. The deep-rooted mesquite and the grabbing catclaw of the desert thrive in these places, but they are secondary in prominence to the large, shade-giving cottonwood, to the box elder, and to thickets of willows and arrowweeds. Down below the trees and larger shrubs, in many places, are clumps of that peculiar jointed plant known as horsetail, whose ancient lineage goes back to the great Coal Age; also there are pretty red or yellow monkey flowers and, in especially sheltered localities, the columbine and maidenhair fern. Trails that follow such water bodies offer many a thrill and surprise for the nature lover, and seldom a dull minute is experienced.

Down among the boulders, wherever permanent streams rush by with white foam and swift current, one is almost sure to encounter the little gray dipper or water ouzel. Continually bobbling up and down on a rock ledge, diving under water and coming into full flight as it reaches the surface, or singing beautifully within narrow canyon confines, this little bird seems strangely out of place to those who have always associated him with high mountain regions. But the dipper appears to be just as much at home and to enjoy life as fully in the depths of hot desert canyons, wherever he has rushing water, as in the highest and boldest mountains. One group of geologists working in the depths of Grand Canyon and climbing about with great effort among granite walls noted the ease with which a little water ouzel flew from cliff to cliff and thereupon adopted as a theme song for their expedition, "If I Had the Wings of an Ouzel . . . "

Equally surprising to most people schooled in less arid regions is the discovery that beaver are perfectly at home in many desert canyons of the Southwest. As a matter of fact, the first Americans to visit Grand Canyon—the Patties, father and son—came up the Colorado River in 1826 trapping beaver. Along the permanent streams, wherever the encroachments of man have not forced a retreat, are found trees and logs cut by these busy animals. Occasionally a dam across some stream or a burrow in a mud bank is discovered, though most of these are temporary and difficult to find.

A multitude of other creatures add distinctive flavor to the setting of a desert canyon oasis. There are the ever present red-spotted toads, with low voices, and the canyon tree toads, with shrill, piping calls. The tree toads continually appear in unexpected places, for suction disks on their toes enable them to climb smooth, vertical surfaces. Down in the water are polliwogs, striders, whirligig beetles, and many of the other interesting animals typical of most bodies of fresh water throughout the country, while in the air may be dragonflies, caddis insects, and damsel flies. The trail through the oasis is a source of unending surprises.

Rocks and Fossils Along the Trail

With wild life and vegetation any facts observed along one trail at a particular altitude apply almost equally well to any other trail located at the same elevation; in contrast, most geologic features do not reflect a control by altitude but by the age and origin of particular rock formations. Accordingly, generalizations concerning the geology of Grand Canyon may be made for all the various trails in that canyon, as every trail passes through the same sequence of formations in the descent, but such generalizations will not fit other trails in the area. On the other hand, canyons of the Navajo country, including the Painted Desert, and of the Zion portion of Utah, contain different rocks, formed during the age of dinosaurs, and other descriptions are necessary for geologic features along the trails in those areas.

In few other parts of the world is the scenery and topography so closely related to the character and structure of the underlying strata as in the canyon region of northern Arizona and southern Utah. The trails of necessity alternate between very steep and more moderate grades according to the relative resistance of the essentially horizontal strata through which they pass. The very fact that, in ascending or descending, one passes through a series of formations, means that with each change a new set of geologic features is encountered. Thus, at one level may be seen deposits and fossils formed under a sea, while at another are relics of ancient desert dunes or of the floodplain of some long-vanished river.

With this in mind we start the descent of Grand Canyon on one of the many trails. Observing closely the rocks at or near the top, we are almost sure to find many sea shells of various types, possibly also some horn corals of conical shape, and other traces of marine life—all petrified yet preserving fine details of ornamentation. On the Kaibab and Bright Angel trails metal labels calling attention to these fossils and their significance have been placed by the National Park Service, but good specimens may be found along nearly all the trails. At some levels, also, great numbers of concretionary spheres of a gray or brown color appear in the limestone, and careful examination shows that these are masses of flint or chert with sponges in the centers. It is a fascinating pastime to see how many varieties of ancient animal remains can be located in this formation and to try to reconstruct in one's mind a picture of the region at that ancient time when the sea washed over its surface.

Farther down in the Grand Canyon one can scarcely avoid noticing the sheer, white sandstone cliff known as the Coconino, for through this formation most of the trails attain maximum steepness. The layers of sandstone within this cliff are seen to slope steeply in several directions and to form great wedges in cross section; these are believed to represent deposits of the lee slopes of ancient dunes, and on them are found, in many places, the tracks left by primitive five-toed reptiles. Here are "footprints on the sands of time." Some were made, apparently, by animals not larger than small lizards, others by creatures having big feet and a stride nearly a yard in length. Some indicate short-limbed, heavy, wide-bodied animals, while others suggest those having long slender limbs and narrow bodies. Many of the tracks are beautifully distinct, and in places they can be traced for considerable distances to where they pass under a covering stratum. These footprints have been found throughout the length of Grand Canyon and along several trails have been uncovered and left exposed in place for visitors to see.

Many another milestone in the record of earth history is encountered as one continues to descend into the depths of Grand

Canyon. At the level of the red shales and sandstones are found the impressions of delicate plants, including ferns that look much like modern types, small cone-bearing trees, and queer jointed-stem species, marking a time of great delta accumulation in this area. At a still lower place one encounters the record in limestone of an inland sea—a sea much older than the one represented by the rim rock referred to before. Some portions of it are seen to be composed entirely of the stems of marine animals known as sea lilies or crinoids, and in other places delicate, lacy skeletons of sea mosses are easy to find on the rock surfaces near the trails. It is fascinating to wander along a canyon path and observe, step by step, unfolding pages in the great panorama of earth history. One is finally forced to the realization that one is passing not only through space but also through time. One is traveling back through the ages. Each fifty or one hundred feet that a traveler descends carries him down in the record of earth history some thousands or even millions of years.

Interesting features to be seen at the many different geologic "horizons" in Grand Canyon are almost numberless. In one place may be found ripple marks in sandstone, looking like a washboard; in another place appear the tiny craters of rain drops impressed in what was mud some hundreds of millions of years ago; elsewhere are seen the water-rounded pebbles of an ancient beach. Even where fossils are absent and the original sediments have been completely changed by terrific heat and pressure, as in the rocks of Archean age that form a large part of the bottom of Grand Canyon, there is no less a wealth of interesting features to be seen. Many of these features are difficult to interpret because of the complexity of processes that have brought about their development, but such things as the varied types of minerals, dikes formed of once-molten granite, and examples of rock layers obviously squeezed and tightly folded give the observer at least a faint concept of the great crustal revolutions and mountain-making movements through which these rocks have gone.

In the colorful canyons of southern Utah and of the Navajo country still other and different glimpses into the spectacular history of the earth may be had. Almost everywhere, in this land of little rain and much erosion, the rock surfaces are well exposed, and the record is clear. It is the record of the great "Age of Reptiles," which, although far younger than that of the eras represented in the walls of Grand Canyon, is still unbelievably old when measured by human standards. In Zion, Tsegi, and numerous other canyons, the lowest strata are those of the Painted Desert, soft in character and brilliant in color, containing much petrified wood and many fragments of ancient amphibian or reptilian life. It is fun to travel the gentle slopes and small hills of this marl or limy clay and find teeth or fragments of bony armor plates from some ancient, bizarre creature, or to stumble upon the log of an extinct type of coniferous tree, either with its cell structure still preserved in delicate detail or with the interior beautifully colored with red, yellow, and brown chalcedony.

At somewhat higher levels in the sequence of rocks through which many of the canyons in the plateau area have been carved, there are red mudstones and lagoonal limestones containing the footprints of mighty dinosaurs. The three-toed tracks, some of them over a foot in length, are found in many places in the region, including a small canyon near Kanab, Utah, Zion Canyon, Dinosaur Canyon, and several localities near Rainbow Bridge. These tracks are impressive when well preserved and when exposed so that a series of evenly spaced prints is in view. Unfortunately, in many localities, they have broken up with weathering or have been stolen, and there is a real need for the development of adequate methods of preservation.

Canyon walls above the dinosaur footprint level are in most places formed of a remarkably uniform sandstone, most red but locally white, known as the Navajo Sandstone. Aside from the color and great cliff-forming character of this sandstone, probably the most noticeable thing about it is the fantastic pattern of swirls and curves formed by the individual layers that make up the rock—it is known to the geologist as cross-bedding. Nearly every visitor to Zion Canyon remarks about this spectacular feature, and the trails in Zion pass through many areas where it is beautifully exposed. In the walls of the Tsegi Canyons where the great cliff dwellings of Betatakin and Keet Seel are located, in Bridge Creek Canyon containing Rainbow Bridge, in Escalante, and many other places, these amazing structures are beautifully exhibited. They are explained by most geologists as the deposits of ancient sand dunes formed when this region was a great desert.

But the records that represent pages of historical geology are by no means the only geologic features of interest to be seen along the trails. There are many others—features formed long since the rock layers themselves and recent as compared with the fossils in the rocks, yet commanding attention and creating interest. The canyons themselves are the products of erosion that is going on even at this moment, and graphic illustrations of erosive processes on a smaller scale, found at every turn along the trails, belong to this category. There are natural bridges and great stone arches in various stages of development in many of the sandstones, especially the Navajo. There are fantastic pinnacles such as those that make Bryce Canyon justly famous and symmetrical buttes and mesas of the type common in Grand Canyon. Then, too, there are exhibits of huge boulders and rounded pebbles along the Colorado River and other permanent streams—the playthings of the water that have been rolled about but at the moment stand idle; tools that have shared in carving the great gorges.

In many places along the trails mute evidence of great crustal disturbances is clearly visible. Here one sees rock strata bent or folded; at another point testimony that the formations were faulted is given in the form of a clear break through the strata or in rock faces polished smooth or grooved when one surface moved past the other. In the Grand Canyon nearly every trail has been made possible because of great faults that have developed lines of weakness through the strata and ultimately resulted in natural passageways through otherwise sheer cliffs. Their effects on the scenery along the trails and the character of these trails is everywhere striking.

John Wesley Powell with a Paiute Indian chief. Smithsonian Institution, photograph by J. K. Hillers.

2

GEOLOGY IS HUMAN KNOWLEDGE

6

The History
And Development Of Geology

BRAINERD MEARS, JR.

From views of the geologic world around us, we now turn inward to examine the nature of geologic endeavor. The next selections present geology as physical adventure and demanding work, moments and methods of discovery, changing and debated ideas. The following synopsis of the history of geology gives a framework for relating various personalities to the state of the science in their times.

The Early Period

ANCIENT TIMES

From Antiquity to about 1800 A.D., there was no organized geology. Worthwhile observations and ideas were intermingled with misinformation and wild speculations. The ancient Greeks did contribute, however, to an understanding of the earth as a planet. Pythagoras (6th Century B.C.) believed the earth was a sphere, and Aristotle (4th Century B.C.) assembled proofs of sphericity. Eratosthenes of Alexandria (3rd Century B.C.) calculated the approximate size of the earth. And, although Greek astronomy was dominated by the earth-centered view represented by the Ptolmaic system, Pythagorus had proposed that the earth and planets revolve around the sun.

In more strictly geologic matters, the Greeks made some creditable observations but had little understanding of external and internal geologic processes. Aristotle, though not the only ancient writer on geologic matters, held characteristic views which were later revived and proclaimed authoritative in Christian theology. The origin of streams he attributed to condensation of vapors, or air, in caverns within mountains, which dripped like sponges. Earthquakes resulted from winds inside the earth; when the internal air was shocked and violently separated into minute fragments, it generated the heat for volcanism.

Aristotle, and others, did recognize that the sea covers former dry land, and predicted that land would emerge from the sea. He realized that streams carried material eroded from the land and deposited it in flood plains and the sea. Some ancients, such as Herodotus (5th Century B.C.), interpreted marine fossils far inland as animal remains indicating former submergence beneath the sea.

The Romans tended to follow Greek ideas. Strabo (54 B.C.-25 A.D.), who travelled widely, observed alluvial deposits and deltas, and recognized the dormant Monte Somma, Vesuvius, as a volcano before its 79 A.D. eruption. Pliny the Elder (23-79 A.D.), a victim of the Vesuvius eruption, wrote a natural history: a compendium of earthquakes and volcanic eruptions of historic times, that also described the plants, animals, and minerals then known (some real and some fictional). The Romans largely compiled observations, and were less interested in speculation than the Greeks.

The Greeks and Romans advanced from explaining geologic features as made by super-human powers, their gods, to the seeking of natural causes—a major step forward. Aristotle said that general principles must arise from and conform to facts, a modern-sounding statement of scientific method. But he adopted speculative theories on inadequate evidence and fitted facts to theories which had not been adequately tested. In general, the Greeks were more interested in logical reasoning from hypotheses and postulates, which they used successfully in deductive mathematics, than in the careful collection of the many observations needed to create and test geological concepts.

MEDIEVAL AND RENNAISANCE TIMES

After the fall of the western Roman Empire, geologic studies in Europe were eclipsed for about a thousand years. In the middle east, Avicenna (973-1037 A.D.), a translator of Aristotle, reflected the Arabic enlightenment during Europe's Dark Ages. He had some concept of geologic time, and he differentiated between mountains of uplift and mountains of erosion. However, he considered fossils as unsuccessful attempts of natural plastic forces, *vis plastica,* to create organic things out of inorganic matter. When his translated works reached Europe, they strongly influenced the revival of thinking based on Aristotle.

The Renaissance and Reformation were marked by an increase in individual observation; a Protestant revolt against the dogmatic theology based on Aristotle, and an unfortunate emphasis on a literal interpretation of the biblical Genesis which had been absent from earlier Catholic teachings.

Leonardo da Vinci (1452-1510), a genius in art, architecture, and engineering, was also a naturalist with an exceptional understanding of the earth. He clearly believed that fossils were once-living organisms, and not relics of the Biblical flood. He appreciated the shifting relations of sea and land and the nature of erosion, transport, and deposition. An original thinker whose mind was not shackled by traditional concepts, Leonardo relied on his own observations and clear reasoning. Unfortunately his advanced geologic notions were lost: he was a man far ahead of his time.

The controversy about whether fossils were sports of nature, unusual concretions, peculiar fermentations, or of organic origin continued for over 200 years after the time of da Vinci. Johannes Beringer, Professor of Natural History at the University of Wurzburg, devoutly believed in Avicenna's *vis plastica.* An ardent collector of fossils, he had little success until his students were enlisted to help. Thereafter an amazing collection was made: queer-looking insects, birds, lizards, frogs in the act of breeding and symbols of the sun and moon in Babylonian and Hebraic script. In 1726 Beringer published a book with many fine illustrations of the fossils. Later when a stone with his name on it appeared, poor Beringer realized he was the victim of a hoax. His students had molded, baked, and planted most of the illustrated "fossils".

Pleased by it all was Johann Scheuchzer (1672-1733), medical doctor in Zurich, who scorned *vis plastica.* He knew fossils were organic, relics of Noah's Flood, Scheuchzer collected many fossil bones including a prize skull and skeleton which he interpreted as the remains of one of the "accursed generation of men" drowned in the Flood. Scheuchzer died famous: the believers in plastic forces were vanquished, and his view that fossils were organic was established. His "sinner's" skeleton; however, was later established as undoubtedly that of a Miocene salamander.

Geology Becomes a Science

A TRUE GEOLOGIST

The beginnings of modern geology are seen in the 18th Century work of Jean Etienne Guettard (1715-1786), although the immediate impact of his work was lost in the sheer volume and cumbersome style of his writings. Guettard emphasized extensive field observation with a minimum of speculative theory. Moreover, he was the first to plot geologic information on maps by using symbols to show the distribution of minerals, rocks, and fossils in the Paris Basin. In paleontology, his detailed notes and drawings clearly demonstrated that the fossils he studied originated as living organisms in the sea. Though far from the first to consider the matter, he presented a most complete picture of the geologic forces tending to level the earth's crust, combining stream erosion and deposition with the activity of the sea in attacking the lands and creating sediments.

In the Auvergne district of southern France, Guettard recognized the evidence of former volcanic activity, although he had never seen an active volcano (and thought they resulted from burning coal and petroleum). However, he attributed polygonal basalts in the district to crystallization from water instead of volcanic activity, thus fathering two contending schools.

VULCANISTS AND NEPTUNISTS

The vehement Vulcanist-Neptunist debate of the late 18th and early 19th Centuries was a major birth pain that greatly affected the development of modern geology. The central figure of the Neptunists was Abraham Gottlob Werner (1749-1817), Professor of Mineralogy at the Mining Academy at Freiberg, Germany. Although Werner published little, he contributed greatly to the systematic organization of mineralogy and his brilliant lectures inspired a generation of his students to become geologists. However, he is best remembered for his Neptunist theory that the earth's crustal rocks were universal formations, all precipitated from sea water. Precipitated as the first universal formation were such rocks as granite, gneiss, and basalt; next, graywackies, slates, and limestones settled out; then another universal layer of sandstones, limestones, rock salt, coal, and more basalts; and finally there were deposited mechanical precipitates including unconsolidated sands, gravels, soils and the like. This grand theory had theoretical objections (for example, what became of the great volume of water in the formative sea?), but its fate finally hinged on the origin of basalt.

Opposition to Werner's theory came quickly from France. Nicolas Desmarest (1725-1804) showed that in the Auvergne, basalts were vesicular, had baked the ground beneath, and could be traced back through continuous lava tongues to source volcanos. By making a geologic map he demonstrated a long volcanic history in the region, marked by three main episodes of eruption interspersed with long intervals of erosion. Old polygonal basalts which erosion had isolated from their sources had the same features as basalts in young flows extending directly from volcanos. Clearly, all the Auvergne basalts were of igneous origin.

In England, James Hutton (1726-1797) led the anti-Wernerian, or Vulcanist school. Perhaps Hutton's greatest contribution to geology involved Uniformitarianism, the doctrine essential for a modern view of geology. Although not the first to entertain the idea, he forced Uniformitarianism into the mainstream of geologic thinking. This doctrine, that "the present is the key to the past," assumes that physical processes are orderly (obey natural laws), but it goes further by stating that these processes have operated in the same way through geologic time. Thus Hutton insisted that the earth's past history must be explained through processes operating today or in the immediate past. No unnatural or extraordinary events account for common geologic features. For example, limestones represent relic shells of organisms in past seas or fragments of older rock, and conglomerates originated as beach or stream gravels like the deposits of today.

Inevitably Hutton's views lead to conflict with the Neptunists. Werner's hypothetical all-enveloping ocean was extraordinary, and the precipitation of basalts and granites is definitely not an observable process in present-day oceans. Moreover, to explain the valleys cut through Werner's original universal rock formations, the Neptunists resorted to Catastrophism. This doctrine assumed global cataclysms that exterminated life and suddenly changed the landscape. Valleys resulted from violent rendings of the earth by earthquake and volcanic action, or from the devastating rush of worldwide floods. The slow erosion of stream valleys and continuous deposition of sedimentary rocks were rejected.

In Hutton's Vulcanist views, basalt was an igneous rock and so was granite (Werner's first precipitate from the universal ocean). In Scotland Hutton found many outcrops with granite dikes intruded into schists. From such evidence, he emphasized the action of the earth's internal heat. Though his views on granite were essentially correct, even Hutton made mistakes. He assumed that sandstone and other sediments were lithified by heat, and he theorized that volcanos were outlets from the earth's completely molten interior. He also suggested that rock deformation and upthrusting of mountains resulted from the expansive force of heat, an idea involved in some recent theories of mountain building.

Hutton's contributions on Uniformitarianism, earth sculpture, the enormity of geologic time, the meaning of unconformities, and the germ of the glacial concept, as well as his vulcanist views, appeared in several short papers and finally in his two-volume work, *Theory of the Earth* (1795). This cornerstone of modern geology was not an immediate triumph, for Hutton, the brilliant conversationalist, wrote in a very involved and complicated style. Fortunately John Playfair (1748-1819) stated Hutton's views clearly in *Illustrations of the Huttonian Theory* (1802). Hutton's work might have attracted little attention but for his friend Playfair's book, and for the violent opposition by Neptunists on the continent of Europe and in the British Isles. In this time of social upheaval, marked by the excesses of the French Revolution, any new ideas were suspect. Moreover Hutton's view of the enormity of geologic time, "no vestiges of a beginning, no propects of an end" seemed heresy to many. Theologically, Werner's concept was more easily reconciled with the Biblical Genesis and story of the Deluge.

Ultimately, the verdict in the Neptunist-Vulcanist affair favored Desmarest and Hutton, yet the issue was long in doubt. Interestingly two of the principals, Werner and Demarest, avoided direct involvement in the clash, which was largely carried on by their associates. The tide finally turned when two of Werner's students, Jean Francois D'Aubisson (1769-1819) and Leopold Von Buch (1774-1852), after visiting the Auvergne were forced to admit the igneous origin of basalt.

Werner's rejected theory, like many others, served some useful purposes. It interested people in geology, gave directions to investigations, and forced contestants to seek field evidence, (Werner, who never left Germany, made very limited observations).

Above all, the Neptunist's collapse and the violence of the debate left geologists wary of any grand theory based on scanty evidence. The conversions of Von Buch and D'Aubisson showed the need for scientific open-mindedness, and the rejection of favorite schemes contradicted by strong evidence.

Werner's discredited theory, although not even the first of the sort, did show clearly the need for a true geologic succession — a time scale missing from Hutton's productive work.

THE GEOLOGIC TIME SEQUENCE

The geologic succession was a major accomplishment in the 19th Century. The establishment of a true succession required the use of Steno's laws of original horizontality and superposition, recognition of faunal succession, and careful tracing of rock units in the field. The break-through resulted from the joint work of Georges Cuvier (1769-1832) with Alexandre Brongniart (1770-1847) in France, and the independent work of William Smith (1769-1839) in England. They all recognized that fossils allow correlation of widely separate strata.

The determination of the lower Paleozoic sequence makes an interesting scientific and human story. Adam Sedgewick (1785-1873), professor at Oxford, established the Cambrian system in 1835. In the same year Sir Roderick I. Murchison (1792-1871), a gentleman of means, established the Silurian system. Working together, Sedgewick and Murchison established the Devonian during a trip into Devonshire. Murchison established the Permian, based on rocks in a Russian province in 1841. Thereafter their friendship degenerated in a dispute over the rocks in Wales.

Sedgewick had started his Cambrian studies on the coast of Wales and worked eastward, while Murchison began his Silurian studies in eastern Wales, where Devonian rocks overlay the Silurian, and worked in the opposite direction. In tracing his rocks westward, Murchison decided that the upper Cambrian rocks claimed by Sedgewick were really his lower Silurian, based on fossil evidence. After this development, the old friends "drifted apart" with Sedgewick becoming increasingly angry and finally vindictive. By 1859 the British Geological Society had sided with Murchison, rejecting a separate Cambrian system. However, after both participants had died, the controversy was settled in a way that was fair to both. Through the work of others, the presence of a basal unit, Sedgewick's Cambrian, was established. Fossils in the disputed overlapping interval were distinctive enough to indicate a separate system, which Lapworth (in 1879) named the Ordovician. From the beginning the problem had been complicated by structurally deformed rocks, a gradational boundary between the Cambrian and Ordovician, and the scarcity of fossils, as well as by human factors arising from a conflict of interests. The rest of the geologic succession was established peacefully before the last quarter of the 19th Century.

ESTABLISHING UNIFORMITARIANISM

While the geologic succession was emerging, Hutton's Uniformitarianism had not yet become doctrine. Cuvier, pioneer in the stratigraphic use of fossils and "father of vertebrate paleontology," maintained that sharp boundaries between strata with differing fossil groups proved wholesale exterminations of animal populations. He died a confirmed Catastrophist. Sedgewick was a Neptunist who grudgingly accepted Hutton's vulcanist views, but never the uniformitarian evolutionary theme of slow change.

Thus, Charles Lyell (1797-1875) faced strong opposition when he began his mission of preaching Uniformitarianism. But the political climate of the times had become more favorable in Britain, where the middle classes liked the social idea of mankind's progressive improvement, an evolutionary concept. Lyell was no great innovator, but his patient pursuit of evidence made him the "high priest of Uniformitarianism." His textbook, *Principles of Geology* (12 editions between 1830 and 1875) firmly established Hutton's concept in the mainstream of geologic thinking. Like many apostles, Lyell went to extremes. Geologists now doubt, for instance, that the intensity of erosional processes has always been constant, note the occurrence of ice ages for example. Yet his enthusiasm was understandable, considering the strong opposition to the valid principle of Uniformitarianism. Lyell's writings on gradual physical change also influenced Charles Darwin (1809-1882).

ORGANIC EVOLUTION

Darwin was not the first to consider organic evolution. Jean Baptiste de Monet Lamarck (1774-1829) had attributed the progressive complexity of the Paris Basin's invertebrate fossils to gradual change from simpler to more complex animals (evolution). Lamarck's views on the causes of evolution, although not generally accepted today, are worth mention. He believed that organisms have an "inner force" tending to improve a species and convert it into higher types, a notion appealing to some philosophers. As a mechanism, Lamarck proposed that characteristics acquired in a lifetime or over several generations would ultimately be inherited. For example, if a born weakling builds a well-muscled body by weight-lifting, his children should have strong natural physiques. The idea of inheriting acquired characteristics attracts certain political theorists who suppose that a new society could cause the evolution of superior people (Recently the Soviet Union had a revival of Lamarkism in biology). But leaving the mechanism aside, the very occurrence of evolution was strongly denied by many scientists well into the 19th Century. In his day, Lamarck was completely overwhelmed by a former protege, the aggressive and flamboyant Cuvier, a devout believer that animal species were unchanging.

Charles Darwin ultimately gained scientific acceptance for the idea of organic evolution, and suggested a reasonable mechanism. His evolutionary views stemmed from geological and biological studies during a five-year voyage starting in 1831 to South America and through the Pacific on H.M.S. Beagle. His observation that closely related species varied slightly along South America, and in the nearby Galapagos Islands led to his concept of evolution, published in *Origin of the Species* (1859).

His recognition of the fact of evolution, Darwin credited to Charles Lyell's influential writings. However, Darwin's proposed mechanism stemmed from reading Thomas R. Malthus's grim view that man's condition could only deteriorate because his unrestrained reproduction would easily outstrip any improvements in food production, a tendancy countered in nature by the struggle for existence. Thus influenced, Darwin suggested that evolution involved the competition of animals for limited food supplies. Individuals born with slight advantages, through chance variations, were those that survived to reproduce future generations. This mechanism of natural selection, sometimes called "the survival of the fittest," was unfortunately adopted in over-simplified form in some social thinking. The outrage Darwin generated in the last half of the 19th Century (and still does in fundamentalist religious circles) is a story in itself.

Interestingly, most of the pioneers who used fossils in assembling the geologic succession did not accept organic evolution. William Smith was unaware of the concept. Murchison saw no merit or evidence for Darwin's theory. Sedgewick argued strongly against it, following Cuvier in proclaiming unchanging species with mass exterminations followed by special creation of entirely new animal populations to explain the different fossil groups in a geologic column. Today the scientific evidence, much of it from the fossil record, overwhelmingly supports the past occurence of organic evolution (the precise mechanism is, admittedly still being studied). Geologically we now recognize that organic evolution accounts for the faunal succession that made fossils an indispensible tool in unravelling and assembling a proper geologic time sequence.

AMERICAN CONTRIBUTIONS

Although north Europeans founded modern geology, there were significant developments in North America. As far back as colonial times, Lewis Evans (1700-1756) understood the origin of fossils, recognized the even crests in the Appalachian Mountains as remnants of an eroded plain, and interpreted raised beaches on Lake Ontario as marking former water levels. In the 19th Century, William Logan (1798-1898) pioneered studies of the Precambrian rocks in Canada. James Hall (1811-1898) of the New York Geological Survey developed the geosynclinal theory, announced in 1856. There were other workers of note in early days, but the great period of American geology followed the Civil War.

With the opening of California, the need for a trans-continental railroad stimulated several explorations for possible routes in the 1850's. The survey parties, which were under army command, generally included naturalists or geologists. The scientific results were modest, but the surveys did generate interest in the geology of the west. After the Civil War, the federal government sent out four geographical and geological parties. Besides providing adventure stories of rugged and competent men in the spectacular scenery of "the Great American Desert", these surveys resulted in remarkable geologic accomplishments.

Clarence King (1842-1901), a man of fine tastes and cultured interests, lead the survey of the 40th parallel across parts of Nevada, Utah, Colorado, and Wyoming. Ferdinand V. Hayden (1829-1887), who had been west before the war, directed the U.S. Geological and Geographical Survey of the Territories—largely in parts of Colorado, Montana, and Wyoming. His reports greatly influenced Congress in the establishment of Yellowstone National Park. Lt. G.M. Wheeler commanded an army-sponsored survey through parts of California, Nevada, Arizona, and New Mexico. Probably reflecting its leader's stiff-necked attitude, this survey was the least productive. John Wesley Powell (1834-1903) led the U.S. Geographical and Geological Survey of the Rocky Mountain Region. A one-armed Union Army veteran, Powell twice took boats down the Colorado River through Grand Canyon, and studied the Colorado Plateau, mainly in Utah and parts of Arizona. The geologic accomplishments of Powell, Clarence E. Dutton (1841-1893), and especially Grove Karl Gilbert (1843-1918) made this the outstanding group. Ultimately, the separate surveys were consolidated, for their operations at about the same time and in the same general region led to overlap, bickering, and competition for federal funds among the separate groups. The U.S. Geological Survey was founded in 1879 with King as the first director, and Powell as the second.

The western surveys produced a mass of necessary description of diverse and well-exposed geology that greatly increased theoretical understanding. Great strides were made in the studies of geologic structure and mountain-making forces, in landforms and the mechanics of subaerial erosion in general, and streams in particular. In Europe at the time, extensive erosion surfaces were attributed to planation by the sea. Streams were credited with cutting individual valleys at the most, and even Lyell doubted that streams excavated major valleys. Powell, Dutton, and others presented evidence from the American west that, given time, streams and associated processes could reduce whole regions to low relief.

GEOLOGY BY 1900

During the 19th Century, geology became an organized science with a considerable base of established facts and such fruitful concepts as a uniformitarian-evolutionary point of view, and a reliable framework for time relations. The Neptunist-Plutonist affair had alerted geologists to the need for a scientific approach suited to the study of the earth, where theories require a vast collection of converging evidence. G.K. Gilbert expressed the outlook in the late 1800's. To avoid being blinded by a "pet" theory, he stressed an open minded approach by considering all significant facts, rather than ignoring contradictory observations.

Perhaps most important, by the end of the 19th Century the times were right for progress. In scientific circles, at least, geology was no longer forced to fit theological notions. Individual workers were no longer isolated from the ideas of others, since many national and regional societies existed for the presentation and discussion of geologic ideas and findings. In the latter half of the 19th Century, general interest in geology was great. Many universities and colleges introduced formal courses in geology. Public lectures by geologists, such as Louis Agassiz on the Ice Age, drew overflow audiences.

A significant development was the growth of a sizeable body of full-time geologists with adequate financial backing which had stemmed from the creation of the U.S. Geological Survey and similar organizations. Early 19th Century geology had been largely a part-time work of people trained in other fields. Hutton, for instance, was trained in medicine, and Sedgewick, an ordained clergyman, had originally taught mathematics. Unless independently wealthy, like Murchison, all of the early geologists had to support themselves by working in some other field or by teaching. In short, geology, founded by talented amateurs, now had a core of professionals.

Modern Geology

INSTRUMENTS

The exciting progress of 20th Century geology reflects the introduction of sophisticated equipment that has greatly expanded observational and experimental techniques. However, the "instrumental revolution" had 19th Century beginnings. In 1828 William Nicol (1768-1851) invented a polarizing prism of calcite, which made possible the petrographic microscope and the technique of making thin-sections of rock for examination. As sometimes happens, the new approach was not accepted immediately. Some 30 years passed until H.C. Sorby stressed its value. Even then, as still happens today, conservatives objected to any new development, and they said that "one couldn't look at mountains through a microscope". The petrographic microscope, long standard laboratory equipment, revolutionized the study of rocks by allowing rapid mineral identification, even in very fine grained rocks, from the patterns and colors produced by polarized light.

ALLIED SCIENCES

Although his close friend Hutton scorned attempts to duplicate grand geologic forces in the crucible, Sir James Hall (1761-1832) melted basalt in an iron foundry's furnace to study how the rate of cooling related to crystal size in igneous rock. Others experimented in the 1800's, but the most productive laboratory studies are recent ones in which experimental geologists have applied principles of physics and chemistry to their work. In such places as the Geophysical Laboratory of the Carnegie Institution, founded in 1904 in Washington D.C., great progress has been made by studying the crystallization history of silicate melts. Such work has allowed the unraveling and precise prediction of the complicated crystallization of igneous minerals. It helps explain, for instance, how a single kind of magma could under certain conditions produce a variety of rocks. By cooling carefully prepared silicate melts to known temperatures, then quickly chilling them, the minerals crystallizing at a particular temperature can be determined.

The dating of radioactive elements in minerals is a laboratory technique that revolutionized our concepts of geologic time. In 1896, Antoine Becquerel, a French physicist, exposed photographic plates by leaving them in a drawer with uranium salts, an accident that revolutionized our understanding of the physical world. Two years later in France, the Curies isolated radium as a prelude to the discovery of the other radioactive elements. Geology is indebted to Bertram Boltwood, an American chemist working with Lord Rutherford's group of physicists in England. When Boltwood discovered, in 1907, that the ratio of lead to uranium was constant in rocks of the same age, the measuring of geologic time in years could begin. A number of different radioactive elements are now being used and techniques have been steadily improved so that today we have a fairly accurate tool for determining the ages of suitable rocks in years.

The application of X-ray techniques to minerals, following Von Laue's work in 1912, was the single most important advance in mineralogy within the 20th Century. The X-ray techniques give a superior way to identify minerals, and they make it possible to locate atoms and thus determine the internal structure of minerals. Since 1912 the internal structure of hundreds of minerals has been determined, and this has contributed a great understanding of minerals and also to an understanding of chemistry and solid state physics. The advances mentioned are involved largely with the expanding field of geochemistry; the accompanying rise of geophysics represents still another cross-cutting field which is also a major 20th Century development.

The seismograph, an instrument originated by John Milne in the late 19th Century, opened the earth's previously inaccessible interior to investigation by the application of physics. Now a battery of sophisticated geophysical devices exists. Their records on dials, paper strips, and flashing lights open a whole new realm of seismic, gravity, heat, and magnetic measurements to supplement field, microscope, and geochemical studies of rocks.

NEW FRONTIERS

Yet another geophysical device, the echo-sounder, has exposed nearly three-quarters of the earth's surface that was formerly inaccessible. So long as depths to the sea floor were known only by reeling out weighted lines, our knowledge was limited largely to shallow waters. Now modern echo-sounders, which continuously record depths beneath ships by bouncing sound pulses off the ocean bottom, have revolutionized our concepts of the earth's submarine surface.

Once thought largely flat, undersea topography is now known to be complex and varied. The continental block's shallowly submerged shelves give way to steeper continental slopes descending to the oceanic depths. The continental slopes grade into continental rises that lead down to abyssal plains, the greatest flat expanses on the Earth's rock crust. Yet the plains are broken by a variety of hills, conical mountains, and plateau-like platforms, as well as by deep, narrow trenches and scarps. The most striking submarine features are the global system of mid-oceanic ridges. Extending from Iceland southward through the length of the Atlantic Ocean, thence into the Indian and Pacific Oceans, they are the earth's greatest continuous system of mountains.

The tops of the ridges are split by axial valleys which may be graben structures.

Besides echo-sounders, and devices for bringing up samples from the ocean bottom, other sea-borne instruments are being used to measure seismic properties, gravity variations, heat loss, and magnetic properties in the ocean floor. As a result of the new techniques and an upsurge of interest in oceanographic research, as well as massive governmental support to finance such operations, a great deal of critical new information has been accumulated. Sifted and related by imaginative minds, it led geologists and geophysicists in the late 1960's to a new view of the earth's global mechanisms.

Essentially, the earth's crust is envisioned as passing through a continuous cycle of regeneration and destruction. New crust forms in the mid-ocean ridges, where magma wells up into their axial valleys and solidifies. Gradually this new-formed crust is displaced outward as still younger material erupts into the ridges. Through time the ocean floors move away from the ridges, and solid plates of crust and uppermost mantle ride on a deeper plastic zone in the mantle. Where opposing plates meet, as along some continental margins and oceanic island belts, one solid plate is deflected downward to be fused and re-incorporated in the deeper mantle.

In a sense, sea-floor spreading is a modified and expanded version of the older theory of continental drift, which postulated last Paleozoic fracturing of super-continents whose fragments, the present continents, drifted apart, leaving an ever-widening Atlantic ocean basin. This idea was strongly supported by Alfred Wegner (1880-1930) and his followers starting in the early 1900's and just as strongly rejected by many other geologists, especially in America. Originally, the "driftists" could not suggest a satisfactory mechanism for "wandering" continents, but they have it now in sea-floor spreading. Although no single person deserves all the credit, H.H. Hess of Princeton University, J.T. Wilson of the University of Toronto, B. Heezen, J.R. Heirtzler, and N.D. Opdyke of Columbia University, F.J. Vine, and D.H. Matthews of Cambridge University, and R.G. Mason, A.D. Raff, and V. Vacquier of Scripps Institute were among those who made significant contributions.

As is often the case in science, the revolution of the late 1960's, when sea-floor spreading upset the established doctrine of the permanent location of continents and ocean basins, is now followed by a more normal period of scientific development in which most workers are supporting, refining, and extending the new concept. Although a great deal of converging evidence seems to support the new theory, the history of geology indicates the need for much testing before accepting any such grand speculative theory of the earth.

The frontier of geology has reached the moon. That geological laboratories would be studying rocks collected there by astronauts in 1969 would have seemed, just a few decades ago, an impossible dream. That the moon was round and had lighter and darker regions was all man could see with his unaided eyes. The first advance, based on technology, came when Galileo used his primitive telescope to see the maria, mountains, and the craters of the moon. Thereafter improved telescopes disclosed much more detail in the diverse lunar landscape marked by many more craters, scarps, groove-like rills, light-colored rays and a host of intriguing features. The next major advance had to await the space age, when orbiting and crashing space vehicles first sent back incredibly detailed pictures of the lunar landscape. Then soft-landing vehicles, culminating with manned landings, allowed experiments and sample collecting on the surface of the moon. Future geologic plans envision more detailed mapping from lunar orbiters, landings in different areas, the use of lunar ground vehicles, and ultimately a base for prolonged geological and geophysical studies.

Studies on the lunar surface should answer many questions and may open a whole new realm of unanticipated problems. The probable origins of lunar rocks may be inferred by comparing their compositions, textures, and micro-structures to the well known rocks on earth. A better understanding of rays, rills, scarps, craters and other surface features should result from lunar field work. Geophysical instruments may reveal whether the moon's internal structure is homogenous or zoned like the earth's. Is it cold and dead or hot and dynamic? Grander problems, such as the origin of the moon, may remain, for after all our study, the origin of the earth is not yet surely known. However, unlike the earth where the incessant attack of air and water long ago obliterated any original surface features, the lunar landscape may still preserve vestiges of the Moon's very early history. If so, the moon may give valuable clues to the formative stages of the earth and other planets in the solar system.

It seems almost inconceivable that just 100 years before astronauts first set foot on the moon, the great geologic adventure was Major Powell's boat trip of 1869 down the Green and Colorado Rivers through Grand Canyon. Yet already a manned landing is planned for Mars. The recent opening of the moon to geologic exploration is a triumph of technology reflecting a national endeavor with the close collaboration of scientists, engineers, and astronauts. Technology, generous support, and teamwork toward directed goals are the new trends in science since the beginning of World War II. Although most dramatic in the development of the atomic bomb and the space program, these trends mark geologic fields, notably submarine geology, with its grand new concept of sea-floor spreading, and astrogeology with its well publicized new findings on the moon.

The rapid and exciting geologic discoveries of the present should not blind us to the past. Present-day geologists are neither more nor less intelligent than their predecessors. Today there are more working geologists than ever before, generally well financed, and living in a culture sympathetic to their work and ideas. The present accomplishments rest on a long history of the development of ideas, of scientific methods, and of techniques. Along with the hard-won successes, the frustrations and failures are also important. Geology, like any science, is not mystical, divine, or ultimate knowledge. It is a collection of images of external reality resulting from the endeavors of very human men.

7

Letters To Tacitus

PLINY THE YOUNGER

The next four selections may be described as non-fictional adventure stories. Their descriptions of the excitement, danger, discomfort, and hard work in dramatic surroundings touch the heart of the challenge that makes some men become geologists.

The first fragments go back to the famous 79 A.D. eruption of Vesuvius. The events surrounding the death of Pliny the Elder, a naturalist and Roman admiral whose fleet lay in the Bay of Naples during the eruption, are eloquently recorded in the letters of Pliny the Younger, his 17-year-old nephew, to the historian, Tacitus.

From Pliny's Letters, Book VI, 16:

Gaius Plinius sends to his friend Tacitus greeting.

You ask me to write you an account of my uncle's death, that posterity may possess an accurate version of the event in your history. . . .

He was at Misenum, and was in command of the fleet there. It was at one o'clock in the afternoon of the 24th of August [A.D. 79] that my mother called his attention to a cloud of unusual appearance and size. He had been enjoying the sun, and after a bath had just taken his lunch and was lying down to read: but he immediately called for his sandals and went out to an eminence from

Excerpts from a translation by Professor J.G. Croswell for Professor N.S. Shaler of Harvard, as published by the latter in his **Aspects of the Earth** (Charles Scribner's Sons, 1896). From Pliny's Letters, in **Causes of Catastrophe,** by L. Don Leet. Copyright 1948 by McGraw-Hill, Inc. Reprinted by permission of McGraw-Hill Book Company.

which this phenomenon could be observed. A cloud was rising from one of the hills which took the likeness of a stone-pine very nearly. It imitated the lofty trunk and the spreading branches... It changed color, sometimes looking white, and sometimes when it carried up earth or ashes, dirty and streaked. The thing seemed of importance, and worthy of nearer investigation, to the philosopher. He ordered a light boat to be got ready, and asked me to accompany him if I wished; but I answered that I would rather work over my books. In fact he had himself given me something to write.

He was going out himself, however, when he received a note from Rectina, wife of Caesius Bassus, living in a villa on the other side of the bay, who was in deadly terror about the approaching danger and begged him to rescue her, as she had no means of flight but by ships. This converted his plan of observation into a more serious purpose. He got his men-of-war under way, and embarked to help Rectina, as well as other endangered persons, who were many, for the shore was a favorite resort on account of its beauty. He steered directly for the dangerous spot whence others were flying, watching it so fearlessly as to be able to dictate a description and take notes of all the movements and appearances of this catastrophe as he observed them.

Ashes began to fall on his ships, thicker and hotter as they approached land. Cinders and pumice, and also black fragments of rock cracked by heat, fell around them. The sea suddenly shoaled, and the shores were obstructed by masses from the mountain. He hesitated awhile and thought of going back again; but finally gave the word to the reluctant helmsman to go on, saying, "Fortune favors the brave. Let us find Pomponianus." Pomponianus was at Stabiae, separated by the intervening bay (the sea comes in here gradually in a long inlet with curving shores), and although the peril was not near, yet as it was in full view, and as the eruption increased, seemed to be approaching, he had packed up his things and gone aboard his ships ready for flight, which was prevented, however, by a contrary wind.

My uncle, for whom the wind was most favorable, arrived, and did his best to remove their terrors. He embraced the frightened Pomponianus and encouraged him. To keep up their spirits by a show of unconcern, he had a bath; and afterwards dined with real, or what was perhaps as heroic, with assumed cheerfulness. But meanwhile there began to break out from Vesuvius, in many spots, high and wide-shooting flames, whose brilliancy was heightened by the darkness of approaching night. My uncle reassured them by asserting that these were burning farm-houses which had caught fire after being deserted by the peasants. Then he turned in to sleep, and slept indeed the most genuine slumbers; for his breathing, which was always heavy and noisy, from the full habit of his body, was heard by all who passed his chamber. But before long the floor of the court on which his chamber opened became so covered with ashes and pumice that if he had lingered in the room he could not have got out at all. So the servants woke him, and he came out and joined Pomponianus and others who were watching. They consulted together as to what they should do next. Should they stay in the house or go out of doors? The house was tottering with frequent and heavy shocks of earthquake, and seemed to go to and fro as if moved from its foundations. But in the open air there were dangers of falling pumicestones, though, to be sure, they were light and porous. On the whole, to go out seemed the least of two evils. With my uncle it was a comparison of arguments that decided; with the others it was a choice of terrors. So they tied pillows on their heads by way of defence against falling bodies, and sallied out.

It was dawn elsewhere; but with them it was a blacker and denser night than they had ever seen, although torches and various lights made it less dreadful. They decided to take to the shore and see if the sea would allow them to embark; but it appeared as wild and appalling as ever. My uncle lay down on a rug. He asked twice for water and drank it. Then as a flame with a fore-running sulphurous vapor drove off the others, the servants roused him up. Leaning on two slaves, he rose to his feet, but immediately fell back, as I understand, choked by the thick vapors, and this the more easily that his chest was naturally weak, narrow, and generally inflamed. When day came (I mean the third after the last he ever saw), they found his body perfect and uninjured, and covered just as he had been overtaken. He seemed by his attitude to be rather asleep than dead. . . .

From Book VI, 20:

Gaius Plinius sends to his friend Tacitus greeting.

You say that you are induced by the letter I wrote to you when you asked about my uncle's death to desire to know how I, who was left at Misenum, bore the terrors and disasters of that night....

My uncle started off and I devoted myself to my literary task, for which I had remained behind. Then followed my bath, dinner, and sleep, though this was short and disturbed. There had been already for many days a tremor of the earth, less appalling, however, in that this is usual in Campania. But that night it was so strong that things seemed not merely to be shaken, but positively upset. My mother rushed into my bedroom. I was just getting up to wake her if she were asleep. We sat down in the little yard, which was between our house and the sea. I do not know whether to call it courage or foolhardiness (I was only seventeen), but I sent for a volume of Livy and, quite at my ease, read it and even made extracts, as I had already begun to do. And now a friend of my uncle's, recently arrived from Spain, appeared who, finding us sitting there and me reading, scolded us, my mother for her patience, and me for my carelessness of danger. None the less, industriously I read my book.

It was now seven o'clock, but the light was still faint and doubtful. The surrounding buildings had been badly shaken, and though we were in an open spot, the space was so small that the danger of a catastrophe from falling walls was great and certain. Not till then did we make up our minds to go from the town. A frightened crowd went away with us, and as in all panics everybody thinks his neighbors' ideas more prudent than his own, so we were pushed and squeezed in our departure by a great mob of imitators.

When we were free of the buildings we stopped. There we saw many wonders and endured many terrors. The vehicles we had ordered to be brought out kept running backward and forward, though on level ground; and even when scotched with stones they would not keep still. Besides this, we saw the sea sucked down and, as it were, driven back by the earthquake. There can be no doubt that the shore had advanced on the sea, and many marine animals were left high and dry. On the other side was a dark and dreadful cloud, which was broken by zigzag and rapidly vibrating flashes of fire, and yawning showed long shapes of flame. These were like lightnings, only of greater extent....

Pretty soon the cloud began to descend over the earth and cover the sea. It enfolded Capreae and hid also the promontory of Misenum. Then my mother began to beg and beseech me to fly as I could. I was young, she said, and she was old, and too heavy to run, and would not mind dying if she was not the cause of my death. I said, however, I would not be saved without her; I clasped her hand and forced her to go, step by step, with me. She slowly obeyed, reproaching herself bitterly for delaying me.

Ashes now fell, yet still in small amount. I looked back. A thick mist was close at our heels, which followed us, spreading out over the country, like an inundation. "Let us turn out of the road," said I, "while we can see, and not get trodden down in the darkness by the crowds who are following, if we fall in their path." Hardly had we sat down when night was over us—not such a night as when there is no moon, and clouds cover the sky, but such darkness as one finds in close-shut rooms. One heard the screams of women, the fretting cries of babes, the shouts of men. Some called their parents, and some their children, and some their spouses, seeking to recognize them by their voices. Some lamented their own fate, others the fate of their friends. Some were praying for death, simply for fear of death. Many a man raised his hands in prayer to the gods; but more imagined that the last eternal night of creation had come and there were now no gods more. There were some who increased our real dangers by fictitious terrors. Some said that part of Misenum had sunk, and that another part was on fire. They lied; but they found believers.

Little by little it grew light again. We did not think it the light of day, but a proof that the fire was coming nearer. It was indeed fire, but it stopped afar off; and then there was darkness again, and again a rain of ashes, abundant and heavy, and again we rose and shook them off, else we had been covered and even crushed by the weight. . . . At last the murky vapor rolled away, in disappearing smoke or fog. Soon the real daylight appeared; the sun shone out, of a lurid hue, to be sure, as in an eclipse. The whole world which met our frightened eyes, was transformed. It was covered with ashes white as snow. . . .

8

The Canyon Of Lodore

JOHN WESLEY POWELL

John Wesley Powell led scientific explorations in the Colorado Plateau and adjacent Rockies. A Union Army veteran who lost an arm at the battle of Shiloh, he twice descended the "white waters" of the Green and Colorado. Powell originally intended publishing only his scientific results. However at the request of then Representative James A. Garfield, he wrote an account of his adventure based on his daily journal, and later a book, Canyons of the Colorado, *which contains the following account of his passage through a canyon of the Uinta Mountains. Powell was the leading figure in the formation of the U.S. Geological Survey and became its second Director.*

June 8.—We enter the canyon, and until noon find a succession of rapids, over which our boats have to be taken.

Here I must explain our method of proceeding at such places. The "Emma Dean" goes in advance; the other boats follow, in obedience to signals. When we approach a rapid, or what on other rivers would often be called a fall, I stand on deck to examine it, while the oarsmen back water, and we drift on as slowly as possible. If I can see a clear chute between the rocks, away we go; but if the channel is beset entirely across, we signal the other boats, pull to land, and I walk along the shore for closer examination. If this reveals no clear channel, hard work begins. We drop the boats to the very head of the dangerous place and let them over by lines or make

From **The Exploration of the Colorado River and Its Canyons,** by John W. Powell.

a portage, frequently carrying both boats and cargoes over the rocks.

The waves caused by such falls in a river differ much from the waves of the sea. The water of an ocean wave merely rises and falls; the form only passes on, and form chases form unceasingly. A body floating on such waves merely rises and sinks—does not progress unless impelled by wind or some other power. But here the water of the wave passes on while the form remains. The waters plunge down ten or twenty feet to the foot of a fall, spring up again in a great wave, then down and up in a series of billows that gradually disappear in the more quiet waters below; but these waves are always there, and one can stand above and count them.

A boat riding such billows leaps and plunges along with great velocity. Now, the difficulty in riding over these falls, when no rocks are in the way, is with the first wave at the foot. This will sometimes gather for a moment, heap up higher and higher, and then break back. If the boat strikes it the instant after it breaks, she cuts through, and the mad breaker dashes its spray over the boat and washes overboard all who do not cling tightly. If the boat, in going over the falls, chances to get caught in some side current and is turned from its course, so as to strike the wave "broadside on," and the wave breaks at the same instant, the boat is capsized; then we must cling to her, for the water-tight compartments act as buoys and she cannot sink; and so we go, dragged through the waves, until still waters are reached, when we right the boat and climb aboard. We have several such experiences to-day.

At night we camp on the right bank, on a little shelving rock between the river and the foot of the cliff; and with night comes gloom into these great depths. After supper we sit by our camp fire, made of driftwood caught by the rocks, and tell stories of wild life; for the men have seen such in the mountains or on the plains, and on the battlefields of the South. It is late before we spread our blankets on the beach.

Lying down, we look up through the canyon and see that only a little of the blue heaven appears overhead—a crescent of blue sky, with two or three constellations peering down upon us. I do not sleep for some time, as the excitement of the day has not worn off. Soon I see a bright star that appears to rest on the very verge of the cliff overhead to the east. Slowly it seems to float from its resting place on the rock over the canyon. At first it appears like a jewel set on the brink of the cliff, but as it moves out from the rock I almost wonder that it does not fall. In fact, it does seem to descend in a gentle curve, as though the bright sky in which the stars are set were spread across the canyon, resting on either wall, and swayed down by its own weight. The stars appear to be in the canyon. I soon discover that it is the bright star Vega; so it occurs to me to designate this part of the wall as the "Cliff of the Harp."

June 9.—One of the party suggests that we call this the Canyon of Lodore, and the name is adopted. Very slowly we make our way, often climbing on the rocks at the edge of the water for a few hundred yards to examine the channel before running it. During the afternoon we come to a place where it is necessary to make a portage. The little boat is landed and the others are signaled to come up.

When these rapids or broken falls occur usually the channel is suddenly narrowed by rocks which have been tumbled from the cliffs or have been washed in by lateral streams. Immediately above the narrow, rocky channel, on one or both sides, there is often a bay of quiet water, in which a landing can be made with ease. Sometimes the water descends with a smooth, unruffled surface from the broad, quiet spread above into the narrow, angry channel below by a semi-circular sag. Great care must be taken not to pass over the brink into this deceptive pit, but above it we can row with safety. I walk along the bank to examine the ground, leaving one of my men with a flag to guide the other boats to the landing-place. I soon see one of the boats make shore all right, and feel no more concern; but a minute after, I hear a shout, and, looking around, see one of the boats shooting down the center of the sag. It is the "No Name," with Captain Howland, his brother, and Goodman. I feel that its going over is inevitable, and run to save the third boat. A minute more, and she turns the point and heads for the shore. Then I turn down stream again and scramble along to look for the boat that has gone over. The first fall is not great, only 10 or 12 feet, and we often run such; but below, the river tumbles down again for 40 or 50 feet, in a channel filled with dangerous rocks that break the waves into whirlpools and beat them into foam. I pass around a great crag just in time to see the boat strike a rock and, rebounding from the shock, career and fill its open compartment with water. Two of the men lose their oars; she swings around and is carried down at a rapid rate, broadside on, for a few yards, when, striking amidships on another rock with great force, she is broken quite in two and the men are thrown into the river. But the larger part of the boat floats buoyantly, and they soon seize it, and down the river they drift, past the rocks for a few hundred yards, to a second rapid filled with huge boulders, where the boat strikes again and is dashed to pieces, and the men and fragments are soon carried beyond my sight. Running along, I turn a bend and see a man's head above the water, washed about in a whirlpool below a great rock. It is Frank Goodman, clinging to the rock with a grip upon which life depends. Coming opposite, I see Howland trying to go to his aid from an island on which he has been washed. Soon he comes near enough to reach Frank with a pole, which he extends toward him. The latter lets go the rock, grasps the pole, and is pulled ashore. Seneca Howland is washed farther down the island and is caught by some rocks, and, though somewhat bruised, manages to get ashore in safety. This seems a long time as I tell it, but it is quickly done.

And now the three men are on an island, with a swift, dangerous river on either side and a fall below. The "Emma Dean" is soon brought down, and Sumner, starting above as far as possible, pushes out. Right skillfully he plies the oars, and a few strokes set him on the island at the proper point. Then they all pull the boat up stream as far as they are able, until they stand in water up to their necks. One sits on a rock and holds the boat until the others are ready to pull, then gives the boat a push, clings to it with his hands, and climbs in as they pull for mainland, which they reach in safety. We are as glad to shake hands with them as though they had been on a voyage around the world and wrecked on a distant coast.

Down the river half a mile we find that the after cabin of the wrecked boat, with a part of the bottom, ragged and splintered, has floated against a rock and stranded. There are valuable articles in the cabin; but, on examination, we determine that life should not be risked to save them. Of course, the cargo of rations, instruments, and clothing is gone.

We return to the boats and make camp for the night. No sleep comes to me in all those dark hours. The rations, instruments, and clothing have been divided among the boats, anticipating such an accident as this; and we started with duplicates of everything that was deemed necessary to success. But, in the distribution, there was one exception to this precaution—the barometers were all placed in one boat, and they are lost! There is a possibility that they are in the cabin lodged against the rock, for that is where they were kept.

But, then, how to reach them? The river is rising. Will they be there to-morrow? Can I go out to Salt Lake City and obtain barometers from New York?

June 10. — I have determined to get the barometers from the wreck, if they are there. After breakfast, while the men make the portage, I go down again for another examination. There the cabin lies, only carried 50 or 60 feet farther on. Carefully looking over the ground, I am satisfied that it can be reached with safety, and return to tell the men my conclusion. Sumner and Dunn volunteer to take the little boat and make the attempt. They start, reach it, and out come the barometers! The boys set up a shout, and I join them, pleased that they should be as glad as myself to save the instruments. When the boat lands on our side, I find that the only things saved from the wreck were the barometers, a package of thermometers, and a three-gallon keg of whiskey. The last is what the men were shouting about. They had taken it aboard unknown to me, and now I am glad they did take it, for it will do them good, as they are drenched every day by the melting snow which runs down from the summits of the Rocky Mountains.

We come back to our work at the portage and find that it is necessary to carry our rations over the rocks for nearly a mile and to let our boats down with lines, except at a few points, where they also must be carried. Between the river and the eastern wall of the canyon there is an immense talus of broken rocks. These have tumbled down from the cliffs above and constitute a vast pile of huge angular fragments. On these we build a path for a quarter of a mile to a small sand-beach covered with driftwood, through which we clear a way for several hundred yards, then continue the trail over another pile of rocks nearly half a mile farther down, to a little bay. The greater part of the day is spent in this work. Then we carry our cargoes down to the beach and camp for the night.

While the men are building the camp fire, we discover an iron bake-oven, several tin plates, a part of a boat, and many other fragments, which denote that this is the place where Ashley's party was wrecked.

June 11. — This day is spent in carrying our rations down to the bay — no small task, climbing over the rocks with sacks of flour and bacon. We carry them by stages of about 500 yards each, and when night comes and the last sack is on the beach, we are tired, bruised, and glad to sleep.

June 12. — To-day we take the boats down to the bay. While at this work we discover three sacks of flour from the wrecked boat that have lodged in the rocks. We carry them above high-water mark and leave them, as our cargoes are already too heavy for the three remaining boats. We also find two or three oars, which we place with them.

As Ashley and his party were wrecked here and as we have lost one of our boats at the same place, we adopt the name Disaster Falls for the scene of so much peril and loss.

Though some of his companions were drowned, Ashley and one other survived the wreck, climbed the canyon wall, and found their way across the Wasatch Mountains to Salt Lake City, living chiefly on berries, as they wandered through an unknown and difficult country. When they arrived at Salt Lake they were almost destitute of clothing and nearly starved. The Mormon people gave them food and clothing and employed them to work on the foundation of the Temple until they had earned sufficient to enable them to leave the country. Of their subsequent history, I have no knowledge. It is possible they returned to the scene of the disaster, as a little creek entering the river below is known as Ashley's Creek, and it is reported that he built a cabin and trapped on this river for one or two winters; but this may have been before the disaster.

June 13. — Rocks, rapids, and portages still. We camp to-night at the foot of the left fall, on a little patch of flood plain covered with a dense growth of box-elders, stopping early in order to spread the clothing and rations to dry. Everything is wet and spoiling.

June 14. — Howland and I climb the wall on the west side of the canyon to an altitude of 2,000 feet. Standing above and looking to the west, we discover a large park, five or six miles wide and twenty or thirty long. The cliff we have climbed forms a wall between the canyon and the park, for it is 800 feet down the western side to the valley. A creek comes winding down 1,200 feet above the river, and, entering the intervening wall by a canyon, plunges down more than 1,000 feet, by a broken cascade, into the river below.

June 15. — To-day, while we make another portage, a peak, standing on the east wall, is climbed by two of the men and found to be 2,700 feet above the river. On the east side of the canyon a vast amphitheater has been cut, with massive buttresses and deep, dark alcoves in which grow beautiful mosses and delicate ferns, while springs burst out from the farther recesses and wind in silver threads over floors of sand rock. Here we have three falls in close succession. At the first the water is compressed into a very narrow channel against the right-hand cliff, and falls 15 feet in 10 yards. At the second we have a broad sheet of water tumbling down 20 feet over a group of rocks that thrust their dark heads through the foam. The third is a broken fall, or short, abrupt rapid, where the water makes a descent of more than 20 feet among huge, fallen fragments of the cliff. We name the group Triplet Falls. We make a portage around the first; past the second and the third we let down with lines.

During the afternoon, Dunn and Howland having returned from their climb, we run down three quarters of a mile on quiet waters and land at the head of another fall. On examination, we find that there is an abrupt plunge of a few feet and then the river tumbles for half a mile with a descent of a hundred feet, in a channel beset with great numbers of huge boulders. This stretch of the river is named Hell's Half-Mile. The remaining portion of the day is occupied in making a trail among the rocks at the foot of the rapid.

June 16. — Our first work this morning is to carry our cargoes to the foot of the falls. Then we commence letting down the boats. We take two of them down in safety, but not without great difficulty; for, where such a vast body of water, rolling down an inclined plane, is broken into eddies and cross-currents by rocks projecting from the cliffs and piles of boulders in the channel, it requires excessive labor and much care to prevent the boats from being dashed against the rocks or breaking away. Sometimes we are compelled to hold the boat against a rock above a chute until a second line, attached to the stem, is carried to some point below, and when all is ready the first line is detached and the boat given to the current, when she shoots down and the men below swing her into some eddy.

At such a place we are letting down the last boat, and as she is set free a wave turns her broadside down the stream, with the stem, to which the line is attached, from shore and a little up. They haul on the line to bring the boat in, but the power of the current, striking obliquely against her, shoots her out into the middle of the river. The men have their hands burned with the friction of the passing line; the boat breaks away and speeds with great velocity down the stream. The "Maid of the Canyon" is lost! So it seems; but she drifts some distance and swings into an eddy, in which she spins about until we arrive with the small boat and rescue her.

Soon we are on our way again, and stop at the mouth of a little brook on the right for a late dinner. This brook comes down from the distant mountains in a deep side canyon. We set out to explore it, but are soon cut off from farther progress up the gorge by a high rock, over which the brook glides in a smooth sheet. The rock is not quite vertical, and the water does not plunge over it in a fall.

Then we climb up to the left for an hour, and are 1,000 feet above the river and 600 above the brook. Just before us the canyon divides, a little stream coming down on the right and another on the left, and we can look away up either of these canyons, through an ascending vista, to cliffs and crags and towers a mile back and 2,000 feet overhead. To the right a dozen gleaming cascades are seen. Pines and firs stand on the rocks and aspens overhang the brooks. The rocks below are red and brown, set in deep shadows, but above they are buff and vermilion and stand in the sunshine. The light above, made more brilliant by the bright-tinted rocks, and the shadows below, more gloomy by reason of the somber hues of the brown walls, increase the apparent depths of the canyons, and it seems a long way up to the world of sunshine and open sky, and a long way down to the bottom of the canyon glooms. Never before have I received such an impression of the vast heights of these canyon walls, not even at the Cliff of the Harp, where the very heavens seemed to rest on their summits. We sit on some overhanging rocks and enjoy the scene for a time, listening to the music of the falling waters away up the canyon. We name this Rippling Brook.

Late in the afternoon we make a short run to the mouth of another little creek, coming down from the left into an alcove filled with luxuriant vegetation. Here camp is made, with a group of cedars on one side and a dense mass of box-elders and dead willows on the other.

I go up to explore the alcove. While away a whirlwind comes and scatters the fire among the dead willows and cedar-spray, and soon there is a conflagration. The men rush for the boats, leaving all they cannot readily seize at the moment, and even then they have their clothing burned and hair singed, and Bradley has his ears scorched. The cook fills his arms with the mess-kit, and jumping into a boat, stumbles and falls, and away go our cooking utensils into the river. Our plates are gone; our spoons are gone; our knives and forks are gone. "Water catch'em; h-e-a-p catch 'em."

When on the boats, the men are compelled to cut loose, as the flames, running out on the overhanging willows, are scorching them. Loose on the stream, they must go down, for the water is too swift to make headway against it. Just below is a rapid, filled with rocks. On the shoot, no channel explored, no signal to guide them! Just at this juncture I chance to see them, but have not yet discovered the fire, and the strange movements of the men fill me with astonishment. Down the rocks I clamber, and run to the bank. When I arrive they have landed. Then we all go back to the late camp to see if anything left behind can be saved. Some of the clothing and bedding taken out of the boats is found, also a few tin cups, basins, and a camp kettle; and this is all the mess-kit we now have. Yet we do just as well as ever.

June 17.—We run down to the mouth of Yampa River. This has been a chapter of disasters and toils, notwithstanding which the Canyon of Lodore was not devoid of scenic interest, even beyond the power of pen to tell. The roar of its waters was heard unceasingly from the hour we entered it until we landed here. No quiet in all that time. But its walls and cliffs, its peaks and crags, its amphitheaters and alcoves, tell a story of beauty and grandeur that I hear yet—and shall hear.

The Canyon of Lodore is 20 3/4 miles in length. It starts abruptly at what we have called the Gate of Lodore, with walls nearly 2,000 feet high, and they are never lower than this until we reach Alcove Brook, about three miles above the foot. They are very irregular, standing in vertical or overhanging cliffs in places, terraced in others, or receding in steep slopes, and are broken by many side gulches and canyons. The highest point on the wall is at Dunn's Cliff, near Triplet Falls, where the rocks reach an altitude of 2,700 feet, but the peaks a little way back rise nearly 1,000 feet higher. Yellow pines, nut pines, firs, and cedars stand in extensive forests on the Uinta Mountains, and, clinging to the rocks and growing in the crevices, come down the walls to the water's edge from Flaming Gorge to Echo Park. The red sandstones are lichened over; delicate mosses grow in the moist places, and ferns festoon the walls.

9

A Perilous Night On Shasta's Summit

JOHN MUIR

John Muir was born in Scotland and emigrated to the United States, where he devoted his life to mountains and mountaineering in the Sierra Nevada, his "Range of Light." An ardent and outspoken conservationist, he was most influential in protecting our natural wonders through the development of governmental parks and forests. He early recognized the glacial sculpturing of the Yosemite Valley and defended his theory against that of J. D. Whitney, a geologist who proclaimed the valley a "dropped block." Though Muir's style seems quaintly Victorian today, his writings remain outstanding among those of naturalists. His description of Mount Shasta and two storms he encountered there are from his book, Steep Trails.

Steep Trails. Copyright renewed 1946 by Helen Muir Sunk. Reprinted by permission of the publisher, Houghton Mifflin Company.

Toward the end of summer, after a light, open winter, one may reach the summit of Mount Shasta without passing over much snow, by keeping on the crest of a long narrow ridge, mostly bare, that extends from near the camp-ground at the timber-line. But on my first excursion to the summit the whole mountain, down to its low swelling base, was smoothly laden with loose fresh snow, presenting a most glorious mass of winter mountain scenery, in the midst of which I scrambled and reveled or lay snugly snow-bound, enjoying the fertile clouds and the snow-bloom in all their growing, drifting grandeur.

I had walked from Redding, sauntering leisurely from station to station along the old Oregon stage-road, the better to see the rocks and plants, birds and people, by the way, tracing the rushing Sacramento to its fountains around icy Shasta. The first rains had fallen on the lowlands, and the first snows on the mountains, and everything was fresh and bracing, while an abundance of balmy sunshine filled all the noonday hours. It was the calm afterglow that usually succeeds the first storm of the winter. I met many of the birds that had reared their young and spent their summer in the Shasta woods and chaparral. They were then on their way south to their winter homes, leading their young full-fledged and about as large and strong as the parents. Squirrels, dry and elastic after the storms, were busy about their stores of pine nuts, and the latest goldenrods were still in bloom, though it was now past the middle of October. The grand color glow—the autumnal jubilee of ripe leaves—was past prime, but, freshened by the rain, was still making a fine show along the banks of the river and in the ravines and the dells of the smaller streams.

At the salmon-hatching establishment on the McCloud River I halted a week to examine the limestone belt, grandly developed there, to learn what I could of the inhabitants of the river and its banks, and to give time for the fresh snow that I knew had fallen on the mountain to settle somewhat, with a view to making the ascent. A pedestrian on these mountain roads, especially so late in the year, is sure to excite curiosity, and many were the interrogations concerning my ramble. When I said that I was simply taking a walk, and that icy Shasta was my mark, I was invariably admonished that I had come on a dangerous quest. The time was far too late, the snow was too loose and deep to climb, and I should be lost in drifts and slides. When I hinted that new snow was beautiful and storms not so bad as they were called, my advisers shook their heads in token of superior knowledge and declared the ascent of "Shasta Butte" through loose snow impossible. Nevertheless, before noon of the second of November I was in the frosty azure of the utmost summit.

When I arrived at Sisson's everything was quiet. The last of the summer visitors had flitted long before, and the deer and bears also were beginning to seek their winter homes. My barometer and the sighing winds and filmy, half-transparent clouds that dimmed the sunshine gave notice of the approach of another storm, and I was in haste to be off and get myself established somewhere in the midst of it, whether the summit was to be attained or not. Sisson, who is a mountaineer, speedily fitted me out for storm or calm as only a mountaineer could, with warm blankets and a week's provisions so generous in quantity and kind that they easily might have been made to last a month in case of my being closely snowbound. Well I knew the weariness of snow-climbing, and the frosts, and the dangers of mountaineering so late in the year; therefore I could not ask a guide to go with me, even had one been willing. All I wanted was to have blankets and provisions deposited as far up in the timber as the snow would permit a pack-animal to go. There I could build a storm nest and lie warm, and make raids up and around the mountain in accordance with the weather.

Setting out on the afternoon of November first, with Jerome Fay, mountaineer and guide, in charge of the animals, I was soon plodding wearily upward through the muffled winter woods, the snow of course growing steadily deeper and looser, so that we had to break a trail. The animals began to get discouraged, and after night and darkness came on they became entangled in a bed of rough lava, where, breaking through four or five feet of mealy snow, their feet were caught between angular boulders. Here they were in danger of being lost, but after we had removed packs and saddles and assisted their efforts with ropes, they all escaped to the side of a ridge about a thousand feet below the timber-line.

To go farther was out of the question, so we were compelled to camp as best we could. A pitch-pine fire speedily changed the temperature and shed a blaze of light on the wild lava-slope and the straggling storm-bent pines around us. Melted snow answered for coffee, and we had plenty of venison to roast. Toward midnight I rolled myself in my blankets, slept an hour and a half, arose and ate more venison, tied two days' provisions to my belt, and set out for the summit, hoping to reach it ere the coming storm should fall. Jerome accompanied me a little distance above camp and indicated the way as well as he could in the darkness. He seemed loath to leave me, but, being reassured that I was at home and required no care, he bade me good-by and returned to camp, ready to lead his animals down the mountain at daybreak.

After I was above the dwarf pines, it was fine practice pushing up the broad unbroken slopes of snow, alone in the solemn silence of the night. Half the sky was clouded; in the other half the stars sparkled icily in the keen, frosty air; while everywhere the glorious wealth of snow fell away from the summit of the cone in flowing folds, more extensive and continuous than any I had ever seen before. When day dawned the clouds were crawling slowly and becoming more massive, but gave no intimation of immediate danger, and I pushed on faithfully, though holding myself well in hand, ready to return to the timber; for it was easy to see that the storm was not far off. The mountain rises ten thousand feet above the general level of the country, in blank exposure to the deep upper currents of the sky, and no labyrinth of peaks and canons I had ever been in seemed to me so dangerous as these immense slopes, bare against the sky.

The frost was intense, and drifting snow-dust made breathing at times rather difficult. The snow was as dry as meal, and the finer particles drifted freely, rising high in the air, while the larger portions of the crystals rolled like sand. I frequently sank to my armpits between buried blocks of loose lava, but generally only to my knees. When tired with walking I still wallowed slowly upward on all fours. The steepness of the slope—thirty-five degrees in some places—made any kind of progress fatiguing, while small avalanches were being constantly set in motion in the steepest places. But the bracing air and the sublime beauty of the snowy expanse thrilled every nerve and made absolute exhaustion impossible. I seemed to be walking and wallowing in a cloud; but, holding steadily onward, by half-past ten o'clock I had gained the highest summit.

I held my commanding foothold in the sky for two hours, gazing on the glorious landscapes spread maplike around the immense horizon, and tracing the outlines of the ancient lava-streams extending far into the surrounding plains, and the pathways of vanished glaciers of which Shasta had been the center. But, as

I had left my coat in camp for the sake of having my limbs free in climbing, I soon was cold. The wind increased in violence, raising the snow in magnificent drifts that were drawn out in the form of wavering banners glowing in the sun. Toward the end of my stay a succession of small clouds struck against the summit rocks like drifting icebergs, darkening the air as they passed, and producing a chill as definite and sudden as if ice-water had been dashed in my face. This is the kind of cloud in which snow-flowers grow, and I turned and fled.

Finding that I was not closely pursued, I ventured to take time on the way down for a visit to the head of the Whitney Glacier and the "Crater Butte." After I reached the end of the main sum-mit ridge the descent was but little more than one continuous soft, mealy, muffled slide, most luxurious and rapid, though the hissing, swishing speed attained was obsecured in great part by flying snow-dust—a marked contrast to the boring seal-wallowing upward struggle. I reached camp about an hour before dusk, hollowed a strip of loose ground in the lee of a large block of red lava, where firewood was abundant, rolled myself in my blankets, and went to sleep.

Next morning, having slept little the night before the ascent and being weary with climbing after the excitement was over, I slept late. Then, awaking suddenly, my eyes opened on one of the most beautiful and sublime scenes I ever enjoyed. A boundless wilderness of storm-clouds of different degrees of ripeness were congregated over all the lower landscape for thousands of square miles, colored gray, and purple, and pearl, and deep-glowing white, amid which I seemed to be floating; while the great white cone of the mountain above was all aglow in the free, blazing sunshine. It seemed not so much an ocean as a *land* of clouds—undulating hill and dale, smooth purple plains, and silvery mountains of cumuli, range over range, diversified with peak and dome and hollow fully brought out in light and shade.

I gazed enchanted, but cold gray masses, drifting like dust on a wind-swept plain, began to shut out the light, forerunners of the coming storm I had been so anxiously watching. I made haste to gather as much wood as possible, snugging it as a shelter around my bed. The storm side of my blankets was fastened down with stakes to reduce as much as possible the sifting-in of drift and the danger of being blown away. The precious bread-sack was placed safely as a pillow, and when at length the first flakes fell I was exultingly ready to welcome them. Most of my firewood was more than half rosin and would blaze in the face of the fiercest drifting; the winds could not demolish my bed, and my bread could be made to last indefinitely; while in case of need I had the means of making snowshoes and could retreat or hold my ground as I pleased.

Presently the storm broke forth into full snowy bloom, and the thronging crystals darkened the air. The wind swept past in hissing floods, grinding the snow into meal and sweeping down into the hollows in enormous drifts all the heavier particles, while the finer dust was sifted through the sky, increasing the icy gloom. But my fire glowed bravely as if in glad defiance of the drift to quench it, and, notwithstanding but little trace of my nest could be seen after the snow had leveled and buried it, I was snug and warm, and the passionate uproar produced a glad exceitement

Day after day the storm continued, piling snow on snow in weariless abundance. There were short periods of quiet, when the sun would seem to look eagerly down through rents in the clouds, as if to know how the work was advancing. During these calm intervals I replenished my fire—sometimes without leaving the nest, for fire and woodpile were so near this could easily be done—or busied myself with my notebook, watching the gestures of the trees in taking the snow, examining separate crystals under a lens, and learning the methods of their deposition as an enduring fountain for the streams. Several times, when the storm ceased for a few minutes, a Douglas squirrel came frisking from the foot of a clump of dwarf pines, moving in sudden interrupted spurts over the bossy snow; then, without any apparent guidance, he would dig rapidly into the drift where were buried some grains of barley that the horses had left. The Douglas squirrel does not strictly belong to these upper woods, and I was surprised to see him out in such weather. The mountain sheep also, quite a large flock of them, came to my camp and took shelter beside a clump of matted dwarf pines a little above my nest.

The storm lasted about a week, but before it was ended Sisson became alarmed and sent up the guide with animals to see what had become of me and recover the camp outfit. The news spread that "there was a man on the mountain," and he must surely have perished, and Sisson was blamed for allowing any one to attempt climbing in such weather; while I was as safe as any-body in the lowlands, lying like a squirrel in a warm, fluffy nest, busied about my own affairs and wishing only to be let alone. Later, however, a trail could not have been broken for a horse, and some of the camp furniture would have had to be abandoned. On the fifth day I returned to Sisson's, and from that comfortable base made excursions, as the weather permitted, to the Black Butte, to the foot of the Whitney Glacier, around the base of the mountain, to Rhett and Klamath Lakes, to the Modoc region and elsewhere, developing many interesting scenes and experiences.

But the next spring, on the other side of this eventful winter, I saw and felt still more of the Shasta snow. For then it was my fortune to get into the very heart of a storm, and to be held in it for a long time.

On the 28th of April (1875) I led a party up the mountain for the purpose of making a survey of the summit with reference to the location of the Geodetic monument. On the 30th, accompanied by Jerome Fay, I made another ascent to make some barometrical observations, the day intervening between the two ascents being devoted to establishing a camp on the extreme edge of the timber-line. Here, on our red trachyte bed, we obtained two hours of shallow sleep broken for occasional glimpses of the keen, starry night. At two o'clock we rose, breakfasted on a warmed tin-cupful of coffee and a piece of frozen venison broiled on the coals, and started for the summit. Up to this time there was nothing in sight that betokened the approach of a storm; but on gaining the sum-mit, we saw toward Lassen's Butte hundreds of square miles of white cumuli boiling dreamily in the sunshine far beneath us, and causing no alarm.

The slight weariness of the ascent was soon rested away, and our glorious morning in the sky promised nothing but enjoyment. At 9 A.M. the dry thermometer stood at 34° in the shade and rose steadily until at 1 P.M. it stood at 50°, probably influenced some-what by radiation from the sun-warmed cliffs. A common bumble-bee, not at all benumbed, zigzagged vigorously about our heads for a few moments, as if unconscious of the fact that the nearest honey flower was a mile beneath him.

In the mean time clouds were growing down in Shasta Valley — massive swelling cumuli, displaying delicious tones of purple and gray in the hollows of their sun-beaten bosses. Extending gradually southward around on both sides of Shasta, these at length united with the older field towards Lassen's Butte, thus encircling Mount Shasta in one continuous cloud-zone. Rhett and Klamath Lakes were eclipsed beneath clouds scarcely less brilliant than their own silvery disks. The Modoc Lava Beds, many a snow-laden peak far north in Oregon, the Scott and Trinity and Siskiyou Mountains, the peaks of the Sierra, the blue Coast Range, Shasta Valley, the dark forests filling the valley of the Sacramento, all in turn were obscured or buried, leaving the lofty cone on which we stood solitary in the sunshine between two skies—a sky of spotless blue above, a sky of glittering cloud beneath. The creative sun shone glorious on the vast expanse of cloudland; hill and dale, mountain and valley springing into existence responsive to his rays and steadily developing in beauty and individuality. One huge mountain-cone of cloud, corresponding to Mount Shasta in these newborn cloud-ranges, rose close alongside with a visible motion, its firm, polished bosses seeming so near and substantial that we almost fancied we might leap down upon them from where we stood and make our way to the lowlands. No hint was given, by anything in their appearance, of the fleeting character of these most sublime and beautiful cloud mountains. On the contrary they impressed one as being lasting additions to the landscape.

The weather of the springtime and summer, throughout the Sierra in general, is usually varied by slight local rains and dustings of snow, most of which are obviously far too joyous and life-giving to be regarded as storms—single clouds growing in the sunny sky, ripening in an hour, showering the heated landscape, and passing away like a thought, leaving no visible bodily remains to stain the sky. Snowstorms of the same gentle kind abound among the high peaks, but in spring they not unfrequently attain larger proportions, assuming a violence and energy of expression scarcely surpassed by those bred in the depths of winter. Such was the storm now gathering about us.

It began to declare itself shortly after noon, suggesting to us the idea of at once seeking our safe camp in the timber and abandoning the purpose of making an observation on the barometer at 3 P.M.,—two having already been made, at 9 A.M., and 12 M., while simultaneous observations were made at Strawberry Valley. Jerome peered at short intervals over the ridge, contemplating the rising clouds with anxious gestures in the rough wind, and at length declared that if we did not make a speedy escape we should be compelled to pass the rest of the day and night on the summit. But anxiety to complete my observations stifled my own instinctive promptings to retreat, and held me to my work. No inexperienced person was depending on me, and I told Jerome that we two mountaineers should be able to make our way down through any storm likely to fall.

Presently thin, fibrous films of cloud began to blow directly over the summit from north to south, drawn out in long fairy webs like carded wool, forming and dissolving as if by magic. The wind twisted them into ringlets and whirled them in a succession of graceful convolutions like the outside sprays of Yosemite Falls in flood-time; then, sailing out into the thin azure over the precipitous brink of the ridge they were drifted together like wreaths of foam on a river. These higher and finer cloud fabrics were evidently produced by the chilling of the air from its own expansion caused by the upward deflection of the wind against the slopes of the mountain. They steadily increased on the north rim of the cone, forming at length a thick, opaque, ill-defined embankment from the icy meshes of which snow-flowers began to fall, alternating with hail. The sky speedily darkened, and just as I had completed my last observation and boxed my instruments ready for the descent, the storm began in serious earnest. At first the cliffs were beaten with hail, every stone of which, as far as I could see, was regular in form, six-sided pyramids with rounded base, rich and sumptuous-looking, and fashioned with loving care, yet seemingly thrown away on those desolate crags down which they went rolling, falling, sliding in a network of curious streams.

After we had forced our way down the ridge and past the group of hissing fumaroles, the storm became inconceivably violent. The thermometer fell 22° in a few minutes, and soon dropped below zero. The hail gave place to snow, and darkness came on like night. The wind, rising to the highest pitch of violence, boomed and surged amid the desolate crags; lightning-flashes in quick succession cut the gloomy darkness; and the thunders, the most tremendously loud and appalling I ever heard, made an almost continuous roar, stroke following stroke in quick, passionate succession, as though the mountain were being rent to its foundations and the fires of the old volcano were breaking forth again.

Could we at once have begun to descend the snow-slopes leading to the timber, we might have made good our escape, however dark and wild the storm. As it was, we had first to make our way along a dangerous ridge nearly a mile and a half long, flanked in many places by steep ice-slopes at the head of the Whitney Glacier on one side and by shattered precipices on the other. Apprehensive of this coming darkness, I had taken the precaution, when the storm began, to make the most dangerous points clear to my mind, and to mark their relations with reference to the direction of the wind. When, therefore, the darkness came on, and the bewildering drift, I felt confident that we could force our way through it with no other guidance. After passing the "Hot Springs" I halted in the lee of a lava-block to let Jerome, who had fallen a little behind, come up. Here he opened a council in which, under circumstances sufficiently exciting but without evincing any bewilderment, he maintained, in opposition to my views, that it was impossible to proceed. He firmly refused to make the venture to find the camp, while I, aware of the dangers that would necessarily attend our efforts, and conscious of being the cause of his present peril, decided not to leave him.

Our discussions ended, Jerome made a dash from the shelter of the lava-block and began forcing his way back against the wind to the "Hot Springs," wavering and struggling to resist being carried away, as if he were fording a rapid stream. After waiting and watching in vain for some flaw in the storm that might be urged as a new argument in favor of attempting the descent, I was compelled to follow. "Here," said Jerome, as we shivered in the midst of the hissing, sputtering fumaroles, "we shall be safe from frost." "Yes," said I, "we can lie in this mud and steam and sludge, warm at least on one side; but how can we protect our lungs from the acid gases, and how, after our clothing is saturated, shall we be able to reach camp without freezing, even after the storm is over? We shall have to wait for sunshine, and when will it come?"

The tempered area to which we had committed ourselves extended over about one fourth of an acre; but it was only about an eighth of an inch in thickness, for the scalding gas-jets were shorn off close to the ground by the oversweeping flood of frosty wind. And how lavishly the snow fell only mountaineers may know. The crisp crystal flowers seemed to touch one another and fairly to thicken the tremendous blast that carried them. This was the bloom-time, the summer of the cloud, and never before have I seen even a mountain cloud flowering so profusely.

When the bloom of the Shasta chaparral is falling, the ground is sometimes covered for hundreds of square miles to a depth of half an inch. But the bloom of this fertile snow-cloud grew and matured and fell to a depth of two feet in a few hours. Some crystals landed with their rays almost perfect, but most of them were worn and broken by striking against one another, or by rolling on the ground. The touch of these snow-flowers in calm weather is infinitely gentle—glinting, swaying, settling silently in the dry mountain air, or massed in flakes soft and downy. To lie out alone in the mountains of a still night and be touched by the first of these small silent messengers from the sky is a memorable experience, and the fineness of that touch none will forget. But the storm-blast laden with crisp, sharp snow seems to crush and bruise and stupefy with its multitude of stings, and compels the bravest to turn and flee.

The snow fell without abatement until an hour or two after what seemed to be the natural darkness of the night. Up to the time the storm first broke on the summit its development was remarkably gentle. There was a deliberate growth of clouds, a weaving of translucent tissue above, then the roar of the wind and the thunder, and the darkening flight of snow. Its subsidence was not less sudden. The clouds broke and vanished, not a crystal was left in the sky, and the stars shone out with pure and tranquil radiance.

During the storm we lay on our backs so as to present as little surface as possible to the wind, and to let the drift pass over us. The mealy snow sifted into the folds of our clothing and in many places reached the skin. We were glad at first to see the snow packing about us, hoping it would deaden the force of the wind, but it soon froze into a stiff, crusty heap as the temperature fell, rather augmenting our novel misery.

When the heat became unendurable, on some spot where steam was escaping through the sludge, we tried to stop it with snow and mud, or shifted a little at a time by shoving with our heels; for to stand in blank exposure to the fearful wind in our frozen-and-broiled condition seemed certain death. The acrid incrustations sublimed from the escaping gases frequently gave way, opening new vents to scald us; and, fearing that if at any time the wind should fall, carbonic acid, which often formed a considerable portion of the gaseous exhalations of volcanoes, might collect in sufficient quantities to cause sleep and death, I warned Jerome against forgetting himself for a single moment, even should his sufferings admit of such a thing.

Accordingly, when during the long, dreary watches of the night we roused from a state of half-consciousness, we called each other by name in a frightened, startled way, each fearing the other might be benumbed or dead. The ordinary sensations of cold give but a faint conception of that which comes on after hard climbing with want of food and sleep in such exposure as this. Life is then seen to be a fire, that now smoulders, now brightens, and may be easily quenched. The weary hours wore away like dim half-forgotten years, so long and eventful they seemed, though we did nothing but suffer. Still the pain was not always of that bitter, intense kind that precludes thought and takes away all capacity for enjoyment. A sort of dreamy stupor came on at times in which we fancied we saw dry, resinous logs suitable for campfires, just as after going days without food men fancy they see bread.

Frozen, blistered, famished, benumbed, our bodies seemed lost to us at times—all dead but the eyes. For the duller and fainter we became the clearer was our vision, though only in momentary glimpses. Then, after the sky cleared, we gazed at the stars, blessed immortals of light, shining with marvelous brightness with long lance rays, near-looking and new-looking, as if never seen before. Again they would look familiar and remind us of star-gazing at home. Oftentimes imagination coming into play would present charming pictures of the warm zone below, mingled with others near and far. Then the bitter wind and the drift would break the blissful vision and dreary pains cover us like clouds. "Are you suffering much?" Jerome would inquire with pitiful faintness. "Yes," I would say, striving to keep my voice brave, "frozen and burned; but never mind, Jerome, the night will wear away at last, and to-morrow we go a-Maying, and what campfires we will make, and what sunbaths we will take!"

The frost grew more and more intense, and we became icy and covered over with a crust of frozen snow, as if we had lain cast away in the drift all winter. In about thirteen hours—every hour like a year—day began to dawn, but it was long ere the summit's rocks were touched by the sun. No clouds were visible from where we lay, yet the morning was dull and blue, and bitterly frosty; and hour after hour passed by while we eagerly watched the pale light stealing down the ridge to the hollow where we lay. But there was not a trace of that warm, flushing sunrise splendor we so long had hoped for.

As the time drew near to make an effort to reach camp, we became concerned to know what strength was left us, and whether or not we could walk; for we had lain flat all this time without once rising to our feet. Mountaineers, however, always find in themselves a reserve of power after great exhaustion. It is a kind of second life, available only in emergencies like this; and, having proved its existence, I had no great fear that either of us would fail, though one of my arms was already benumbed and hung powerless.

At length, after the temperature was somewhat mitigated on this memorable first of May, we arose and began to struggle homeward. Our frozen trousers could scarcely be made to bend at the knee, and we waded the snow with difficulty. The summit ridge was fortunately wind-swept and nearly bare, so we were not compelled to lift our feet high, and on reaching the long home slopes laden with loose snow we made rapid progress, sliding and shuffling and pitching headlong, our feebleness accelerating rather than diminishing our speed. When we had descended some three thousand feet the sunshine warmed our backs and we began to revive. At 10 A.M. we reached the timber and were safe.

Half an hour later we heard Sisson shouting down among the firs, coming with horses to take us to the hotel. After breaking a trail through the snow as far as possible he had tied his animals and walked up. We had been so long without food that we cared but little about eating, but we eagerly drank the coffee he prepared for us. Our feet were frozen, and thawing them was painful, and had to be done very slowly by keeping them buried in soft snow for several hours, which avoided permanent damage. Five thousand feet below the summit we found only three inches of new snow, and at the base of the mountain only a slight shower of rain had fallen, showing how local our storm had been, notwithstanding its terrific fury. Our feet were wrapped in sacking, and we were soon mounted and on our way down into the thick sunshine—"God's Country," as Sisson calls the Chaparral Zone. In two hours' ride the last snow-bank was left behind. Violets appeared along the edges of the trail, and the chaparral was coming into bloom, with young lilies and larkspurs about the open places in rich profusion. How beautiful seemed the golden sunbeams streaming through the woods between the warm brown boles of the cedars and pines! All my friends among the birds and plants seemed like *old* friends, and we felt like speaking to every one of them as we passed, as if we had been a long time away in some far, strange country.

In the afternoon we reached Strawberry Valley and fell asleep. Next morning we seemed to have risen from the dead. My bedroom was flooded with sunshine, and from the window I saw the great white Shasta cone clad in forests and clouds and bearing them loftily in the sky. Everything seemed full and radiant with the freshness and beauty and enthusiasm of youth. Sisson's children came in with flowers and covered my bed, and the storm on the mountaintop vanished like a dream.

10

On The Vema In Southern Waters

LEONARD ENGEL

The Lamont Geological Observatory of Columbia University, under the inspiration of Maurice Ewing, a geophysicist, has been a leader in the remarkable developments that have marked submarine geology since World War II. The next selection by Leonard Engle, a writer on scientific subjects for "The New Yorker," "Harper's. and several other magazines, suggests not only the sophistication of modern technological research, but also the realities of life aboard an ocean-going research vessel.

About seven-tenths of the earth's surface is covered by the oceans. Consequently, man can hardly claim to know the earth or understand its history or structure until he knows what lies within the sea and in the huge fraction of the earth's crust concealed by the sea.

In the past seven or eight years, an impressive volume of new knowledge of the oceans and the ocean bottom has come from the voyages of a single ship, a 37-year-old three-masted iron schooner named *Vema. Vema* has found evidence that the earth's crust is thinner below the sea than below the land and that there are differences, extending well into the earth's interior, between the land and sea provinces. *Vema* has amassed much of the data establishing the existence of one of the world's most majestic geographical features— the Mid-Ocean Ridge, a massive 40,000-mile-long mountain range

Reprinted from the **Columbia Forum,** Spring, 1961, Volume IV, Number 2. Copyright 1961 by the Trustees of Columbia University in the City of New York.

with a puzzling crack running down its crest, meandering through all the oceans excepting the North Pacific. She has recorded direct evidence of the drop in sea level during the Ice Ages, and of the rise in the ocean's temperature that set in with the retreat of the ice.

This is but part of what has been accomplished with the research ship of the Lamont Geological Observatory, the Columbia University center for research in the earth sciences. A few months ago, I joined the scientific party aboard *Vema* for the second time, to learn more about how the oceanic 70 per cent of the world is probed and what it discloses. My choice was fortunate; I was on board during part of *Vema's* remarkable 1959-60 cruise, when the Columbia vessel became the first ship in history to circle the globe and sail both the Arctic and Antarctic on a single voyage.

As I knew, sea research demands all the ingenuity of modern science—plus a strong back and a dash of iron in the soul. Wherever else the tasks of ocean science may be performed, their focus is a rolling, tossing ship's deck. On *Vema's* deck—almost always wet with spray if not with green water cascading over the rail—the work went on literally around the clock. In recent decades most oceanographic work has been carried out with small vessels capable of short cruises only; few oceanographic institutions have been able to operate bigger craft. The Lamont Geological Observatory is one of the few institutions that has made long cruises a regular part of its sea research program. *Vema's* 1959-60 cruise, which lasted a few days less than a year, was her sixth of six months or longer in as many years. While *Vema* is hardly large, her basic operating cost (exclusive of the direct cost of the experiments and studies she carries out) is more than $1,000 a day. Lamont justifies this formidable expenditure—which is furnished mainly by government grants—by sending her out to distant waters beyond the reach of a smaller ship, and by utilizing her for a multiplicity of simultaneous projects and so making every hour of ship-time count—an arrangement responsible in no small measure for the ship's and Lamont's unusual productiveness.

Thus, when *Vema* left New York at the start of the 1959-60 voyage, more than two dozen distinct research projects were included in her cruise plans. Among them: exploration of those sections of the Mid-Ocean Ridge that lie beneath the Indian Ocean, to see, in particular, whether they showed the rift previously found in other parts of the Ridge. (The crack was there, in all parts of the Indian Ocean Ridge that the men from Lamont examined.) In the Antarctic, *Vema* was to collect water samples for radio-carbon studies to help settle a long-standing question: how long it takes cold water from the Antarctic to reach the equator. The effect of the circulation of the oceans on the world's climate is, of course, enormous. Also on the ship's schedule was extensive charting of undersea crustal strata in several parts of the world by the ingenious technique of seismic shooting—firing explosives in the sea and timing the travel of the sound waves through the layers of sediment and rock below the ocean floor. And deep-water microbes would be collected for investigation as possible sources of new antibiotics.

Several key instruments were to record and measure wherever the schooner went. The precision depth recorder: for there are still immense areas of the ocean that have never been plumbed accurately. The magnetometer: for measuring variations in the earth's magnetic field. The piston corer: a device for punching out and bringing up samples of the ocean floor with sedimentary layers undisturbed. Withal, a sharp watch was to be kept for the unexpected discovery—and there was one.

I joined *Vema* when she called at Ushuaia, a port in the Argentine half of Tierra del Fuego and one of the most southerly inhabited points in the world. I flew down with Maurice Ewing, director of Lamont, who was to be ship's chief scientist for the next two months. *Vema* had already been on cruise for seven months, in all three major oceans—Atlantic, Indian and Pacific—and in the Antarctic.

As a rule, each leg of a *Vema* cruise has one or more special missions, to be carried out in addition to the work she does wherever she goes. On this leg, the main mission was to be a heavy program of seismic shooting in cooperation with an Argentine navy vessel, the *Capitan Canepa*. The two ships were to plumb the structure of the continental shelf at the lower end of South America and along the east coast of the continent as far north as Rio de Janeiro.

Vema was no stranger to these waters. In fact, she had been instrumental in opening them to modern oceanographic study. When the Columbia vessel first went far south in 1956, it had been widely held that the windswept waters of the extreme South Atlantic and South Pacific were too rough for sustained oceanographic work. *Vema* had proved that a stout ship and determined men can carry on systematic sea research almost anywhere — and thereby added greatly to knowledge of one of Earth's farthest corners. This was to be *Vema's* fifth campaign in or near Cape Horn.

The seismic shooting program on this leg of *Vema's* cruise had several specific objectives. One was to seek geological links between the Falkland Islands and South America. Some geologists have suggested that the Falklands are not, geologically speaking, part of South America, but were left behind when (as these geologists believe) South America and Africa drifted apart eons ago; the presence or absence of a structural connection between the Falklands and South America would provide a test of their theory. Another aim of the seismic program was to clear up uncertainties in data from previous South Atlantic seismic surveys. A third was to chart sedimentary basins. The basins of the South American continental shelf are of wide interest not only because several probably contain oil, but also because they differ in important geological respects from a number of the basins off North America.

Maurice Ewing calls seismic shooting the geophysicist's "X-ray eye into the earth's crust." Essentially, it is a scheme for using sound waves, generated by explosions, for probing deeply hidden structures man can scarcely hope ever to see for himself. Seismic shooting has been used as an aid to geophysical research and in oil prospecting since the end of World War I; the technique was first taken to sea by Ewing, Albert P. Crary, and H. M. Rutherford before World War II.

Seismic shooting can be carried out in a variety of ways. Thus, in "reflection shooting," explosion echoes are simply bounced off crustal strata. *Vema* and *Canepa,* however, were to employ "seismic refraction," a more sophisticated procedure which yields more information. In seismic refraction, the aim is to pick up sound waves that have entered and been bent or refracted along crustal strata before they return to the surface of the earth. Sound velocities are calculated as sound waves travel through several layers of sediment and rock. This procedure not only provides clues to the composition of the earth's crust (different classes of sediment and rock have characteristic velocities for the transmission of sound), but it also makes it possible to deduce the depth and thickness of strata.

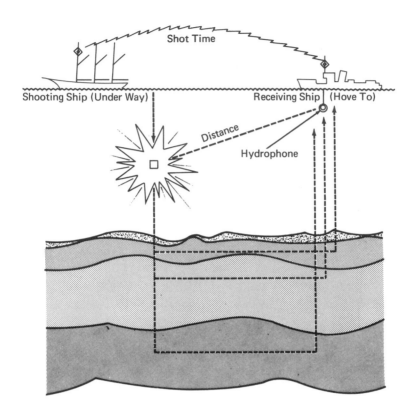

Fig. 10-1 HOW SEISMIC REFRACTION WORKS: The shooting ship drops a shot. The shot goes off, sending waves in all directions. Radio signal to the receiving ship gives the shot time. Those sound waves from the shot which go by the shortest path through the water to the hydrophone gives the position of the shot, its distance from the shooting ship. Other sound waves enter various layers of the ocean bottom and are refracted back. The time of arrival of these waves at the hydrophone tells the speed of the waves through each layer and the depth of the layer, the speed indicating the material penetrated. The record of the arrival times of all of the shock waves sent out by the explosion thus indicates the materials to be drawn in a geological cross-section.

The minimum requirements for refraction shooting at sea are two ships and a fund of skill in coordinating tricky details. Two ships are necessary because, to pick up refraction shots, the shooting and listening stations must be some distance apart. The ships take turns shooting and listening.

The shooting ship must steam along a prescribed course through the area to be studied, firing charges ranging in size from half-pound TNT blocks to 300-pound Navy depth charges at carefully timed intervals. As each charge goes over the side, the time and the ship's position and the water depth must be noted, to permit computation of the shot's location when it explodes. The shooting vessel must pick up the sound of the explosion it has set off and transmit it by radio to the listening ship, to give the listening ship "zero time" for its calculations.

Meanwhile, the listening ship is hove to, and has suspended hydrophones (usually two) in the water. Since the charges dropped by the shooting ship ordinarily go off well above the ocean floor, only a very small part of the sound energy generated by the explosion actually penetrates the ocean floor and travels through it toward the listening ship. As a result, in order to hear these "ground waves" at all, the listening ship must be on "silent ship," its engines, generators and other noise-making machinery turned off. The hydrophones must also be well away from the ship and below the surface waves, floating free, no tension in the cables connecting them to the ship. The sensitive recording gear aboard must be set in motion at precisely the right moment.

We made the first seismic shoot of this particular cruise leg just two days after leaving Ushuaia. *Vema* left the Tierra del Fuego port a day and a half ahead of *Capepa,* using the headstart time to explore the continental shelf off Tierra del Fuego with one of *Vema's* newest instruments, the sub-bottom depth recorder. The SDR sounds the bottom with loud electric sparks generated at the end of a cable trailed behind the ship. In water up to 1,000 feet in depth, the "sparker" is uncanny at picking out ancient shorelines, one-time river mouths, old beaches, buried lagoons—all under sediment now, and to be further explored with still better instruments some day.

The first shoot, *Canepa* firing and *Vema* receiving, took place in the small hours of the morning, as soon as the two ships were in close radio contact. Over the next four weeks, seismic work went on for up to twenty hours a day. More than 100 profiles (the geophysicist's term for a series of shots over a given line, generally Twenty-five to forty miles long) were completed. Excepting brief halts for repairs of equipment or to steam to new positions, there were only three breaks in the nearly ceaseless round of shooting and receiving. One was due to a storm; the shooting ship could not make enough speed in the prescribed direction to get away from its own charges. Another occurred when *Canepa* had to go to port for fuel and ammunition—and to take a member of *Vema's* scientific party, stricken with appendicitis, to hospital. The third interruption came when *Vema* discovered a gigantic undersea canyon, perhaps the largest canyon on either land or sea. But more of that later.

The best way to tell what seismic shooting at sea is like is to relate what happened one Saturday night, when we were not far from the Falklands. It was a bad night, with a fifty-knot gale blowing. I choose it, not as typical of ocean research—gales don't blow all the time—but to show what can be accomplished even under trying conditions by a determined scientific party.

That afternoon, the weather had been bad enough for William Ludwig, a Lamont seismic specialist aboard *Canepa,* to call over the radio, as *Vema* shot to *Canepa,* that he hoped our "shooter" was lashed to the deck. The wind was blowing harder still when *Vema* hove to and prepared to receive. Three men brought aboard the heavy magnetometer "fish," which is towed at the end of a 500-foot cable when *Vema* is under way. The laboratory watch put the bathythermograph over the side, another metal fish that records water temperatures down to 900 feet. (Knowledge of water temperatures is essential to interpretation of marine seismic records.) Two other members of the scientific party broke out the hydrophones. Tom De Witt, the young engineering student who was to keep the record book and operate the recording gear for the first part of the station, pressed the buzzer for "silent ship."

Vema was ready just after 7 P.M. *Canepa's* executive officer, Lt. Federico Aragno—who had had two years of training at the Scripps Institution of Oceanography and who did the bookkeeping on *Canepa's* shooting—called on the radio: "*Canepa* to *Vema*. This will be shot No. 402. Three hundred pounds. Six-foot fuse. Estimated burning time, 90 seconds. One minute warning *now.*" Then, a minute later, "Over the side, the charge is over the side."

Aboard *Vema,* De Witt watched the chronometer. Twenty seconds before the charge was due to explode, he sang out, "Slack." Out on deck, two figures hooded against the spray threw loops of hydrophone cable over the rail and the hydrophones floated free in the black, frigid water. Five seconds before, De Witt had switched on the recording gear. Then, almost exactly on the expected second, a hiss was heard on the radio—and was automatically put on the hydrophone record. It was the "shot instant." Forty seconds later, the ground waves and the direct wave had all come in. De Witt shut off the recording camera. A member of the party stationed in the darkroom quickly developed the record. In hardly more than a minute, Ewing—who does nearly all the record reading himself when he is aboard—was studying the film to make sure a good record had been obtained and to determine the size and schedule of subsequent shots. Meanwhile, the "slackers" were hauling in the hydrophone cables to be ready to slack for the next shot.

When I came on watch at midnight, *Canepa* had not yet finished shooting in toward the stationary *Vema*. She had still to shoot past *Vema* and then on out some thirty-five miles—a task that might take, in that sea, six or seven hours. I would be slacking throughout my watch and, if we were short of men, perhaps longer.

The ship was rolling heavily. In the laboratory, Ewing and the men at the camera and in the darkroom had had to wedge themselves into place. Outside, even *Vema's* high stern deck, where the slackers worked, was periodically swept by water as the schooner slapped her stern down, raising a heavy spray, or rolled her rail under. It was hard, wet work hauling the hydrophone cables in after every shot—shots were coming at five-minute intervals now—and keeping the coils of cable from swimming around the deck.

It seemed to me that the noise of the sea and the slap of the ship must be drowning out many sound wave arrivals. A while later, when the interval between shots had become longer, my slacking partner and I were able to duck into the laboratory between shots. We found Ewing smiling broadly. We were obtaining good records, with unambiguous "shot instant" traces and clean sound wave arrivals. Tonight's records would be translated into stratigraphic charts of the earth's crust—some 1,200 feet below the ship—after *Vema's* return to New York.

For the next nine days, the round of seismic work continued without interruption. The weather improved as we zigzagged our way north out of the sub-Antarctic Zone. We also trawled frequently for bottom life, did bottom photography, and "cored" at least once a day.

When the two ships reached the latitude of Argentina's main naval base, Puerto Belgrano, *Canepa* headed for port to refuel and restock with ammunition, and to take the ailing man from

Vema to the Puerto Belgrano naval hospital. *Vema,* meanwhile, headed for deep water in order to test some new equipment for deep-ocean bottom studies. Twelve hours later, we were crossing the continental slope, where the continental shelf plunges abruptly to the true ocean floor—a region where submarine canyons are particularly apt to be found. A sharp watch on the precision depth recorder was ordered.

A small canyon turned up early the next morning. Then, the following day, there it was—a whopper. Its outline, as picked out by the probing sound beam of the echo fathometer and traced on the depth recorder's moving roll of paper, was unmistakable.

Plans had previously been made for *Vema* and *Canepa* to rendezvous and resume shooting that evening. The arrangements were changed forthwith. The Argentine ship—a fully equipped oceanographic vessel and the only one registered in the southern hemisphere equipped with a depth recorder comparable to *Vema's* —would join the Columbia schooner to survey the canyon as far and as thoroughly as possible.

Over the next forty-eight hours, the two ships maneuvered back and forth across the hidden abyss, with Ewing (a veteran collector of submarine canyons) signaling the turns for both. Aboard *Vema,* the laboratory watch marked the time, the log reading and the ship's course and speed every few minutes on the depth recorder record. Another young man plotted the readings on a chart spread out on a laboratory bench. (Others were doing the same aboard *Canepa;* later, the Argentine ship's operations officer would come aboard to work with *Vema's* navigator to re-concile the two sets of charts.)

Slowly, the outline of the canyon grew. At first, it ran parallel to the Argentine coast, some 300 miles away. We began to wonder what we had found; submarine canyons are usually associated with ancient river beds and generally run out to sea. All of a sudden, the canyon made a sharp turn to the east—toward the sea.

The two ships made nearly forty carefully-plotted crossings of of the canyon altogether. They revealed an astonishing prize, a canyon dwarfing the Grand Canyon of the Colorado and perhaps the biggest in the world: half a mile to a mile deep, five to twelve miles wide at the top, nearly a mile wide at the bottom, and at least several hundred miles long. Seventy miles of the canyon— which is centered about halfway between Buenos Aires and the

Falklands Islands—were explored in detail and another fifty-five miles were less completely surveyed.

Vema halted several times to put her corer down into the bottom of the canyon—an operation that took two hours or more, since the bottom of the canyon was in some places nearly 15,000 feet below the ocean's surface. The time was well spent, for the corer brought back smooth, round pebbles of the kind found in river beds. There could be no doubt. The canyon (which has not yet been given a name) owed its existence to an ancient river, probably dating to the Ice Ages, when the Argentine coast extended farther seaward than at present. But it is doubtful that the river had carved it, for the canyon had been under water even then. Ewing believes it must have been carved out by powerful "turbidity currents," generated when heavy loads of sediment, deposited by the river at the edge of the continental shelf, slid down the edge of the continental slope.

Ewing and a number of his colleagues hold that turbidity currents generated by undersea avalanches have been a primary force in the shaping of many prominent features of the ocean floor. Their theory was first put forward about a dozen years ago, not long after the discovery of the Abyssal Plain, a remarkably flat region of the ocean bottom in the western half of the North Atlantic. Ewing, who had discovered the Abyssal Plain during his first long research voyage, found it hard to believe so flat an expanse of ocean bottom could have been formed by the slow settling of sediment from above. He felt that some sort of sediment-bearing current must have flowed over the plain, depositing mud and filling every hollow, as a muddy river does when it overflows its banks and spreads over the surrounding land. Further evidence of ocean currents capable of moving large masses of material over great distances was added piece by piece through the next several years. The Hudson Canyon, off New York, was followed out into an area that was certainly never dry land and to which no river current could have reached; layers of sand were discovered in the deep ocean bottom, hundreds of miles from where they could have originated. The peculiar pattern of transatlantic cable breaks following the famous 1929 earthquake in the Grand Banks off Newfoundland seems now most thoroughly explained as a consequence of a high-speed turbidity current. Not all marine geologists and geophysicists are convinced of the reality of these currents.

But now Ewing had a powerful new argument in the huge slash *Vema* had uncovered in the ocean floor.

I left *Vema* when she called at Buenos Aires. Her cruise, however, was far from over. She had yet to make the long run from Rio de Janeiro to Nova Scotia for a seismic campaign in the Arctic with the Canadian research vessel *Slackville*. She did not return to New York until fall.

And her stay in New York was brief—just long enough for refitting and the installation of a new engine. Early in December, she was off again, to add not only to knowledge of Earth's oceanic provinces, but to a cruising record without parallel in modern oceanography.

As this went to press, word was received of the death of John F. Hennion, one of the Lamont Observatory's specialists in marine seismic work. Mr. Hennion was killed in the premature explosion of a TNT charge aboard Vema *during a seismic study off Chile. This was the first fatal accident in more than 25 years of seismic exploration of the ocean bed. The loss is a grievous one.*

11
Dinosaurs and Victorians

EDWIN H. COLBERT

The next two selections deal with men and events in the history of two major geologic fields, vertebrate paleontology and radiometric dating.

Dinosaurs, which still excite the lay public, were at the forefront of geologic research during the 19th Century. Edwin H. Colbert, of the American Museum of Natural History, is one of the leading modern students of these extinct reptiles and the author of several popular books. He takes us behind the scenes of the early discoveries in this chapter from his book, Dinosaurs. *His concluding discussion of the affair between Professors Marsh and Cope is not in the least overplayed. Geology has had its professional feuds, but this was one of the most acrimonious.*

Dinosaurs are Discovered

It does seem strange to think that a mere century and a half ago dinosaurs were quite unknown. So far as man was concerned, at that time there were no dinosaurs—there was not even the concept of a dinosaur. Dinosaurs were still to be discovered, studied, and described.

Without doubt people had seen dinosaur bones in the rock, not only 150 years ago but far back in time, probably back into the days before recorded history when throughout the world men were still primitive savages. But until the early days of the last century even the most knowledgeable men had looked at dinosaur bones with unseeing eyes; they had looked but they had not understood what they had seen. In early historic times, and even through the centuries of medieval history in Europe, people were prone to attribute large fossil bones to giants—a most natural deduction to folks who believed in mythological tales. Indeed, some of the ancient legends of giants may have been based in part on the discoveries of large fossil bones; usually the bones of extinct elephants, but perhaps now and then the bones of other large extinct vertebrates, including dinosaurs.

But by the beginning of the nineteenth century rational explanations for natural phenomena were being sought. No longer were intelligent people satisfied with tales that taxed their credulity; things in nature must be explained in accordance with uniform natural laws. It was the age of reason.

The first dinosaur to be properly described was discovered on a morning in March, 1822, in Sussex, England. Dr. Gideon Mantell, a physician in the city of Lewes, was an unusual man of wide interests, one of that band of devoted amateurs in the Western world who did so much to found the natural sciences during the eighteenth and nineteenth centuries. He was particularly interested in fossils, and for some years had spent his spare hours in the countryside of southern England, searching for the petrified remains of extinct animals. His young wife often accompanied him on many of his excursions into the country, and thus she had gained some experience in looking for fossils in the field.

On this particular day Dr. Mantell drove some miles outside Lewes to attend to one of his patients, and his wife went along for the ride. While he was in the house of his patient, Mrs. Mantell walked up and down to pass the time, and in the course of her walk she saw some rocks that had been piled by the road, to be used for repairs of the surface. As she was looking at the pile of rubble she noticed an object in a rock that looked most unusual, something that had the definite form and the shining surface of a fossil. She picked up the piece of rock containing the specimen, and saw that the object that had attracted her attention was a fossil tooth. Of course, she showed it to her husband.

It was a tooth quite different from any fossil that Dr. Mantell had seen before, and it aroused his curiosity. So, during the ensuing weeks he returned often to the country near where it had been found, and located not only more teeth like it, but also a number of fossil bones as well. All of these fossils were unfamiliar to him.

In those early days of paleontology there was one great authority to whom other students of fossil vertebrates might turn for help, namely Baron Georges Cuvier of Paris. Cuvier was a great leader in the study of animals with backbones, and with good reason he has been called the founder of comparative anatomy and the father of vertebrate paleontology. He was a man of immense erudition and of extraordinary energy, who because of his preeminent position as a scientist had lived safely through all of the troubled events that befell France during his long lifetime, from the Revolution, through Napoleonic times, into the days of the Bourbon Restoration. In the 1820's Cuvier was nearing the end of his wonderfully productive life, and he had behind him a vast amount of experience resulting from long years of research on the vertebrates.

Mantell sent his fossils to Cuvier, who identified the teeth as those of an extinct rhinoceros and the bones as belonging to an extinct hippopotamus. In spite of the high authority of Cuvier, Mantell could not accept these identifications because he was certain that the fossils had come from rocks far too ancient to have contained the remains of advanced mammals.

He therefore sought the advice of Dean William Buckland of Oxford University, one of the founders of paleontology in England, a man with whom Mantell had worked. Buckland urged Mantell to do nothing at the time; above all, to publish nothing. Buckland was ready to accept Cuvier's identifications, and this led him to the conclusion that the fossils were remains that had been found in superficial deposits (the *diluvium,* as such deposits were then called) of comparatively late geological age.

But Mantell was not convinced, either by Cuvier or by Buckland, so he decided to make a detailed study of the fossils on his own behalf. At first he made little progress, for he could locate nothing ancient or recent that resembled the fossils in his possession. Then one day in 1825, when Mantell was comparing his fossils with various specimens in the museum of the Royal College of Surgeons in London, he met a naturalist named Samuel Stutchbury, who had just been studying an iguana, a large recent lizard inhabiting Mexico and Central America. Stutchbury immediately saw that Mantell's fossil teeth resembled the teeth of an iguana, but on a much magnified scale. Here was the clue that Mantell needed, and in that same year he published a description of the fossil teeth under the name of *Iguanodon,* which can be translated literally as "iguana tooth".

When Mantell's description appeared, Cuvier very handsomely acknowledged his own previous error of identification in a statement that may very freely be translated as follows:

"Do we not have here a new animal, an herbivorous reptile? And even as among the modern terrestrial mammals it is within the herbivores that one finds the species of the largest size, so among the reptiles of another time, then the only terrestrial animals, were not the greatest of them nourished on vegetables? Time will confirm or reject this idea. If teeth adhering to a jaw would be found, the problem might then be solved."

In the meantime, in 1824, Dean Buckland had described another large reptile, upon the basis of a lower jaw and various parts of the skeleton which had been found at no great distance from Oxford. This was clearly a carnivorous, or meat-eating, reptile, with large blade-like teeth, quite in contrast to the rather leaf-shaped teeth of *Iguanodon.* Buckland gave the name of *Megalosaurus* to the fossil remains.

Megalosaurus was obviously a very large reptile. Cuvier, with whom Buckland had consulted before publishing the description of *Megalosaurus,* advanced the opionion that, judging by the size of the bones, this reptile must have been over forty feet long. Actually, the skeleton of *Megalosaurus* is about half as long as Cuvier had estimated it.

Dinosauria

Iguanodon and *Megalosaurus* were thus recognized as large reptiles that had lived upon the land, but what kind of reptiles they may have been was a question as yet unanswered. More fossils were needed.

In 1833 Mantell moved from Lewes to Brighton, because he felt that in the fashionable resort then frequented by royalty he might get adequate support for his work on fossils. (By now he was thoroughly committed to the acquisition and study of fossils, and medicine was to him a necessary money-grubbing profession and little more.) Just before he moved to Brighton he had described a partial reptilian skeleton he had found, and he called it *Hylaeosaurus.*

Two years later a discovery was made near Maidstone, Kent, that at last was to give something of a clue as to the nature of *Iguanodon.* In a quarry a slab of rock was found that contained *Iguanodon* teeth in *association* with various bones of the skeleton. The discovery so prophetically hoped for by Cuvier a decade earlier had finally come to light. *Iguanodon* now began to take form as an animal, rather than as disassociated parts, in the minds of Mantell and other early students of fossils.

By the middle of the 1830's, therefore, three large fossil reptiles had been discovered in the south of England, and named, but beyond that their respective places in the tree of animal life were unknown. These three reptiles, *Iguanodon, Hylaeosaurus,* and *Megalosaurus,* were in truth dinosaurs, but no one realized it. It was to be several years before another Englishman, Richard Owen, would create the concept of dinosaurs.

Richard Owen was the first great comparative anatomist and vertebrate paleontologist of England, a worthy successor of Cuvier. He had a long and distinguished career. In 1842 much of this career was still ahead of him, yet even then, as a comparatively young man, he was Hunterian Professor of Anatomy at the Museum of the Royal College of Surgeons, and his word on the anatomical characters and the relationships of the vertebrates carried a great deal of weight. It was in that year, in one written sentence, that Owen brought the dinosaurs into being, in a report on British fossil reptiles, published by the British Association for the Advancement of Science. Here is what Owen said:

"The combination of such characters, some, as the sacral ones, altogether peculiar among Reptiles, others borrowed, as it were, from groups now distinct from each other, and all manifested by creatures far surpassing in size the largest of existing reptiles, will, it is presumed, be deemed sufficient ground for establishing a distinct tribe or suborder of Saurian Reptiles, for which I would propose the name of *Dinosauria.*"

In a little more than ten years after this, dinosaurs were to be widely known among Englishmen in many walks of life.

Owen, Hawkins, and Iguanodon

The first world's fair was the Crystal Palace exposition of 1851, a new kind of fair to celebrate the technology of steam power and the Industrial Revolution that was rapidly making the Western world into something different from what it had been a few decades earlier. In London a great building of iron and glass was erected in Hyde Park, in which many exhibits were displayed and in which concerts and other programs were performed. The Crystal Palace was so popular during the time of the exposition that in 1854 it was removed to a spacious site away from the center of the city, and in the landscaped grounds around the relocated Crystal Palace there were constructed life-sized restorations in concrete of *Iguanodon* and of other animals that had inhabited southern England in the geologic past.

These models were reconstructed by a technician having the unusual name of Waterhouse Hawkins, and they were done under the eagle eye of Professor Owen. The project was carried through with all the scientific exactitude that could be brought to bear upon it at the time.

In those days it was thought that *Iguanodon* was a sort of rhinoceros-like reptile with a horn on its nose, and in this shape it was restored for the Crystal Palace grounds. We now know that *Iguanodon* was a large bipedal dinosaur that walked on strong hind legs, and that the supposed horn placed upon the nose of the beast by Mr. Hawkins is actually a large spike that formed the end of the thumb. But we need not criticize the Crystal Palace models because of our superior modern knowledge; they were laudable and sincere efforts for their time, and they were most effective. A century later they are still standing.

The construction of these restorations was a long and exacting task, and we may well understand the feeling of real accomplishment that came over Hawkins and his assistants as the job neared its end. To celebrate the completion of the *Iguanodon* model, a party was given at which Professor Owen and other distinguished scientists of the day sat down to what must have been a rather crowded dinner *inside* the model, while the lesser lights banqueted on tables set up around the reconstructed dinosaur. It was a memorable affair.

According to a contemporary report, "Cards of invitation were issued as follows:— 'Mr. D.W. Hawkins solicits the honor of Professor————————'s company at dinner in the *Iguanodon,* on the 31st December, 1853, at 4 o'clock, p.m.' This incredible request was written on the wing of a Pterodactyle, spread before a most graphic etching of the Iguanodon with his socially-loaded stomach, so practically and easily filled as to tempt all, to whom it was possible, to accept the invitation. Mr. Hawkins had one-and-twenty guests around him in the body of the Iguanodon—before the upper part of the back and head was completed. In the chair, most appropriately, sat Professor Owen, occupying part of the head of the gigantic animal; he was supported by the late Professor Forbes; Mr. Prestwich, the eminent geologist; Mr. Gould, the celebrated ornithologist; with other eminently scientific men. This model, it must be remembered, represents the natural size of the animal during life. In building the model the following was consumed as its material:—600 bricks, 650 two-inch half round drain tiles, 900 plain tiles, 38 casks of cement, 90 casks of broken stone, together with 100 feet of iron hooping, and 20 feet of cube inch bar."

Leidy, Hawkins, and Hadrosaurus

During those early decades of the nineteenth century, when Gideon Mantell and William Buckland and Richard Owen were discovering and describing the first dinosaurs to be found in England, the remains of these interesting reptiles were as yet unknown across the Atlantic. More than a decade was to pass after Owen had defined and given substance to the dinosaurs before any dinosaurian fossils were to be recognized in North America, the land that ultimately was to yield the most abundant and the most spectacular dinosaurs to be found in the wide world.

It all began in a very modest way with the discovery of dinosaur teeth near the confluence of the Judith River and the Missouri River in Montana, and of two vertebrae and a toe bone in South Dakota. These fossils were picked up in 1855 by one of the early scientific explorers of our West, Dr. Ferdinand V. Hayden. He took the fossils east with him and gave them to Dr. Joseph Leidy of the Academy of Natural Sciences in Philadelphia, for study and description. Dr. Leidy's account of these fossils was published in the following year.

Dr. Leidy was a remarkable man. Like Mantell in England, Leidy in America was a medical doctor who devoted part of his time to natural history. But Leidy was much less of an amateur than was Mantell; in many respects he was much like the professional scientist of today. In addition to his association with the Academy of Natural Sciences, he was Professor of Anatomy in the Medical School of the University of Pennsylvania, which means that he was thoroughly familiar with the details of human anatomy. Beyond this, Leidy was a man of broad knowledge who spoke with authority on the anatomy of many different vertebrates. Indeed, he ranged so widely in scientific interests that he was also an authority on parasites. Those were the days, a century ago, when a man could become a master in more than one field of science; they were the days of the all-around naturalist, a type that has by now almost completely disappeared. As a pleasant adjunct to his wide-ranging professional competence, Leidy was possessed of a calm and friendly nature that made him well beloved to all who knew him. He was a man of great dignity.

His descriptions in 1856 of the dinosaurian fragments that had been found by Hayden in Montana and South Dakota prepared him for participation in the next discovery of a North American dinosaur, a discovery that was spectacular and significant. This was the recovery in 1858 of a partial dinosaur skeleton in Haddonfield, New Jersey, immediately across the Delaware River from Philadelphia.

In the summer of that year a Mr. W. Parker Foulke of Philadelphia was spending some time in Haddonfield. While there he heard stories of bones that had been found several years before in a marl pit on the farm of his neighbor, Mr. John E. Hopkins. Foulke's interest was immediately aroused by the accounts of the bones, and he tried to locate some of the fossils. In this he was doomed to disappointment— the bones had all disappeared. Nevertheless Mr. Foulke made arrangements with Mr. Hopkins to reopen the pit where the fossils had been discovered, and as a result he unearthed various additional bones—vertebrae, the front and hind limbs, parts of the pelvis, some jaw fragments, and nine teeth. There was no skull, which was most unfortunate. Even so, he had a considerable portion of a skeleton, and all of this material was turned over to Dr. Leidy at the Academy in Philadelphia. Leidy described the bones and teeth very carefully and showed that they belonged to a hadrosaur, or trachodont, one of the so-called duck-billed dinosaurs related to *Iguanodon*. He named the new dinosaur *Hadrosaurus folkei.*

Hadrosaurus was discovered at a time when most Americans were preoccupied with the great dispute that was growing ever more acrimonious and ever more ominous between the northern and southern states. Within three years after the finding of this first dinosaur skeleton in America, the country was torn by its most fearful trial, and there was little time for fossils. Consequently a lapse of several years occurred during which the bones of the new dinosaur gathered dust on the shelves of a case in the Philadelphia Academy, while outside in the streets columns of blue-uniformed men marched off to war.

After the war was over who should appear in New York but Mr. Waterhouse Hawkins, to begin work on some life-sized reconstructions of ancient animals that were to be displayed in a large iron and glass museum in the newly created Central Park. Mr. Hawkins intended to add a touch of local color by constructing a full-scale model of *Hadrosaurus,* as a companion to the other "antediluvian monsters" he was bringing to life in such a substantial fashion. The work was well under way when the "iconoclastic Central Park Commission," as Mr. Hawkins bitterly designated the men in charge of the Park, forced him to end his labors. The reason for this change of plans was complex, and in part was a result of some unsavory politics of the Tweed era. Poor Hawkins was the innocent victim of all of this, and he had to see his partially constructed models, including *Hadrosaurus,* unceremoniously buried in the park, where presumably they are still interred, awaiting the attention of some future archaeologist.

Waterhouse Hawkins then went to Philadelphia — the year was 1868 — to set up the skeleton of *Hadrosaurus* for the Academy there. This task was carried to its successful conclusion, owing partly to the skill of Mr. Hawkins and partly to the liberal use of plaster in reconstructing missing bones. Certainly an impressive-looking plaster and bone skeleton was erected, but its resemblance to *Hadrosaurus* as we now know that dinosaur was anything but absolute. His effort remained visible for many years, to be seen by several generations of museum visitors. It was only shortly before the Second World War that the Hawkins restoration of the hadrosaur skeleton was finally dismantled.

Even though he had labored long, and had successfully restored the skeleton of *Hadrosaurus,* Waterhouse Hawkins still was not satisfied to rest upon his laurels; he wanted to fulfill his frustrated ambition of making a full-scale restoration of this dinosaur as it had appeared in life. In a report published by the Philadelphia Academy of Natural Sciences in 1874, we read that "the opportunity was afforded by the trustees of the New Jersey College at Princeton, who desired to possess for their museum one of Mr. Hawkins's restorations of an extinct animal of New Jersey. For this purpose they selected *Hadrosaurus...*" But alas! it would appear that even in Princeton Mr. Hawkins was unable to bring to reality his plans for a model of *Hadrosaurus;* at least there seems to be no record of his ever having completed this work. Consequently, *Hadrosaurus* for the public remained a name, a collection of bones, and a reconstructed skeleton in Philadelphia.

Dinosaur Tracks in the Connecticut Valley

Although the *Hadrosaurus* specimen unearthed at Haddonfield in 1858 was the first dinosaur *skeleton* to be described from North America, indications of dinosaurs on this continent came to light at a much earlier date—at the beginning of the nineteenth century, as a matter of fact. These were dinosaur footprints, found in the valley of the Connecticut River, but since Mantell and Buckland and Owen were not to describe and define dinosaurs for some years to come, the true nature of the footprints was not then realized.

In 1802 a New England farm boy named Pliny Moody found fossil tracks in sandstones exposed near his home in South Hadley, Massachusetts. These footprints looked as if they had been made by gigantic birds, for which reason they were regarded by some people as the fossil tracks of ground-living birds, perhaps similar to modern ostriches. By other people they were designated as the tracks of "Noah's raven." Within the next two or three decades more and more of these "bird tracks" were found up and down the length of the Connecticut River Valley, in rocks of Triassic age.

The tracks had a particular fascination for one man—the Reverend E.B. Hitchcock, president of Amherst College. He ferreted out stone footprints, wherever they might be, spending years and all the money he could scrape together on the project, and in 1858 published his monumental tome *Ichnology of New England,* a volume brought out by the Commonwealth of Massachusetts. In this work he described the tracks, most of which he considered to have been made by ancient birds, and figured them in beautiful full-page lithographic illustrations. He gave them names. And he established for himself a reputation as a solid man of science.

Professor Hitchcock envisaged a special building at Amherst to house and exhibit the tracks that he had collected through the years. About this time Mr. Samuel Appleton of Boston had willed a large sum to be used for benevolent and scientific purposes, and Hitchcock tapped this fund for ten thousand dollars, a sum sufficient in those days to construct a very adequate and suitable museum building. Thus there came into being the "Appleton Cabinet," the entire lower floor of which was devoted to the display of Triassic footprints.

Even as late as 1863, when a supplement to Hitchcock's *Ichnology of New England* was edited by his son, the tracks were still considered as having been made by ancient birds. But already in 1861 the oldest unmistakable fossil bird had been found, the famous skeleton of *Archaeopteryx,* a fossil many millions of years later in age than the Connecticut Valley tracks. Therefore it soon became apparent that the New England tracks were too ancient to have been made by birds; in short, there were no birds on the earth when these tracks were formed. The Connecticut Valley tracks clearly had been made by small primitive dinosaurs, the bones of which were turning up in sediments contemporaneous with those containing the tracks.

So the tracks took their rightful place as the first evidences of dinosaurs to be *found* but certainly not the first to be *recognized* in North America. It required a considerable advance of knowledge about the ancient life of the earth before the tracks were finally identified correctly. As the nineteenth century drew to a close, man's knowledge about past life of the earth had become established on a solid foundation of facts, the result of many discoveries of fossils. Dinosaurs were no longer known from mere isolated fragments; there were now respectable if not abundant collections of bones in various institutions. The pioneer days of discovery were past.

Professor Marsh and Professor Cope

The history of the discoveries and the early studies of dinosaurs is a continuing story, as any history is bound to be, and its division into stages or phases is an artificial, man-made pigeonholing of events, as are any such divisions of history. Yet these artifices help us to remember things. Let us, therefore, regard the first half-century or so of work on the dinosaurs as the "pioneer" period—a span of time ending with the beginning of the last decade of the nineteenth century. This pioneer period, so defined, was a time of groping, rather haphazard efforts in the field, and of intellectual groping in the laboratory, a period when the depths and limits of our knowledge of dinosaurs were first being realized.

In Europe the first period of early, undirected investigations of the dinosaurs gradually merged into a more modern period of field work and research, based upon a greatly augmented knowledge not only of the dinosaurs but also of their relatives, of all extinct life, and of the past events of earth history. In North America the first phase of work on the dinosaurs was brought to an exciting close by the unrivaled discoveries and researches of two men, and by the ensuing acrimonious dispute between them.

The collecting of dinosaurs in North America during the seventies and eighties was very colorful and romantic, to put it mildly, and studies on the fossils so collected were far removed from the quiet, objective activities that are supposed to encompass scientific research. The reason for all of the excitement is to be found in the conflicting personalities and ambitions of Professor Marsh and Professor Cope.

Othniel Charles Marsh was born during the year 1831 in Lockport, New York. He grew up in a region where there were woodlands to explore and where there were many cliffs and outcrops of rocks containing fossils that would be of interest to a boy having a larger than usual bump of curiosity and an interest in nature. When he was a young lad his mother died, and subsequently he came under the patronage of his uncle, George Peabody, a man of wealth and influence. Marsh attended Yale College, graduating in 1860, after which he studied in Germany for several years. To the great good luck of Marsh, his uncle left him an ample income for life. Thus Marsh became independent, and devoted all his time to fossils. He returned to Yale in 1866 as Professor

of Paleontology, and immediately launched upon an active career
of research on fossil vertebrates at Yale. He persuaded his uncle
to found the Peabody Museum at Yale, which under the direction
of Marsh was to become one of the outstanding natural-history
museums of North America. He never married. He died in 1899.

Edward Drinker Cope was born in Philadelphia in 1840, the
son of a wealthy Quaker philanthropist and the grandson of a
shipowner. His mother died when Cope was a very small boy, but
he had the good fortune to acquire a loving and understanding
stepmother. He also was fortunate in having a very sympathetic
father. Cope was amazingly precocious, and at the age of seven
was making detailed and sophisticated notes concerning the fine
points of the anatomy of an ichthyosaur that was on display at the
Academy of Natural Sciences in Philadelphia. At nineteen he
was publishing scientific papers. He also went to Europe during
the early sixties, and returned to become a Professor at Haverford College. But he did not remain at Haverford for long; he had
too independent a personality to be confined within the bounds
of any institutional organization. Consequently, for much of his
life he was an independent worker, financing his own field work
and research with the fortune that had been left to him by his
father. In the course of time he became an outstanding world
authority in at least three fields—the study of recent fishes, the
study of recent amphibians and reptiles, and the study of fossil
vertebrates. He published prolifically and stands as one of the
great men of American science. During his middle age he invested
heavily in mines and virtually lost the fortune that had been his.
In the last years of his life (he died in 1897) he served as Professor at the University of Pennsylvania, and this appointment
helped him through the last difficult, and in a sense tragic, years.
He married his cousin, Annie Pim, and they had one daughter.

Cope and Marsh were able men; both were brilliant (Cope
was a true genius), both were inordinately ambitious, both had
money, and both were very independent people, accustomed to
getting what they wanted. What both wanted was to collect and
describe extinct vertebrates of all kinds. They put their wishes
into actions, and in so doing they literally discovered a vast new
world of dinosaurs.

Cope and Marsh began their scientific careers immediately
after the Civil War, during an expansionist period of American

history. They became acquainted with each other in the early seventies, and at that time they had some pleasant trips together in eastern North America, searching for early dinosaurs and other fossils. Then they turned their attentions to the great open country west of the Mississippi. They began to collect in the Great Plains and in the Rocky Mountain region, and in a very short time all of western North America was not big enough to accommodate the two of them. Their rivalry and their jealousy of each other grew by leaps and bounds, and exceeded all reasonable limits. Each man became a monomaniac with regard to the other, and their activities assumed almost the nature of open warfare. They spied upon each other, and the collectors they hired likewise spied on each other. They tried to keep secret their fossil localities. Upon at least one occasion their collectors engaged in a free-for-all fight among the lonely ridges of Wyoming. They engaged in vituperative attacks upon each other in the newspapers, and they promulgated all sorts of machinations in scientific societies, in the government surveys on which they both served, and in the universities and museums with which they were connected. Indeed, the story of Cope and Marsh would make a swashbuckling novel.

Yet all of this had its good side: their intense rivalry stimulated them to collect and describe fossils on a grand scale. They sent back tons of bones to New Haven and to Philadelphia, to be cleaned by their technicians, drawn by their artists, and measured by their assistants. In doing so they laid the groundwork for our modern methods and techniques of vertebrate paleontology. Marsh and Cope did not rely upon the chance discovery of fossils. They planned well-integrated campaigns of fossil hunting and collecting. They developed methods for taking large bones out of the ground that are still in use. They maintained laboratories, staffed by assistants. They published their research studies in two veritable blizzards of scientific reports.

Of course, Cope and Marsh ranged far and wide in their search for dinosaurs, but each made certain outstanding discoveries. Cope discovered and described primitive Triassic dinosaurs from New Mexico. He opened a large quarry in the Jurassic Morrison deposits near Canyon City, Colorado, and there excavated some excellent skeletons. One of his specimens, the skeleton of the meat-eating dinosaur *Allosaurus,* was still in boxes, unopened and unstudied, when the American Museum of Natural History purchased the Cope collection. This dinosaur now occupies a prominent place in Brontosaur Hall in New York. Cope also described Cretaceous dinosaurs, especially from along the Judith River in Montana.

The outstanding dinosaurian discovery by Marsh was the great deposit of skeletons in the Morrison sediments along Como Bluff, in eastern Wyoming. From this place Marsh and his collectors excavated a veritable treasure-trove of skeletons, some of which are finely displayed in the Yale Peabody Museum. This is one of the most prolific and famous dinosaur localities in the world, and incidentally one that Cope's men tried to invade when it was being actively worked by Marsh collectors. Marsh also collected Cretaceous dinosaurs, especially from the region near Denver. It was from this locality that he first made known the existence of horned dinosaurs.

The work of Marsh and Cope amazed and overwhelmed paleontologists in Europe and in other parts of the world. (Because the feud between Marsh and Cope reached such intensity, Joseph Leidy retired from the field and turned his attention to other things. Leidy was a gentle man and he could not put up with the fireworks of the Cope-Marsh affair.) In short, these two men expanded the knowledge of the dinosaurs from scattered descriptions, based largely upon parts of skeletons or isolated bones, to a large body of knowledge founded upon an imposing array of skulls and relatively complete skeletons.

As far as dinosaurs are concerned, Marsh was clearly the winner. He collected more and described more than did Cope—perhaps largely because Cope's interests were wider than Marsh's. Cope had a greater variety of fossils to study than did Marsh, and in addition he was at the same time studying and publishing on recent fishes, amphibians, and reptiles. The box score: Marsh described 19 genera of dinosaurs; Cope trailed behind with only 9.

With the passing of Marsh and Cope, shortly before the turn of the century, the pioneer period of work on dinosaurs was truly ended. In this country and abroad, partly as a result of the great work done by these two men, the modern period of dinosaurian field and laboratory studies was inaugurated.

12

Kulp: The Age Of The Earth

RUTH MOORE

Of the major geologic developments of the 20th Century, none has been more significant than the determination of the earth's true age and the construction of a time scale in approximate years. Ruth Moore is a jounalist with the ability to instill life into complex scientific developments. In this chapter from "The Earth We Live On", she vividly presents some of the people, technical methods and problems in the history of radiometric dating.

Lord Rutherford—then Ernest Rutherford—was strolling across the campus of McGill University in Montreal, where he had been serving for several years as professor of physics. In his hand he tossed a small piece of pitchblende, the natural ore of uranium. When he met one of his faculty colleagues, a professor of geology, he stopped him and asked: "How old is the earth supposed to be?" The geologist told him that the weight of opinion favored an age of about a hundred million years.

"I *know,*" said Rutherford quietly but with full effect, "that this piece of pitchblende is seven hundred million years old."

Here perhaps was the first collision—for the year was 1904—between the old estimates of the age of the earth and the vastly larger figures that would grow out of a wholly new kind of dating. Rutherford was just beginning to apply the recent discovery of radioactivity to the determination of the age of uranium ore, and hence—in part, at least—to the determination of the age of the earth.

The announcement soon afterward of his findings went a long way toward upsetting all that the Western world then believed about the age of the earth. The brilliant New Zealand-born physicist administered a severe shock to geological thought.

And yet this increase in the estimate of the age of the earth was only the first in a long series of upward revisions. Within the next fifty years scientists working with radioactivity would not only begin the dating of the major events of the earth's past; they would bring about a startling change in our very concept of time.

Some of the most dramatic of these discoveries, making it possible to pinpoint happenings a billion years and more in the past, have come in the decade following the close of World War II. Science now has learned to read the past with uranium and thorium, and with the radioactive isotopes of rubidium, potassium, carbon, oxygen, and other similar materials whose very existence was unsuspected not long before.

Whole laboratories, like the geochemistry laboratory of Columbia University's Lamont Geological Observatory, have been organized to work with the new methods of dating and to focus them on the primary problems of the earth. And young men like Lamont's J. Laurence Kulp have brought the new atomic physics and chemistry to bear in these studies of the earth. But their striking findings, revealing many a long-hidden secret of the earth, are an outgrowth of much that went before.

Man long has wanted to know how old the earth is. The sages of Greece probably were not the first to try to devise a chronology. Their unknown predecessors may have worked at the same baffling problem. But of all the ancient prophecies which have come down to us, that of the Hindus was, as Arthur Holmes points out in his *Age of the Earth,* the most remarkable. The Manusmitri, the Hindu sacred book, fixed the whole existence of the earth, past and future, at 4.32 billion years or one "day" in the life of Brahma. Beyond, it was believed, would come the night of Brahma where the finite would merge into the infinite. Brahma's "day" was divided into fourteen great cycles, the seventh of which, it is said, has now been reached. According to this reckoning, the earth is now slightly less than 2 billion years old.

The Western world had no similar concept of time. When Archbishop Ussher, early in the seventeenth century, worked out the age of the earth from his interpretation of the ages recorded in the narratives of the Old Testament, he fixed the date of the earth's creation at 4004 B.C. It was printed in the margins of many Bibles, and was largely accepted by Christian peoples as the authoritative and incontrovertible date of the earth's beginning. The West was certain that the earth was thus only a few thousand years old.

Geologists of the late nineteenth century were not deterred. They could not accept this constrictive limitation when they saw the changes the earth had undergone, for such changes could not have come about in a few thousand years. Joly, who in 1899 attempted to measure the age of the earth by computing the amount of salt washed into the seas, arrived at a figure of 80 million to 90 million years. Another scientist set an age of 57 million years, based on the separation of the earth from the moon; and others, studying the thickness of the sediments, estimated it at about 100 million years.

It was at about this time, in 1897, that the scientific attempts to establish an age for the earth received the famed setback from Lord Kelvin. Kelvin's estimate that the earth could not be less than 20 million years old or more than 40 million was carefully calculated. It could be held invalid, he accurately declared, "only if sources of heat now unknown to us are prepared in the great storehouse of creation."

Almost at the moment that Kelvin was speaking, exactly such new sources of heat were being discovered. Professor Henri Becquerel of France had followed with great interest Rontgen's discovery of X rays in 1895. Was it possible, Becquerel wondered, that the phosphorescence that he was studying might also be produced by invisible rays? Perhaps certain substances absorbed light from the sun and later reflected it. Becquerel exposed a large number of substances to the sun and then laid them on photographic paper to see if they would fog it as X rays did.

None of them produced the slightest effect until one day he tried a salt of the rare metal uranium, the heaviest of all the elements.

Then he found that the uranium salts blurred the paper in the same way whether or not they were previously placed in the sun. Some form of radiation was being given off by the uranium, Becquerel saw, and it was being given off continuously and spontaneously. Here was a new form of radiation. He announced his discovery in 1896.

Marie and Pierre Curie, who were interested in the same general problems, at once began to work on the mystifying radiation. By the laborious process of testing all the elements, they learned that thorium also gave off the strange radiation. But even more puzzling was the behavior of a piece of the natural uranium ore called pitchblende which came from a mine at Joachimsthal in Bohemia.

The intensity of the radiation it emitted far surpassed the combined amount that should have come from both the uranium and the thorium it contained. Only one conclusion was possible: some unknown material in the ore also was radioactive (the term Marie Curie had coined for the phenomenon). For four years the Curies labored in their drafty, shed-like laboratory, trying to isolate the elusive material. In the end, as all the world knows, they found radium—only a tenth of a gram of it, but a substance a million times more radioactive than uranium. It was one of the classic achievements of science, and for it the Curies and Becquerel shared a Nobel Prize.

Such discoveries have their own radiance. Ernest Rutherford, then a young science student at the Cavendish Laboratory at Cambridge, was inspired by the reports coming from France, and in his turn went to work with the new phenomenon.

Before the year 1902 had ended, Rutherford was ready to announce that radioactivity was nothing less than the spontaneous disintegration of radioactive atoms and their conversion into wholly different atoms. The atom, up to that time, had been considered the eternal and unchangeable base of all matter. But here was a transmutation of matter, almost as fabulous as the transmutations the alchemists had dreamed of. A new era in science was suddenly opened.

Rutherford, in the meanwhile, had gone to Canada, to McGill University. He had realized almost at once that if the particles were given off at a fixed rate (and every sign indicated that they were), the loss in radioactivity would afford a new kind of measurement of the age of minerals and of the earth of which they were a part.

Later it was to be learned that this disintegration of radio-active substances is one of the steadiest, most undeviating things in existence; the particles are thrown off at the same rate regardless of changes in temperature, time, or condition. Research revealed, however, that each radioactive material disintegrates at its own distinctive rate, and that the decay of uranium is almost inconceivably slow. At the end of 4.5 billion years only half of its radioactivity is gone; only half of it has changed into another substance. At the end of a second 4.5 billion years another half would be altered, and so it continues in ever diminishing halves. Rutherford understood enough of this "half-life" system of measurement to estimate that the uranium he was studying was at least 700 million years old, and that the earth also had to be that old.

In 1904 Rutherford was invited to discuss the whole exciting question of radioactivity and its indications about the age of the earth at a London meeting. He tells of what happened in his own ebullient way:

"I came into the room which was half dark, and presently spotted Lord Kelvin in the audience and realized that I was in for trouble at the last part of my speech dealing with the age of the Earth where my views conflicted with his. To my relief Kelvin fell fast asleep, but as I came to the important point, I saw the old boy sit up, open an eye and cock a baleful glance at me! Then a sudden inspiration came and I said that Lord Kelvin had limited the age of the Earth *provided no new source* (of heat) *was discovered.* 'That prophetic utterance refers to what we are considering tonight, radium!' Behold! the old boy beamed upon me!"[1]

Discoveries continued to come in a great surge, for scientists everywhere were working intensively at the new findings. In 1906, B.B. Boltwood, professor of chemistry at Yale University, noted that lead was nearly always present in uranium-bearing rocks. Could lead, he asked, be the final disintegration product of the radioactive uranium?

To test the point, Boltwood collected ore samples from all parts of the world and measured their lead and uranium contents.

[1] A.S. Eve: **Rutherford.** New York: MacMillan and Co.; 1939.

Both varied widely. That suggested another idea; he would line them up according to their lead-uranium ratios.

When he did so, the results were highly enlightening. The greater the amount of lead, the older was the geological formation from which the mineral had come. His hunch that there was a relationship between lead and age was thus supported by the geological evidence. With this encouragement Boltwood calculated the time that would have been required for the uranium to deteriorate into the lead in each sample. For one specimen he obtained the startling figure of 2.2 billion years. The earth certainly was as old as its oldest rocks, but the suggestion that the earth could have had a life of more than 2 billion years seemed fantastic.

The time had come to test and consolidate the new discoveries. As this was done, it became apparent that many serious errors could creep into the computation of ages by a simple lead-uranium ratio. A better method was needed, and it was found. Science learned that uranium is in fact a mixture of two isotopes, and that each of them decays into a different kind of radiogenic or radioaction-derived lead. For uranium and thorium the atomic transformation may be set out in this way:

URANIUM 238 *decays to* LEAD 206
URANIUM 235 *decays to* LEAD 207
THORIUM 232 *decays to* LEAD 208

As soon as the complex nature of uranium was recognized, scientists saw that three age-determinations might be made from one specimen. In a specimen containing uranium alone the decay of uranium 238 to lead 206 would provide one measurement, and the disintegration of uranium 235 to lead 207 a second. The ratio of lead 206 to lead 207 provided a third, for the amount of lead 207 increases proportionately with time—uranium 235 decaying more rapidly than uranium 238. In effect, therefore, each piece of uranium ore had three clocks built within it, or at least two clocks and a continuous record of their performance. Age could be read by each of the three and the answers compared. Still, many problems of technique remained to be solved.

When Alfred O. Nier, then of Harvard University, published the first isotopic dates in 1939, it was certain that a new phase in dating was beginning. The war, however, intervened and many of the physicists who had been working with these problems joined the Manhattan Project.

One of the young scientists who gained experience in radioactive work during the war was J. Laurence Kulp. The director of geochemistry at the Lamont Observatory was born at Trenton, New Jersey, on February 11, 1921. He went to the midwest to college, graduating from Wheaton, near Chicago, in 1942. After taking a master's degree in science at Ohio State University he returned east, to Princeton, for his Ph.D. He received the latter degree in 1945. Work with the Manhattan Project had fitted in, and at the close of the war Kulp joined the faculty of Columbia University.

Kulp had been trained as a physical chemist, but after four years in the laboratcry he thought longingly of work that would get a man out of doors at least once in a while. More than this, the possibilities of applying the new physics and chemistry to the study of the earth and its large unsolved problems offered an auspicious and exciting prospect. Geochemistry, which obviously was on the threshold of a rapid development, offered exactly the combination that Kulp wanted.

Columbia University at the time was expanding its work in this and related fields. Mrs. Thomas W. Lamont had presented the university with a beautiful estate on the west bank of the Hudson about thirteen miles above the city, for a geological observatory. Under the direction of Dr. W. Maurice Ewing it was dedicated to the application of the methods of physics, chemistry, mathematics, and engineering to the study of the earth. In 1955 a special building, clean-lined, modern in design, and fitting well into the wooded Palisades, was completed for the geochemistry laboratory. The latter is in fact something quite new in the world: a center for radioactive dating, for studying the chronology of the earth.

Potentially any long-lived radioactive isotope may be used for what might be called a time clock, if that term may be recovered from industry. The three isotopes of uranium were only the first to be employed. Indeed, the scientists of the world are in something of a race to identify and develop others.

Immediately after the war Dr. Willard F. Libby, of the University of Chicago and later a member of the Atomic Energy

Commission, demonstrated that radioactive carbon, Carbon-14, may be used to determine the age of recent organic remains, like charcoal from a fire of prehistoric man. "Recent" at first meant up to twenty-five thousand years. It was then extended to fifty thousand.

Kulp set up extensive programs at Lamont both to carry on lead and Carbon-14 dating and to improve these techniques. The laboratory also went to work on the promising new rubidium and potassium dating and on other possible dating materials.

In all of these studies, extremely small amounts of the radioactive materials and of their end products must be extracted from the raw ores. Analyses must be made, and radioactivity measured in supersensitive Geiger counters. Elaborate computations also are necessary.

To cope with this intricate chemistry, Lamont has a vacuum fusion laboratory, an emission spectograph and X-ray diffraction laboratory, a natural radioacarbon laboratory, and laboratories of general geochemistry, electrochemistry, analytical chemistry, mass spectrometry, and low-level counting.

In these rooms is a forest of equipment. Miles of glass tubing, it seems, turn and twist, enter flasks of all sizes and shapes, and emerge from them only to intertangle with pumps and other equipment confounding and awesome to the non-scientist. In other rooms are batteries of computing machines, their dial-studded faces alight with the flickering of red and green bulbs. And there are recording instruments whose pens trace back and forth across fine-lined graph paper. Here is all the panoply of science at its most imposing.

To these laboratories come specimens from all parts of the world. At this point, though, the formal terminology of science is thoroughly snubbed. Many of the specimens that are fed into the spectrometers, the diffraction apparatus, and all the rest of it, originate in such non-scientifically named mines as the Hoot Owl, Brown Derby, Deep Creek, and Happy Jack's. So are they known as they go through the laboratory, and so are they reported in the journals of science.

Lamont has one other unique measuring implement. It is a handsome brass balance that stands on Kulp's desk. One of its shining pans holds a perfect white crystal, sharp-planed, symmetrical, and clear. The other bears a chunk of grayish black, slag-like rock, pitted, rough, and contorted. This burned, fused piece with its small twisted inclusions came from zero point at Hiroshima.

As this laboratory and others ran uranium specimens through their apparatus, they sometimes found that all three clocks showed remarkable agreement. Nier dated one sample of pitchblende from Joachimsthal—the same pitchblende that Mme Curie had used—which revealed an age of 244 million years on the lead-206-uranium-238 test, an age of 249 million years on the 207/235 test, and one of 242 million years on the 207/206 ratio. Another ore from Connecticut showed, respectively, ages of 250 million years, 266 million, and 280 million.[2]

Such an agreement is remarkable. What makes it one of the astonishing facts of science is that both uranium and thorium decay at different rates. They also have different chemical characteristics, and occur in different concentrations. It is as if three entirely different clocks, each of which had been running for hundreds of millions of years, pointed in the present to the same hour and minute. Even though the ages recorded by these natural clocks go back several billion years there is virtually no doubt, when all three agree, that the ages are correct.

"Agreement between them is strong evidence of the reliability of the calculated age," said Kulp. Adolph Knopf, former chairman of the Committee on the Age of the Earth, has made the same point: "If all three determinations agree, assurance is rendered trebly sure."

But occasionally there is a wide discrepancy in some of the dates. A pitchblende from northern Canada had an age of 337 million years on the 206/238 ratio and one of 389 million years on the 207/235 determination, but jumped to an unacceptable 705 million years on the third. More disturbing was the discrepancy in a radioactive mineral from Sweden known as kolm. It is the one known radioactive material that contains fossils. Nearly all other uranium ores occur in masses that have intruded into unstratified rocks, and afford no evidence of where they stand in the geological sequence. But the fossils embedded in the kolm are trilobites and other forms that

---

[2] All dates are presented with the scientist's familiar ± although it is omitted here. If a date is given as, say, 2,400 ± 200 years, it means that from the evidence of the measurement alone, the chance is 68 per cent that the true value is between 2,200 and 2,600 years.

trace back to the early stages of life on this earth, or at least to its first appearance in the rocks. That definitely helps to place the kolm among the oldest of the fossil-bearing rocks. On the 206/238 test the kolm gave an age of 380 million years, on 207/235 an age of 440 million years, and on the 207/206 determination one of 800 million years. Something unquestionably had gone wrong.

For a while the investigators were baffled. As the work went on, however, and more was learned about the complex way in which uranium decays, the source of the trouble was spotted. Uranium does not disintegrate directly into lead. According to the latest tables, uranium 238 goes through fifteen transformations before it emerges as the plain grub lead.

At one of the stages on its diminishing course the uranium 238 transmutes very briefly into an inert gas, called radon. At this stage it is fairly easy for some of the gas to leak away. If this occurs, the uranium-238 stock is reduced, and ultimately less lead 206 is formed. Both the 206/238 and the 207/206 ratios would be thrown into error.

But all is not lost, as far as dating goes. In the first place, any radon lost in the uranium-238 chain does not affect thorium decomposition. And even the value of 238 time is not destroyed by what might figuratively be called the loss of a fraction of a second. The loss can be taken into account, and allowance made for it.

Kulp began to study radon leakage in the laboratory. At room temperatures, he found, uranianites and pitchblendes lose from 0.1 to 10 per cent of their radon. At higher temperatures the leakage may be larger. But if a loss of about this amount is allowed for at the sixteen-hundred-year-long radon stage, the U-238 dates of very young minerals often are brought closely into line. It is like knowing that the clock is a minute slow and allowing for it. In that way, the discrepancy is overcome.

Other alterations that affect the radioactive clocks are being studied in much the same way. Lead, for example, is commonly lost if the mineral in which it is contained has been reheated since it was first formed.

As the trouble spots were singled out and dealt with, improvements also were made in the apparatus that does the measuring. A mass spectrometer has been developed that on one runthrough of a single sample supplies the data on all three types of radio-

genic lead. The amounts of each present in the sample are registered
by a pen that makes red zigzags on green graph paper. Kulp also has
worked out charts that make it possible to obtain the age directly
from the figures shown on the graphs. A sample can now be prepared
for testing in about a week, and after half a day in the spectrometer
its age may be read. Formerly, weeks of work were involved in
making one determination.

With the greater number of tests the improved techniques
have made possible, and with the solution of some of the major
problems of making measurements, Kulp and his associates have
reported that there are at least eight localities in the world where
the age of the rocks can now be established to within five per cent
of the true and absolute figure. The geochemists thus are saying
that science at last can accurately time the last several billion
years. It is a striking achievement—to reach back through time and
fix the birthday of a rock in an unknown and almost unimaginable
era.

The following ages—which science now regards as "nearly
sure"—have been set for areas in two continents:

Southeast Manitoba, Canada	2,650 MILLION YEARS
Southern Rhodesia, Africa	2,650 MILLION YEARS
Great Bear Lake, Canada	1,400 MILLION YEARS
Parry Sound, Ontario, Canada	1,040 MILLION YEARS
Wilberforce, Ontario, Canada	1,030 MILLION YEARS
Katanga, Belgian Congo, Africa	610 MILLION YEARS
Bedford, New York	350 MILLION YEARS
Middletown, Connecticut	260 MILLION YEARS
Gilpin County, Colorado	60 MILLION YEARS

In addition, there is a long list of dates that the scientists are
willing to say are correct to within twenty per cent.

But the uranium-lead methods made possible the dating of
only a scanty part of the rocks of the earth. In many parts of the
world there are no uranium ores suitable for dating. Technical
difficulties also block the dating of uranium-derived lead that
is less than a million years of age.

Carbon-14 dating also is limited. Even with the latest improvements in method, it can reach back into the past no further
than fifty thousand years.

Thus, there was no quantitative way of dating rocks too old
for the Carbon-14 method and too young for the uranium-lead
system, and an undatable gap was left between fifty thousand

years and a million. It was a crucial gap, for therein lies the larger part of human history.

Additional methods of dating that could reach more rocks and more eras were badly needed. Scientists concentrated on the problem, and research already is well advanced on two additional new dating methods that hold great promise.

One of the two methods is based on rubidium, a soft, silvery metal that explodes if it comes in contact with water. This metal consists of two isotopes: radioactive rubidium 87 and rubidium 85. Over a period of billions of years the former decays to strontium 87—about seven per cent of ordinary strontium is made up of this end-product strontium. It follows that any mineral relatively rich in rubidium and low in ordinary strontium can be dated.

Goldschmidt was the first to see the possibilities of the rubidium-strontium decay chain. That was in 1937. Ten years later, scientists had sufficiently developed the new method to date a number of rubidium specimens from areas where lead-uranium dates had already been obtained. That provided a nice check on the uncertain new method, and at first both sets of dates seemed to agree closely. It looked as though an accurate new dating tool was at hand.

Later, however, as new mass spectrometers were used for additional rubidium studies, it was discovered that the rubidium dates ran regularly fifteen- to twenty-per-cent higher than the best uranium dates. Again something was wrong. Apparently defective analyses had produced the early agreement. The disagreement therefore had to be explained.

Scientists at Lamont and other laboratories dug into the problem. They found that the trouble probably lay in the lack of an exact "half-life" for rubidium. It was assumed in the tests that half of any rubidium 87 percent would decay to the next product in the chain in 6.3 billion years. But this had not been positively determined, and Kulp pointed out that if a possible twenty-five-per-cent error had been made, the half-life would become 4.9 billion years. If the latter figure turns out to be the right one, it would bring the uranium-lead and the rubidium-strontium ages into agreement.

As Kulp notes, the decay of radioactive rubidium is immensely slow, even slower than the protracted decay of uranium, and when the technical difficulties are solved, it will be an extremely valuable new clock for dating the most ancient of rocks.

The other new dating tool on which research is far along also holds great promise. When the work with radioactivity was in its early stages in 1907, scientists saw that a rare form of potassium is radioactive. Not until 1943, however, was it possible to track down its end product. Von Weizsächer had earlier predicted that potassium 40 might decay under some circumstances to calcium 40 and under others to the colorless, odorless gas argon 40; this proved to be the case.

Excitement ran high in scientific circles, for potassium is one of the most common elements in the rocks. If it could be used for dating, vast areas would be opened to research. Uranium, thorium, and rubidium all are relatively rare, but there are rich concentrations of potassium in the basalts that underlie a large part of the oceanic areas of the globe. Their study might reveal much about the age of the oceans and about some of the deep movements of the crust of the earth. And besides, there is potassium aplenty in such hard ancient rocks as the schist and gneiss that often form part of the continent's "basement complex" and crop up in the shields.

Extremely difficult technical problems at first blocked the use of the new potassium clue. The exact steps in the decay chain that converts the radioactive potassium to argon had to be worked out. It also seemed quite likely that some of the end-product argon might leak from the rocks, and that would further complicate the measurements.

Kulp and the Lamont staff were among many who labored steadily at this problem. Step by step, using radioactive tracer materials from the piles of the Atomic Energy Commission, they were able to demonstrate that argon in the rocks may be released and measured to an accuracy of within three per cent.

Some samples of rock from the highly metamorphosed southern Appalachians were among the first measured. The argon/potassium clock showed that the rocks had undergone two periods of metamorphism, the first about 600 to 700 million years ago and the second about 350 million years ago. The argon gas imprisoned in the rocks also revealed that the "Storm King" granite at Bear Mountain, in New York, had pushed its way up through the older rocks of the area about 700 to 800 million years ago. The "Canada Hill" granite gneiss of the same area showed an age of about 1,200 to 1,400 million years.

Although much more research must be done before the argon-potassium clock is perfected, Kulp predicts its ultimate success. "If the branching ratio and retentivity can be established with high precision," he said, "this method may become the most useful of all isotopic chronometers."

As precise ages were established for a significant number of rock areas of the earth, science gave its renewed attention to that venerable question: how old is the earth itself?

Prior to 1946, estimates, as the radioactivity scientists politely say, were not based on "quantitative reasoning." They were rather informed guesses. Such guesses had then placed the age of the oldest rocks at about 2 billion years, which meant that the earth was somewhat older.

But the new isotopic data on the age of the rocks provided a basis for some exact computations. Arthur Holmes assumed that when the earth began it contained no lead 207, and that all the lead 207 present in the modern crust had therefore been generated from uranium 235. Calculating the time it would have taken the lead to form, he arrived at an age of 3.3 billion years for the earth. A number of other scientists working with the same data also found that the earth was from 3 to 3.5 billion years old. The figures showed notable agreement.

"But," Kulp warns, "the ages obtained should probably be considered a minimum, particularly if the continental areas are becoming increasingly granitic as well as expanding." Wilson has made the point that they are.

The first isotopic calculations of the age of the earth were thus made uncertain by doubts as to the accuracy of the figures on the amount of lead 207 in the crust today. Lead, it is known, is not concentrated heavily in the outer part of the crust, as is uranium. The age-old remaking of the crust, with the upward movement of the granites and the growth of the nucleus areas, may have affected the distribution of lead in the samples on which the figures were based. If some lead was being overlooked, the figures on the age of the earth would be low.

The search went on for a more exact way to determine the age of the earth. The scientists saw that if they could obtain the isotopic *composition* of the lead that was present when the earth was formed, and compare it with the isotopic *composition* of the lead in the crust today, simple subtraction would supply a measure of what lead had been added by the decomposition of uranium. This might perhaps be compared to saying that if we started with two lead coins and ended with three, it would be easy to tell what had been added during the life of the earth. And if the added coin could have been produced only by the decay of uranium and if the rate of that decay was known, the exact time that it had taken to produce the extra coin could be calculated. The age of the earth could be fixed absolutely.

But how was the composition of primeval lead, of the earth's starting quota, to be found? One possibility suggested itself to C.C. Patterson of the California Institute of Technology: the answer, he suspected, might lie in meteorites. There was a strong likelihood that the chunks of stone and metal which occasionally rain down on the earth from outer space have the same composition as the original earth. Many scientists believe that they are the debris of a shattered planet that came into existence at the same time as the earth and the other planets. Patterson and his associates obtained samples from metal meteorites that had fallen in Iowa, Mexico, Kansas, and, later, in Australia.[3]

The metal meteorites contained virtually no uranium or thorium, but they were rich in lead. Because such quantities of lead could not have been produced by uranium decay after the consolidation of the meteorites, Patterson concluded that the lead was present in the materials from which the meteorites were formed. It was, then, primeval lead.

Another problem had to be solved simultaneously. Patterson and his associates needed to know the average isotopic composition of the earth's lead today, and they had to be certain it was the average composition of the lead in the crust as a whole, and not of that in some particular locality. The scientists found what they were seeking in the fine red clay that accumulates endlessly in the great depths of the Pacific Ocean. By the time this extremely fine "rock flour" reaches the bottom of the ocean, so much mixing has occurred that it represents an average of the rocks of all the continents. As an additional check in determining the composition of the earth's lead, the Patterson group wanted data on the lead in the lower parts of the earth's crust. They obtained samples of such rock and lead

[3]Meteorites are the solid bodies that fall to earth. Meteors or shooting stars burn themselves out as they flash across the heavens.

from the basalts in some of the deeper, undisturbed lava flows in the Snake River area.

When analyses were made of the isotopic composition of the lead in the red clay, in the basalt, and in the meteorites, dramatic results appeared:

Lead	206	207	206/207 AGE
Oceanic	18.93	15.73	
Meteoritic	9.41	10.27	
Radiogenic	9.52	5.46	4.5 BILLION YEARS
Basaltic	18.12	15.45	
Meteoritic	9.41	10.27	
Radiogenic	8.71	5.18	4.6 BILLION YEARS

In the meteorites, in which almost no lead could have been formed by radioactive decay, there was strikingly less lead than in the clay and the basalt, where lead had been added by the breakdown of uranium. In fact, the figures vividly showed that the earth today contains almost twice as much lead 206 as the meteorites, and about a third more lead 207. And this was true whether the earth's lead came from the bottom of the ocean or from the lower part of the crust. The effect of the decay of uranium was definitively established: radioactive decay was adding large amounts of lead to the earth.

Knowing how much radiogenic lead the earth had gained and knowing the rate of uranium decay, Patterson could calculate the time that had gone into its formation. It had taken 4.5 billion years to accumulate the earth's uranium-derived lead; some 4.5 billion years had passed since the earth had consolidated out of the dusts of space. This, then, was the age of the earth: 4.5 billion years!

Remarkable additional proof that this vast age is in truth the correct age of the earth or very close to it was soon to come.

If meteorites were formed at the same time as the earth, and if their age could be measured, the scientists working in this field quickly saw that they would have an accurate check on the age revealed by the isotopic comparison. By 1956 several groups had succeeded in making three distinct measurements of the age of meterorites.

Studying certain stony types of meteorites which contained uranium that had partially decayed into lead 206 and 207, Patterson, Harrison Brown, and Mark Inghram, all three then at the University of Chicago, and George Tilton, of the Carnegie Institution, separated enough lead to obtain a 206/207 ratio. It showed that the meteorites they tested were 4.5 billion years old!

One of the newest methods of dating, rubidium-strontium, was used for another determination. Ernst Schumacher found that the metal meteorites held no rubidium, and that their strontium therefore had to be primeval strontium, present when the black piece of metal was formed. On the other hand, the stony meteorites, which crash into the earth more frequently than their metal counterparts, contained rubidium, and its decay had produced measurable amounts of radiogenic strontium. By simple subtraction Schumacher could tell how much strontium had been created by radioactive disintegration, and by calculating how long it had taken that strontium to form, he could obtain the age of the meteorite. It was 4.5 billion years!

Still more confirmation was to come from a third and different procedure, a potassium-argon measurement. G. J. Wasserburg and R. D. Hayden of the University of Chicago found that the stony meteorites they studied contained argon which had come from the decay of radioactive potassium. By measuring the time required for its formation, they calculated the age of the meteorites. It was 4.6 billion years!

Three entirely different methods thus placed the age of the meteorites at very close to 4.5 billion years. Worlds apparently were in the making in that vast recess of time, and among them was the body from which the meteorites came, and probably the earth as well. For the evidence of that most certain of all clocks, radioactivity, said that the earth, too, had come into existence at that time, 4.5 billion years ago. The agreement was overwhelming.

So strong and convincing was this converging of proofs that most scientists in the field concluded that the age of the earth had at last "been reasonably well defined." After the fluctuation of centuries, after the earth's age had been variously estimated at less than 6,000 years, at 100 million, at 2 billion, at 3.5 billion, it is now expected that the true and absolute age will be found to be very close to the 4.5 billion years recorded in the laboratories in 1955 and 1956. Man at last, it seems, has reached back to the very beginnings of this earth; we know at the middle of the twentieth century approximately where our earth stands in the endless panorama of time.

Only one major doubt remains. There is, as Patterson has reminded the scientific world, no absolute proof that the newly born earth had the same composition as the meteorites. That still is an assumption, though it is one that most scientists make and it is supported by powerful evidence.[4] But it is a doubt that is not likely to be completely resolved until we learn finally and fully the still-veiled story of the origin of the solar system.

An age of 4.5 billion years at once created havoc in the traditional geologic time scale, the scale developed to indicate the time and order of the formation of the rocks and of the life of the earth.

In nearly all of the scales, the time from the present back to the first appearance of the fossils in the rocks is generally set at 500 million years. The new dating does not change this span allotted to fossil life; it confirms it. The time set for the three divisions in this great period, the Paleozoic, the Mesozoic, and the Cenozoic, is thus unchanged.

Before life came upon the scene, however, or before the tiny creatures swarming in the early seas developed shells or bony structures that could survive as fossils, the time scales assumed two great, vague preceding periods totaling about 1.5 billion years. These periods—pre-Cambrian periods—were called the Proterozoic and the Archeozoic. A few recently corrected scales even set this opening age of the earth at 2.5 billion years.

It is here that a great change must now be made. With the revision of the earth's time and perhaps with its definition, this period—which man has traditionally regarded as prelude—has been extended to 4 billion years. No longer prelude but the major portion of the earth's time, it has become a span eight times longer than that of fossil life upon the earth. The age of recorded life has now diminished to a relatively brief one, the occupancy of humans to a mere minute, and the tenancy of homo sapiens, or men like ourselves, to a fleeting moment in the long history of

[4] The meteorites also contain diamonds. The pressures necessary for their formation could only have been secured in a "large object." Calculations indicate that this parent body from which the meteorites came must have been about the size of the moon.

the earth.[5] Even the scientist who ordinarily speaks only in restrictive terms cannot restrain his wonder when he regards this lengthening of the earth's time. "It is somewhat shocking," said Kulp, to find such an enormous age going before.

A re-study of the oldest rocks has underwritten the new time scale growing out of the comparison of meteoritic and earthly lead. The oldest rocks measured so far, the 2,650-million-year-old uranium ores of Canada and Africa, occur in what are called pegmatites. They are minerals that intruded into an older complex of greatly altered sedimentary and volcanic rocks. Thus, they pushed their way in among rocks that had already been on the earth long enough to have crumbled into sand and dust, to have been washed down into primordial waters, and to have re-formed and undergone great change.

"There was a great deal of earth history before the regional metamorphism which culminated in the southern Manitoba pegmatites," Kulp points out. The Columbia University geochemist expects that new refinements in lead-uranium dating, particularly with zircons, will make it possible to date some of the ancestral rocks, the rocks whose detritus made the rocks the pegmatites invaded. The identification and dating of still older rocks seems certain to come, and the expansion of this very early period in the earth's history promises to fit well into a total 4.5-billion-year time for the earth.

And the new dating is fixing the time and order of earth-shaping events that followed. In North America, Kulp points out, a series of events now appears well defined. Around the 2,650-million-year-old rocks that formed part of the nucleus of the continent—the inner Canadian shield—rose younger rocks. Work at Lamont has already timed the Lake Athabasca mineralization in north central Canada at about 2 billion years, the metamorphism in the Black Hills at 1.5 billion years, and the Great Bear Lake mineralization—in the Canadian northwestern territories—at 1.4 billion years.

If the Swedish kolm—the earliest fossil-bearing rock that can be dated by radioactive techniques—is finally timed at about

[5] If the 4.5-billion-year age of the earth is thought of as one year, the 500 million years of the fossil record is equivalent to about forty days. The time of humans, assuming that it is about 1 million years, would amount to less than two hours, and that of modern man, giving him 50,000 years, to a little more than five minutes.

440 million years, as the Lamont studies have shown it may well be, another highly significant point in the earth's history will have been confirmed. When the simple, minute animals entombed in the kolm swarmed in the early seas, life had already taken a long step forward. The 500-million-year date long set for the beginning of such life would appear correct.

And further along in the earth's development, radioactive dating has shown the formation of the southern Appalachians at about 340 million years ago, and the buckling up of the Rocky Mountains about 60 million years ago. Events whose beginning man could only imagine are being dated with as much precision as modern historians might bring to setting the birth date of some early king or dynasty. And the pattern is consistent, consistent with itself and with all that the layering of the earth and the order of the fossils reveals about the earth.

But the dates so far recorded are only the opening figures. They are only a foretaste of the filling in of the record which is certain to come. "Geochronometry should produce many details of geologic history in the next decade," says Kulp.

Even now, however, man knows more about his earth than the most sanguine might have expected just a few decades ago. We know, as none of our human predecessors could have known with any assurance, that this is a tremendously old earth that we live on, a 4.5-billion-year-old earth. We know that it probably came together out of the dusts of space and that it has sorted itself into a crust, a mantle, and a core. We know that continents grew from island-like nuclei and are still growing. We know that this is an earth whose orderly, inevitable breakdown is matched only by its endless regeneration. Many utterly unsuspected answers have been supplied to the questions that Hesiod began to ask and that man has asked through his brief centuries.

It is an amazing, an ever changing, a never wholly revealing earth we live on.

13

Rational And Empirical Methods Of Investigation In Geology

A significant new trend in geology is the introduction of much more numerical data, treated with mathematical and statistical techniques. The following two fragments of a longer article cannot do full justice to the opposition case, but they do give a lively glimpse inside modern geology.

Most of us are concerned, and some of us have strong feelings, pro or con, about what has been happening to geology in the past 25 years: greatly increased use of nongeologic techniques in the solution of geologic problems, such as dating by radioisotope methods; the tendency for what were special fields of interest to become nearly or wholly independent disciplines, with separate journals and jargon; and most of all, because it penetrates every field, what may be called the swing to the quantitative.

At meetings of our societies, when the elder brethren gather together in hotel rooms after the technical sessions, the discussion usually comes around to these changes. There are apt to be sad postmortems for certain departments, once powerful, which are now, owing to the retirement or flight of their older stalwarts, largely staffed by dial twisters and number jugglers. It is stated, as

From **The Fabric of Geology,** Claude C. Albritton, Editor. Copyright 1963 by the Geological Society of America. Reprinted by permission of the Geological Society of America.

a scandalous sign of the times, that in certain departments geologic mapping is considered to be, not research, but a routine operation —something like surveying from the point of view of an engineer— and therefore not suitable as a basis for the doctoral thesis. There is almost always at least one sarcastic remark per evening along the line of what our equation-minded youngsters think is the function of the mirror on a Brunton compass; a comment or two on their ignorance or disregard of the older literature; some skepticism as to whether the author of a new monograph on the mechanism of mountain building had ever been *on* a mountain, *off* a highway; and so on. This is partly banter, because we are aware that these are merely the usual misgivings of every older generation about the goings-on of every younger generation. But sometimes there is evidence of real ill-feeling, which in part at least reflects a defensive attitude; and there may be a few who seem to think that the clock ought to be stopped—that nothing new is good.

Though I am one of the elders, I often cross the hall to a concurrent session of another group, our avant-garde, where there is an almost evangelical zeal to quantify, and if this means abandoning the classical geologic methods of inquiry, so much the better; where there are some who think of W.M. Davis as an old duffer with a butterfly-catcher's sort of interest in scenery; where there is likely to be, once in a while, an expression of anger for the oldsters who, through their control of jobs, research funds, honors, and access to the journals, seem to be bent on sabotaging all efforts to raise geology to the stature of a science; where, in the urgency for change, it seems that nothing old is good.

This picture is not overdrawn, but it applies only to a small number: the blacks and the whites, both sure of their ground. Most geologists are somewhere in the gray between, and are beset with doubts. As for myself, I have sometimes thought that the swing to the quantitative is too fast and too far, and that, because a rather high percentage of the conclusions arrived at by certain methods of manipulating numerical data are superficial, or wrong, or even ludicrous, these methods must be somehow at fault, and that we do well to stay with the classical geologic methods. But at other times I have been troubled by questions; why the swing has been so long delayed in geology as compared with physics and chemistry; and whether, with its relative dearth of quantitative laws, geology is in fact a sort of subscience, as implied by Lord Kelvin's pronouncement that what cannot be stated in numbers is not science. (For original wording, and a thoughtful discussion, see Holton, 1952, p. 234.) Even more disturbing is the view, among some of my friends in physics, that a concern with cause-and-effect relations merely confuses the real issues in science; I will return to this matter later. If only because of the accomplishments of the scientists who hold these views, we must wonder whether our accustomed ways of thinking are outmoded, and whether we should not drastically change our habits of thought, or else turn in our compasses and hammers and fade away quietly to some haven reserved for elderly naturalists.

Preparation for a talk on quantitative methods in geomorphology, as a visiting lecturer at the University of Texas last year, forced me to examine these conflicting appraisals of where we stand. I suggest that two changes, quite different but closely interlocking, are occurring at the same time and have become confused in our thinking.

One of these changes includes an increase in the rate of infusion of new ideas and techniques from the other sciences and from engineering, an increase in precision and completeness of quantitative description of geologic features and processes of all kinds, and an increased use of statistics and mechanical methods of analyzing data. This change fits readily within the framework of the classical geologic method of investigation, the most characteristic feature of which is dependence on reasoning at every step; "Quantitative Zoology," by Simpson, Roe, and Lewontin (1960) shows the way. In so far as it merely involves doing more completely, or with more refinement, what we have always been doing, it is evolutionary; and it is axiomatic that it is good. Some of us may find it hard to keep abreast of new developments, but few oppose them even privately, and even the most reactionary cannot drag his feet in public without discredit to himself.

The other change is the introduction, or greatly increased use, of an altogether different method of problem-solving that is essentially empirical. In its purest form this method depends very little on reasoning; its most characteristic feature, when it functions as an independent method, is that it replaces the reasoning process by operations that are largely mechanical. Because in this respect and others it is foreign to our accustomed habits of thought, we are inclined to distrust it. By "we" I mean, of course, the conservatives of my generation.

At least a part of the confusion in our thinking comes from a failure to distinguish between the evolutionary quantification, which is good, and the mechanical kind of quantification, which I think is bad when it takes the place of reasoning. It is not easy to draw a line between them because the empirical procedures may stand alone, or they may function effectively and usefully as parts of the classical geologic method; that is, they may replace, or be combined in all proportions with, the reasoning processes that are the earmarks of that method. When this distinction is recognized it becomes evident that the real issue is not qualitative versus quantitative. It is, rather, rationality versus blind empiricism.

Although the timing has been influenced by such leaders as Chayes, Hubbert, Leopold, Krumbein, and Strahler, we are now in the swing to the quantitative because of the explosive increase in the availability of numerical data in the last few decades (Krumbein, 1960, p. 341), and because basic descriptive spadework has now advanced far enough in many fields of geology to permit at least preliminary formulation of significant quantitative generalizations. The quantification of geology will proceed at a rapidly accelerating rate no matter what we do as individuals, but I think the rate might be quickened a little, and to good purpose, if the differences between the two groups on opposite sides of the hall, at least those differences that arise from misunderstanding, could be reduced. An analysis of certain quantitative methods of investigation that are largely empirical will, I hope, serve to bring out both their merits and limitations, and may convince some of our oldsters that although disregard of the limitations may produce questionable results, it does not follow that there is anything wrong with quantification, as such, nor with blind empiricism, as such. But this is not very important — time will take care of the oldsters, soon enough. This essay is for the youngsters — the graduate students — and its purpose is to show that as they quantify, which they are bound to do, it is neither necessary nor wise to cut loose from the classical geologic method. Its message is the not very novel proposition that there is much good both in the old and the new approaches to problem-solving. A brief statement of what I am calling the rational method will point up the contrast between it and the empirical method, with which we are principally concerned.

The Rational Method

I'm sure that most American geologists are acquainted with our three outstanding papers on method: G.K. Gilbert's "Inculcation of the Scientific Method by Example," published in 1886; T.C. Chamberlin's "Method of Multiple Working Hypotheses," published in 1897; and Douglas Johnson's "Role of Analysis in Scientific Investigation," published in 1933. I do not need to describe the so-called scientific method here; for present purposes I need only remind you that it involves an interplay of observation and reasoning, in which the first observations suggest one or more explanations, the working hypotheses, analysis of which leads to further observation or experimentation. This in turn permits a discarding of some of the early hypotheses and a refinement of others, analysis of which permits a discarding of data now seen to be irrelevant to the issue, and a narrowing and sharpening of the focus in the search for additional data that are hidden or otherwise hard to obtain but which are of special diagnostic value; and so on and on. These steps are spelled out in formal terms in the papers just mentioned, and it was useful to do that, but those who use the method all the time never follow the steps in the order stated; the method has become a habit of thought that checks reasoning against other lines of reasoning, evidence against other kinds of evidence, reasoning against, evidence, and evidence against reasoning, thus testing both the evidence and the reasoning for relevancy and accuracy at every stage of the inquiry.

It now seems to be the vogue to pooh-pooh this method, as differing in no essential way from the method of problem-solving used by the man in the street. I've been interested in watching the way in which men in the street, including some medical doctors — practitioners, not investigators — arrive at conclusions, and I can only suggest that the scientists who insist that all persons arrive at conclusions in the same way should reexamine their conviction. There are, of course, rare intellects that need no disciplining, but for most of us with ordinary minds, facility in the operations that I have just outlined must be acquired by precept, example, and practice.

The objective of the scientific method is to understand the system investigated—to understand it as completely as possible. To most geologists this means understanding of cause and effect relations within the system (Garrels, 1951, p. 32). Depending on the nature of the problem and its complexity, quantitative data and mathematical manipulations may enter the investigation early or late. In general, the larger the problem, the more many-sided it is, the more complicated by secondary and tertiary feedback couples, and the more difficult it is to obtain the evidence, the more essential it is to the efficient prosecution of the study that the system first be understood in *qualitative* terms; only this can make it possible to design the most significant experiments, or otherwise to direct the search for the critical data, on which to base an eventual understanding in quantitative terms.

A problem—any problem—when first recognized, is likely to be poorly defined. Because it is impossible to seek intelligently for explanations until we know what needs explaining, the first step in the operation of the scientific method is to bring the problem into focus. This is usually accomplished by reasoning, i.e., by thinking it through, although we will see shortly that there is another way. Then, if it is evident that the problem is many-sided, the investigator does not blast away at all sides at once with a shotgun; he shoots at one side at a time with a rifle—with *the* rifle, and *the* bullet, that he considers best suited to that side.

This means that the investigator admits to his graphs, so to speak, only items of evidence that are relevant to the particular matter under investigation, and that are as accurate as practicable, with the probable limits of sampling and experimental error expressed graphically. In reading answers from the graph, he does no averaging beyond that required to take those limits into account. And once an item of information has been admitted to the graph, it cannot be disregarded; as a rule, the items that lie outside the clusters of points are at least as significant, and usually much more interesting, than those that lie within the clusters. It is from inquiry as to why these strays are where they are that most new ideas—most breakthroughs in science—develop.

The scientific method tries to visualize whole answers—complete theoretical structures—at the very outset; these are the working hypotheses that give direction to the seeking-out and testing of evidence. But one never rushes ahead of the data-testing process to a generalization that is regarded as a conclusion. This is not because there is anything ethically wrong with quick generalizing. It is only that, over a period of 500 years, investigators have found that theoretical structures made in part of untested and ill-matched building blocks are apt to topple sooner or later, and that piling them up and building on them is therefore not an efficient way to make progress. The need to test the soundness of each building block *before* it gets into the structure—to determine the quality and the relevance of each item of evidence *before* it gets onto the graph—is emphasized by Douglas Johnson (1933). His approach was the antithesis of that to which we may now turn.

The Empirical Method

What I have long thought of as the engineering method or the technologic method (we shall soon see that it needs another name) deals almost exclusively with quantitative data from the outset, and proceeds directly to a quantitative answer, which terminates the investigation. This method reduces to a minimum, or eliminates altogether, the byplay of inductive and deductive reasoning by which data and ideas are processed in the scientific method; this means that it cannot be critical of the data as they are gathered. The data are analyzed primarily by mathematical methods, which make no distinction between cause and effect; understanding of cause and effect relations may be interesting, but it is not essential, and if explanations are considered at all, there is usually only one, and it is likely to be superficial. All of the reasoning operations that characterize the so-called scientific method depend on a fund of knowledge, and on judgment based on experience; other things being equal, the old hand is far better at these operations than the novice. But the operations of the "engineering method" are much less dependent on judgment; in applying this method the sharp youngster may be quicker and better than the experienced oldster. For this reason and because of its quick, positive, quantitative answers, it makes a strong appeal to the younger generation. I would like now to explain the logic of this method, as it operates in engineering.

Many engineers feel that unless a relation can be stated in numbers, it is not worth thinking about at all. The good and sufficient reason for this attitude is that the engineer is primarily a doer—he designs structures of various types, and supervises their building. In the contract drawings for a bridge he must specify the dimensions and strength of each structural member. Nonengineers may be able to think of a drawing that indicates the need for a rather strong beam at a given place in the bridge. But a young man who has spent five years in an engineering school is incapable of thinking seriously of a "rather strong" beam; all of the beams of his mind's eye have numerical properties. If the strength of a beam cannot be put in numerical terms, thinking about it is mere daydreaming.

The matter of stresses in a steel structure is fairly cut and dried. But the engineer is confronted with many problems for which there are no ready answers; he must deal with them—he must complete his working drawings—against a deadline. If he is charged with the task of designing a canal to carry a certain flow of irrigation water without either silting or erosion of the bed, or with the immensely more complex task of developing and maintaining a 10-foot navigable channel in a large river, he cannot wait until he or others have developed a complete theory of silting and scouring in canals and rivers. It may be 50 or a 100 years before anything approaching a complete theory, in quantitative terms, can be formulated; and his drawings, which must be entirely quantitative, have to be ready within a few days or weeks for the contractors who will bid the job. So he has to make certain simplifying assumptions, even though he realizes that they may be wide of the mark, and he has to make-do with data that are readily available, even though they are not entirely satisfactory, or with data that can be obtained quickly from experiments or models, even though the conditions are significantly different from those existing in his particular canal or river.

He is accustomed to these expedient operations, and he is not much concerned if, in plotting the data, he mixed a few oranges with the apples. In fact, he wouldn't worry much if a few apple *crates* and a few orange *trees* got onto his graph. He cannot scrutinize each item of evidence as to quality and relevancy; if he did, none but the simplest of structures would ever get built. He feels that if there are enough points on a scatter diagram, the bad ones will average out, and that the equation for the curve drawn through the clustered points will be good enough for use in design, always with a goodly factor of safety as a cushion. And it almost always is. This method is *quantitative, empirical,* and *expedient.* As used by the engineer, it is logical and successful.

It is of course used by investigators in many fields other than engineering. Friends in physics and chemistry tell me that it accounts for a large percentage of the current research in those sciences. A recent paper by Paul Weiss (1962) with the subtitle "Does Blind Probing Threaten to Displace Experience in Biological Experimentation?" calls attention to its increasing use in biology. The approach and examples are different, but the basic views of Dr. Weiss correspond so closely with those expressed in this essay that I am inclined to quote, not a passage or two, but the whole paper. Because this is impracticable, I can only urge that geologists interested in this phase of the general problem—whither are we drifting, methodologically?—read it in the original.

In view of its widespread use in science, what I have been calling the engineering or technologic method certainly should not be identified, by name, with engineering or technology as such. And on the other side of the coin, the so-called scientific method is used more consistently and effectively by many engineers and technologists than by most scientists. Besides being inappropriate on this score, both terms have derogatory or laudatory connotations which beg some questions. So, with serious misgivings that will be left unsaid, I will from here on use the term "rational method" for what we are accustomed to think of as the scientific method, and what I have been calling the engineering method will be referred to as the empirical method.

Actually, the method that I am trying to describe is *an* empirical method; it is shotgun or scatter-diagram empiricism, very difdifferent from the one-at-a-time, cut-and-dry empiricism of Ehrlich who, without any reasoned plan, tried in turn 606 chemical substances as specifics for syphilis. The 606th worked. Both the scatter-diagram and the one-at-a-time types can be, at one extreme, purely empirical, or, if you prefer, low-level empirical. As Conant (1952, pp. 26-30) points out, the level is raised—the empirical approaches the rational—as the gathering and processing of the data are more and more controlled by reasoning.

* * *

Whither Are We Drifting, Methodologically?

I would like now to return to some of the questions asked at the outset. Must we accept, as gospel, Lord Kelvin's pronouncement that what cannot be stated in numbers is not science? To become respectable members of the scientific community, must we drastically change our accustomed habits of thought, abandoning the classic geologic approach to problem-solving? To the extent that this approach is qualitative, is it necessarily loose, and therefore bad? Must we now move headlong to quantify our operations on the assumption that whatever is quantitative is necessarily rigorous and therefore good?

Why has the swing to the quantitative come so late? Is it because our early leaders, men such as Hutton, Lyell, Agassiz, Heim, Gilbert, and Davis, were intellectually a cut or two below their counterparts in classical physics? There is a more reasonable explanation, which is well known to students of the history of science. In each field of study the timing of the swing to the quantitative and the present degree of quantification are largely determined by the subject matter: the number and complexity of the interdependent components involved in its systems, the relative ease or difficulty of obtaining basic data, the susceptibility of those data to numerical expression, and the extent to which time is an essential dimension. The position of geology relative to the basic sciences has been stated with characteristic vigor by Walter Bucher (1941) in a paper that seems to have escaped the attention of our apologists.

Classical physics was quantitative from its very beginning as a science; it moved directly from observations made in the laboratory under controlled conditions to abstractions that were quantitative at the outset. The quantification of chemistry lagged 100 years behind that of physics. The chemistry of a candle flame is of an altogether different order of complexity from the physics of Galileo's rolling ball; the flame is only one of many types of oxidation; and oxidation is only one of many ways in which substances combine. There had to be an immense accumulation of quantitative data, and many minor discoveries—some of them accidental, but most of them based on planned investigations—before it was possible to formulate such a sweeping generalization as the law of combining weights.

If degree of quantification of its laws were a gage of maturity in a science (which it is not), geology and biology would be 100 to 200 years behind chemistry. Before Bucher (1933) could formulate even a tentative set of "laws" for deformation of the earth's crust, an enormous descriptive job had to be well under way. Clearly, it was necessary to know what the movements of the crust *are* before anybody could frame explanations of them. But adequate description of even a single mountain range demands the best efforts of a couple of generations of geologists, with different special skills, working in the field and the laboratory. Because no two ranges are alike, the search for the laws of mountain growth requires that we learn as much as we can about every range we can climb and also about those no longer here to be climbed; the ranges of the past, which we must reconstruct as best we can by study of their eroded stumps, are as significant as those of the present. Rates of growth and relative ages of past and present ranges are just as important as their geometry; the student of the mechanics of crustal deformation must think like a physicist and also like a historian, and these are very different ways of thinking, difficult to combine. The evidence is hard to come by, it is largely circumstantial, and there is never enough of it. Laboratory models are helpful only within narrow limits. So it is also with the mechanism of emplacement of batholiths, and the origin of ore-forming fluids, and the shaping of landforms of all kinds, and most other truly geologic problems.

It is chiefly for these reasons that most geologists have been preoccupied with manifold problems of description of geologic things and processes—*particular* things and processes—and have been traditionally disinclined to generalize even in qualitative terms. Because most geologic evidence cannot readily be stated in numbers, and because most geologic systems are so complex that some qualitative grasp of the problem must precede effective quantitative study, we are even less inclined to generalize in quantitative terms. Everybody knows the story of Lord Kelvin's calculation of the age of the earth.

These things are familiar, but they are worth saying because they explain why geology is only now fully in the swing to the quantitative. Perhaps it would have been better if the swing had begun earlier, but this is by no means certain. A meteorologist has told me that meteorology might be further ahead today if its plunge to the quantitative had been somewhat less precipitous— if there had been a broader observational base for a qualitative understanding of its exceedingly complex systems before these were quantified. At any rate, it is important that we recognize that the quantification of geology is a normal evolutionary process, which is more or less on schedule. The quantification will proceed at an accelerating pace, however much our ultraconservatives may drag their feet. I have been trying to point out that there is an attendant danger: as measurements increase in complexity and refinement, and as mathematical manipulations of the data become more sophisticated, these measurements and manipulations may become so impressive in form that the investigator tends to lose sight of their meaning and purpose.

This tendency is readily understandable. Some of the appealing features of the empirical method have already been mentioned. Moreover, the very act of making measurements, in a fixed pattern, provides a solid sense of accomplishment. If the measurements are complicated, involving unusual techniques and apparatus and a special jargon, they give the investigator a good feeling of belonging to an elite group, and of pushing back the frontiers. Presentation of the results is simplified by use of mathematical shorthand, and even though nine out of ten interested geologists do not read that shorthand with ease, the author can be sure that seven out of the ten will at least be impressed. It is an advantage or disadvantage of mathematical shorthand, depending on the point of view, that things can be said in equations, impressively, even arrogantly, which are so nonsensical that they would embarrass even the author if spelled out in words.

As stated at the outset, the real issue is not a matter of classical geologic methods versus quantification. Geology *is* largely quantitative, and it is rapidly and properly becoming more so. The real issue is the rational method versus the empirical method of solving problems; the point that I have tried to make is that if the objective is an understanding of the system-investigated, and if that system is complex, then the empirical method is apt to be less efficient than the rational method. Most geologic features—ledges of rock, mineral deposits, landscapes, segments of a river channel—present an almost infinite variety of elements, each susceptible to many different sorts of measurement. We cannot measure them all to any conventional standard of precision—blind probing will not work. Some years ago (1941) I wrote that the "eye and brain, unlike camera lens and sensitized plate, record completely only what they intelligently seek out." Jim Gilluly expresses the same thought more succinctly in words to the effect that most exposures provide answers only to questions that are put to them. It is only by thinking, as we measure, that we can avoid listing together in a field book, and after a little while, averaging, random dimensions of apples and oranges and apple crates and orange trees.

Briefly, then, my thesis is that the present swing to the quantitative in geology, which is good, does not necessarily and should not involve a swing from the rational to the empirical method. I'm sure that geology is a science, with different sorts of problems and methods, but not in any sense less mature than any other science; indeed, the day-to-day operations of the field geologist are apt to be far more sophisticated than those of his counterpart—the experimentalist—in physics or chemistry. And I'm sure that anyone who hires out as a geologist, whether in practice, or in research, or in teaching, and then operates like a physicist or a chemist, or, for that matter, like a statistician or an engineer, is not living up to his contract.

The best and highest use of the brains of our youngsters is the working out of cause and effect relations in geologic systems, with all the help they can get from the other sciences and engineering, and mechanical devices of all kinds, but with basic reliance on the complex reasoning processes described by Gilbert, Chamberlin, and Johnson.

14

Debate About The Earth

J. TUZO WILSON AND V. V. BELOUSSOV

The new concept of sea-floor spreading with continental drift may seem a revolution in geologic thinking, but it is not yet universally accepted. Professor J.T. Wilson, of the University of Toronto, and Academician V.V. Beloussov, of Moscow, hold different opinions. Both are leading geologists in the study of the earth's global structures and the causes of crustal deformation. Although some technical points of the discussion will be difficult to follow, the exchange gives a good idea of the skeptical approach and critical spirit that characterize scientific method.

A Revolution in Earth Science

In recent years the hypothesis of continental drift has become accepted by more earth scientists. Indeed, many now believe that the hypothesis has attained the status of a theory. Further, as Dr. Wilson explains in the next few pages, the implications for geology are far greater than a question of whether the continents were once a single mass.

This article was first presented at the 1967 meeting of the Canadian Institute of Mining & Metallurgy, held in Ottawa. Then it was printed in the February 1968 issue of the *Canadian Mining & Metallurgical Bulletin,* and its appearance there led to the open letter by V.V. Beloussov that follows. In order to provide context for that letter,

Dr. Wilson and the editor of the *Bulletin* have given permission to reprint this article in *Geotimes*. Together with Dr. Wilson's reply, they are published here as a debate about the Earth.

J. Tuzo Wilson
Principal, Erindale College
University of Toronto

At any time, it is a pleasure to me to be invited to address the Canadian Institute of Mining & Metallurgy because it represents an industry and men with whom I have been happily, if peripherally, associated continuously for 40 years. This time it is a particular pleasure because I believe that an important subject has arisen which I propose to discuss.

This is a major discovery in the earth sciences, first fully revealed in the winter of 1966-67, but already widely accepted. The basis of this revolution, for it is no less, is that measurements of 3 different features of the Earth all change in exactly the same ratios. These ratios are the same in all parts of the world. The results from one set are thus being used to make precise numerical predictions about all the sets in all parts of the world. No such accurate correlations and predictions have ever been found before in geology or areal geophysics. The whole subject of earth science has thereby been radically altered.

The first of these measurements is that of the direction of magnetic polarity in lava flows. When piles of young lavas are examined with the aid of a pocket compass, it is noted that some flows are magnetized in the direction of the earth's field and some in the reverse direction (Fig. 1). By accurately dating enough flows, a time scale of the dates of reversals has been established. During the past 4 million years, 9 reversals have occurred synchronously all over the world (Cox, Dalrymple & Doell 1967, McDougall & Chamalaun 1966). It appears that the scale can be extended back to Precambrian times (McMahon & Strangway 1967). This time scale is the first of the 3 identical ratios.

The second group of measurements is that of the widths of successive strips of magnetic anomalies measured over ocean basins.

TIME (Years)	PILE OF LAVA FLOWS	DIRECTION OF MAGNETIZA-TION	TIME (Years)	PILE OF LAVA FLOWS	DIRECTION OF MAGNETIZA-TION
Present				R	
	N			R	REVERSED
	N			R	
	N	NORMAL		R	
	N		2,400,000	R	
700,000	N			N	
850,000	R	REVERSED		N	
	N	NORMAL		N	NORMAL
950,000	N			N	
	R		3,000,000	N	
	R			R	
	R	REVERSED		R	REVERSED
	R		3,100,000	R	
	R			N	
1,800,000	R		3,400,000	N	NORMAL
	N			N	
	N	NORMAL		R	
	N			R	REVERSED
2,000,000	N			R	
	R				

Fig. 14.1 Diagram of a pile of lava flows, showing that during some periods the flows had the same magnetic orientation as at the present day, although during other periods the polarity is reversed.

In 1956, Heezen & Ewing (1963) suggested that a world-encircling system of mid-ocean ridges exists. Their colleagues soon showed that magnetic anomalies occur in patterns of strips parallel to the crest of these ridges. Vine & Matthews (1963) predicted, and Vine & Wilson (1965) demonstrated, that the widths of successive anomalies were the same as the times between successive reversals of the earth's magnetic field. Heirtzler et al. (1966) and Vine (1966) have shown that, although the absolute widths vary in different places, the ratio of the widths is constant everywhere. This is the second of the 3 identical ratios.

The third group of measurements has been made by Opdyke & colleagues (1966) on deep-sea cores. They discovered that the directions of the feeble magnetization of samples taken at intervals along a deep-sea core could be measured. This magnetization is either parallel to the present field or is exactly reversed. The depths at which successive reversals take place is in a constant ratio for all cores. This is the third of the identical ratios.

The essence of the new revolution lies in this identity of the ratios of 3 independent groups of measurements. The proved coincidences are already too great to be due to chance, and are all the more remarkable because one measurement is of time in millions of years, one is of horizontal distance in hundreds of kilometers and one is of vertical distance in centimeters.

The only explanation proposed for Opdyke's discovery is that the Earth's magnetic field is reversing and imprinting the sediments as they are uniformly deposited on the sea floor.

If the Earth's field is reversing in a known time scale and that scale reappears in distances, the relationship must take the form of a velocity. This suggests that the ocean floor is being generated along mid-ocean ridges, that it is being magnetically imprinted as it forms, and that it is being carried away from the ridges at uniform velocities. This is the theory advanced by Vine & Matthews (1963) and independently by Morley & Larochelle (1964). It is a development of the convection current hypothesis of Holmes (1928-29), as modified by Hess (1962), and likened by Dietz (1966) to a system of conveyor belts carrying the ocean floors about.

What distinguishes this theory from all past hypotheses is its precision. For example, its application to a magnetic map of the sea floor off Vancouver Island gives the following interpretation for one point chosen at random. At lat. 47 15' N and long. 130 W, the sea floor is calculated to be moving away from the United States coast in a N 72 W direction at a rate of 6 cm per year. The rate has been steady for several million years, and the rock samples dredged from the vicinity should have been erupted between 2.0 and 2.5 million years ago and should have an average polarity exactly opposite to the present direction of the Earth's field. This information was read from plate 1 in Raff & Mason (1961).

As magnetic surveys are extended, it seems likely that it will be possible to make accurate statements, such as the above, for any point on the ocean floors. Such forecasts can be checked, and this is being done. So far, the theory is proving satisfactory; if this continues, earth science will have entered a new era.

The mechanism which causes the surface to move about is considered to be upwelling and outward flow from mid-ocean ridges (Fig 2). By this process, large segments of the surface layers of the Earth are spread apart. They appear to be carried about at rates of a few centimeters a year by currents in the plastic white-hot mantle. Estimates based upon melting-point curves, isostatic rebound after ice-sheet melting and the propagation of seismic waves all agree that this plastic asthenosphere begins at a depth of about 50 km. Above it, the cooler and brittle lithosphere includes some of the uppermost mantle and all of the crust except deep roots of mountains (Fig. 3).

Upwelling and extension in some places is matched elsewhere by downward flow and compression. These segments of the litho-

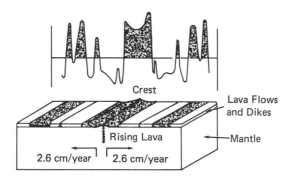

Fig. 14-2   Pattern of upwelling and reversals, showing how the strips of anomalies are believed to be due to the spreading of the ocean floor from mid-ocean ridges.

sphere are forced together, one side over-riding the other and forcing it down to be reabsorbed under young mountains and beneath deep ocean trenches, such as exist off Chile, Japan or the Aleutians (Coats 1961).

This theory is one form of continental drift, but it is not the same as Wegener's version, for he visualized the continents as rafts moving through the ocean floors; the present concept visualizes the continents as being carried about along with ocean floor like logs frozen in ice. Elsasser (1966), Orowan (1964) and Tozer (1965) have examined this theory and find convection in a shallow layer to be physically acceptable.

The theory also explains many other observations favoring drift. These include paleoclimatic and paleomagnetic evidence; the excellent fit of the two sides of the Atlantic Ocean; the increasing age of islands and cores away from mid-ocean ridges; the apparent youth of all oceans; the new class of transform faults proposed to fit the geometry of a spreading or absorbed crust (Wilson 1965); the demonstration by Sykes (1967) that earthquakes on the mid-ocean ridge do have the direction of motion required by that theory; and other reasons summarized by Blackett, Bullard & Runcorn (1965), Runcorn (1962) and Munyan (1964).

Effect on Earth Science

It must be obvious that the discovery of a series of precise, worldwide coincidences between different measurements will have a great effect upon earth science. To understand what this may be, consider the manner of progress in any science. There are 4 stages. The collection of data, the discovery of a precise theory, its use to make forecasts, and the checking of these predictions. Repetition of the cycle often leads to improved theories.

The difficulty about earth science has been that because the Earth is so complex and has to be studied in so much detail, the science has never progressed beyond the first stage. Geologists and solid-earth geophysicists amass data, but this has been of limited use because of the lack of any good theories upon which to base predictions. Prospectors are very well aware of the need to make predictions about the occurrence of ore and the difficulty of doing this.

Because the data which they had was inadequate and failed to lead to any sound theories, the very idea of prediction, which is the powerful essence of the scientific method, fell into disrepute among geologists. Instead, they quite rightly tried to improve their observations and techniques. This they have preferred to do by making more precise measurements in their customary fields (paleontology, mineralogy, petrology, geochemistry, structural geology, geomorphology, etc.) rather than by studying other fields. In particular, most geophysical subjects have traditionally been studied by physicists. Most members of both groups have been so absorbed in techniques and data collection that the search for principles has been neglected. Neither group found satisfactory theories of the Earth's behavior. Geologists' concepts applied to the real Earth, but were vague; geophysicists' ideas were more precise, but, because of their usual lack of knowledge of geology, they were applied to models too simple to bear much resemblance to the Earth.

The development of the present revolutionary theory has de-depended upon the contributions of both groups. These include geologists from all parts of the world, too numerous to mention, but especially those who have recently studied the petrology of ocean islands and floors, geochemists who have provided useful chemical guides to the probable nature of the Earth's interior and physicists who have devised instruments for studying the sea floor and for making new physical measurements of the Earth's interior and of the age and history of rocks.

However, it seems fair to suggest that the men who have most clearly understood what was happening and who have done most to produce this revolution have been geologists with two characteristics in common: a broad training, including an unusually good grasp of physics, and an interest in worldwide problems rather than small areas. This description applies, I believe, to such men as Holmes, Du Toit, Carey, Hess, Heezen, Menard, Dietz, Vine, Matthews, Irving, Morley, Doell, and McDougall. Wegener, Vening Meinesz, Bullard, and Runcorn are like-minded physicists.

As so often happens when something really new is discovered in science, the establishment, if we may so describe those larger institutions most closely concerned with the subject, had little to do with it. The petroleum and mining industries paid little attention to the idea of continental drift. The majority of the members of geological surveys and departments of geology were not expecting any such revolution. All these organizations were concentrating upon limited objectives. They were so absorbed in improving techniques, in amassing data and in planning computer codes by which to store the information that they forgot that other sciences have simplified their problems by discovering principles. Was not Tycho Brahe followed by Kepler and Newton?

The isolation in which some scientists live is well illustrated by a letter I received on March 14, 1967, from the director of a distinguished geological survey, which read 'The opinion of the National Committee is that the subject of continental drift, attractive and stimulating as it is, is not of priority interest to geologists in . . .'

In large measure, the revolution can be said to have come about because of defense spending, which provided, for the first time, abundant knowledge of hidden places on the ocean floors, of the Moon and Mars, and of the Earth's interior (through new seismic arrays built to detect atomic bomb explosions). These discoveries opened men's minds to new possibilities.

This outlook promises well for the future for the two following reasons. Science, like clothing, moves from one fashion to another. Earth science is not forever destined to be a poor relation. A century ago, geology was the leading science. The names of Sir William Logan, Sterry Hunt and Sir William Dawson remind us of how true that was in Canada. Geology did not decline, but the faster rise of organic chemistry, engineering, atomic physics, molecular biology, electronics, computers and other sciences eclipsed it. However, these in turn lose their freshness and there is no reason why earth science, having made an important break-through, should not again rise to preeminence.

The other reason is that one discovery often leads to others. The excitement in earth science today lies in new ideas and the search for powerful new principles—no longer are we limited to the blind collection of better data.

EFFECT ON UNIVERSITIES

It is well known that the teaching of earth science in universities and schools is in a state of flux. A very few departments teach combined courses in geology, geophysics and geochemistry with a suitable measure of math, physics and chemistry. Others have scarcely enlarged their teaching of geology at all. Some departments have even narrowed their interests. The commonest change has been to introduce new instrumentation into the traditional geological subjects. This has been an excellent start, and now new ideas are being introduced. Geophysics has traditionally been taught separately, as an offshoot of physics, with insufficient attention to the complexities of geology.

Attitudes toward the new discoveries are likely to be equally diverse. Integrated departments will be in a position to appreciate and test the new ideas. Others will want to have little to do with them and will say that the proposed theory is still unproved. This is quite true. One can never reconstruct the distant past with mathematical certainty, and in that sense continental drift has not been proved and, by its nature cannot ever be proved. Nevertheless, the coincidence of the three ratios is now so well established that it cannot be ignored. This coincidence must be underlain by some principle. Drift is the only explanation yet offered, and, for that reason, it may become widely accepted.

If, in this way, the precise pattern of continental drift in later geological time has been revealed, this discovery will have a great effect upon traditional subjects.

As R. T. Chamberlin said in 1928, 'If we are to believe Wegener's hypothesis we must forget everything which has been learned in the last 70 years and start all over again.' This, fortunately, is an exaggeration, but the standard textbooks assume that drift has not occurred. If it has, in fact, been occurring at a relatively rapid rate, much of our teaching must have been nonsense. Isn't it time we changed? Is there not evidence of much weakness in methods, both geological and geophysical, which were too feeble to distinguish between drift and non-drift?

Studies of minerals, rocks and fossils as intrinsic objects are legitimate fields of science. If they are to be pursued, this should be done in the most efficient ways possible. The properties of these objects suggest links with crystallography, chemistry and biology, respectively.

On the other hand, if the object of a department is to study the Earth, then it is clear that mineralogy, petrology and paleontology are tools and that they have not been very efficient ones. Clearly, less instruction should be given to geologists of the details of these and similar subjects in order that time can be found to study other techniques, principles of the Earth's behavior and a general view of world geology.

Consider, for example, paleontology. It now seems evident that magnetic reversals provide a succession of dates, synchronous all over the world, which were frequent during the Tertiary, and less so before. This can test and aid paleontological correlations and extend them to lava flows and other nonfossiliferous rocks. Many paleontologists have tackled the problem of continental drift and they have come to such diverse and opposite views that one is forced to conclude that the study of fossils alone has not been a powerful enough method to resolve the matter. On the other hand, these new techniques promise to make some precise reconstructions possible. As the former relations of continents become known, the entire field of paleontology will need reworking. The new proposals, far from eliminating the need for paleontology, promise to aid this subject and place it on a much sounder basis, but this cannot be done without understanding the new theory. One should learn the powerful new theory first and paleontology second. Paleontology should increasingly be moved to graduate study.

In petrology, it is clearly important to be able to decide whether a volcano or other outlet for igneous rocks is remaining fixed over a source which it would tend to exhaust or whether it is moving about and thus continually drawing on fresh material. Conclusions about the origin of andesite, the common rock in island arcs, will surely be affected if it is decided that the ocean floors are not static, but are constantly being pushed down in ocean trenches to pass under the arcs where the andesite is generated (Coats 1962). However, these important ideas have not been adequately considered. The chemistry of the Earth cannot be decided in the laboratory alone, but only when the laboratory is used as an aid to field work and new ideas.

Again, one cannot properly discuss structural geology without considering the implications of continental drift on the subject. Most of the existing textbooks do not. It has been suggested that many of the largest faults in the world are due to drift and hence are of a special type—transform faults—not previously recognized or described. There has been no time yet for any book on structural geology to even mention them.

If the continents are moving about they must react on one another. This can now be studied in a precise way for Tertiary and Cretaceous time. Once the idea of relatively rapid drifting becomes accepted, its earlier history can be traced back using paleomagnetic and geological methods. This will modify most of what has been written about historical geology. For example, the only modern book on world geology does not seriously entertain the idea of drift (Kummel 1961). Precambrian history can be sensibly interpreted only when many age determinations are combined with ideas about drift.

In geophysics, the patterns of magnetic anomalies over the oceans now make sense, gravity observations can be more easily interpreted and the processes that cause earthquakes can be better understood. Practically all courses and textbooks in stratigraphy, physics of the Earth, paleontology, petrology, and structural and historical geology have been rendered somewhat out of date and need revising. Under the circumstances, our ideas about economic geology and prospecting could not help but be vague and of little use.

It is thus apparent that the traditional methods in geology and in geophysics have not been as effective or as powerful as they could be in combination on a broader scale. The new discoveries have been made by utilizing all types of information on a worldwide basis—over oceans as well as over land. It is evident that the Earth is a single system, with all subjects and all regions reacting upon one another.

Heretofore, geophysics has been taught as one group of subjects and geology has been taught to other men as another. To too great an extent, each subject has been treated as an isolated set of methods of data collecting and as a package of accummulated facts. Descriptive geology has been confined to a discussion of local areas and continents. To many, geophysics is only one specialized way of prospecting.

Clearly, all this is due to change. As fast as the details of the new story can be worked out and as fast as new books can be written, it is evident that earth science can be put on a similar basis as other sciences, with a discussion of the principles of the behavior of the Earth as the framework and the essential parts of the traditional subjects used to illustrate this. Regional geology must be discussed on a worldwide basis, not solely in terms of local details. Great adjustments in the curriculum are needed and room can only be found by removing, to the graduate school, the greater part of the work in those subjects which have proved to be less effective and less powerful.

The change that is needed is similar to the removal of much classical physics from undergraduate courses to make room for modern physics. No one has ever suggested that mechanics or acoustics were wrong. They are still used by specialists to track satellites and to design concert halls. On the other hand, everyone now agrees that it was wise to drop these undergraduate courses so that electronics, quantum mechanics, computer technology and other more powerful and more generally useful subjects could be taught. It is clear that the same pattern must be followed in geology and geophysics. I am not suggesting that the traditional subjects of geology are wrong. What I believe is that they can be better taught and more easily understood after the principles of Earth behavior are mastered. Most of the new techniques they employ are developments of physics, chemistry and mathematics, and geologists are correct in placing more stress on these courses. This change will take time and arouse opposition, but it is necessary if earth science is to rise to its opportunities. I can attest to the difficulty of conversion from my own experience.

This new approach is necessary if earth sciences are to attract students. Why should bright young scientists elect to study subjects mulled over for a century, shown to be ineffective in coping with major problems of the Earth and often of only local interest when a more modern undergraduate curriculum can introduce him to the whole spectrum of Earth and planetary science? The full scope and the greatest appeal of the subject lie in pointing out that now is the time when other planets are being explored, when the sea floor is yielding exciting discoveries, when the interior of the planets is becoming understood and when jet airplanes, helicopters and flying sensors are carrying out prospecting work all over the world.

This is the introduction that will appeal to, and be a sound basis for, our undergraduates. There may be a question of who, with that training, would do the necessary routine jobs, such as mapping. I believe the answer is that earth science is not now as effective as it might be and is not attracting many students. If modernized, it might be more useful and would attract more students who could do a better job.

EFFECT ON INDUSTRY

The mining and petroleum industries want men in the earth sciences who can find useful deposits and they want more of them. Industry has always provided the greatest rewards and the highest praise to the successful prospectors and the great oil-finders. To find an orebody or a petroleum reservoir requires a theory in order to know where to drill. Today, that theory is more likely to be based upon an assessment of combined geological, geophysical and geochemical data than on a wildcatter's hunch. The broad training proposed is what is needed in industry. Geology has, in the past, lacked theories and this has been a serious defect as far as industry is concerned. A precise theory has now been proposed. It may have immediate benefits, but the most promising aspect is that discoveries in other sciences have rarely come singly. Witness the great succession of revolution in physics between 1895 and 1920. There is the same promise in the earth sciences.

Some immediate effects can, however, be stated. For the petroleum industry, these effects will be least in young strata in the heart of continents. In older rocks, and especially those which lie near coasts, the acceptance of drift will serve to explain why very great volumes of sediments seem in some places to have been derived from areas that are now ocean. Borderlands, as envisioned by Barrell and Schuchert, are not only possible but likely. Continental blocks have apparently been drifting together and breaking apart repeatedly (Wilson 1966). This may explain thrusting and rifting.

The movements of continents can explain changes from marine to continental to evaporite conditions. Thus, Belmonte, Hirtz & Wenger (1965) have suggested that the salt domes and oil deposits of Gabon and the opposite coast of Brazil were formed as those continents started to break apart and the sea first entered the rift opening in mid-Cretaceous time.

The largest effects are to be expected in offshore oil basins. Continental drift can explain why the Gulf of Mexico may have been a small evaporite basin during Jurassic time and hence why there seem to be salt domes on its floor under 12,000 ft of water. Drift would make possible very large strike-slip faults along the south side of the Grand Banks and off the equatorial coast of West Africa. If these are transform faults, they need never have penetrated the continents, a situation that would be impossible without drift and not previously envisioned in interpretations.

Arguments based on geophysical studies over oceans can reinforce geological arguments on land. Thus, Hamilton & Myers (1966) concluded from a long study of the geology of California that the San Andreas fault is moving at the rate of 6 cm per year. This happens to be precisely the value obtained by Vine & Wilson (1965) from a study of offshore magnetics.

In mining, the present theory could have a marked effect upon the interpretation of one type of large ultrabasic body with associated copper and nickel deposits. Gass & Masson-Smith (1963) and Maxwell suggest that these bodies in Cyprus, and presumably also in Cuba, Italy, Turkey, Greece, New Caledonia and New Guinea, have been thrust up from the sea floor over the land. This interpretation has not been previously placed on these areas and the change could make a large difference in prospecting.

Again, it is well known that no deposits have been found along some large faults, like the Great Glen in Scotland, whereas others, like the Kirkland Lake break or certain faults in British Columbia and California, are well mineralized. Perhaps the new theory can explain the difference and draw attention to new and favorable localities. It may do the same for batholiths and their associated ores. It is becoming evident that the building of the Andes and the Cordillera, with their important ore deposits, is likely to have been a consequence of the movements which opened the Atlantic Ocean. The whole subject of economic geology is ripe for a deeper understanding.

DEMAND AND NEED

As I have said, the proposals I have made for changing university curriculums are certain to meet with opposition. What amazes me is that the opposition is not greater. This winter, the leaders of several large and formerly very conservative university departments have radically altered their views. At the Geological Society of America's regional meeting in Boston in March 1967, I discovered that new editions of some of the leading American textbooks will soon appear, incorporating the new ideas. Earth scientists knew that their subject was ripe for a change, and their conversion to the new ideas has often been rapid.

The remaining opposition is chiefly based on the view that there is still a good demand for conventional geologists, so why change? Of course, all science graduates get jobs today, because there are not enough of them. Many conventional geologists are now going into school teaching, which is desirable, but this does nothing to meet the needs of industry, which wants men trained in all methods of prospecting.

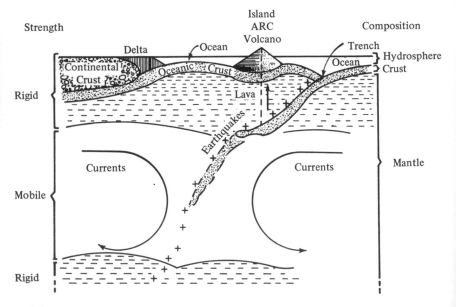

Fig. 14-3 Diagram of reabsorption of the oceanic crust into trenches on the convex side of island arcs.

The Geological Association of Canada suggests 'that the tendency for many graduates to continue research in areas not applicable to mineral exploration . . . considerably limits the number entering the industrial field.' (Canada 1966). The Association also mentions the demand for teachers. The American Association of Petroleum Geologists made a review which states that the supply of graduates in geology is about equal to demand, while the supply of geophysicists is only one-tenth of the demand (Royds et al. 1965).

The concept that every professor should be free to do whatever research he wishes and should be supported is a popular one and has its merits. The chief claim is that only in this way can new ideas freely start. I believe that, for modest grants, the system is sound.

Anyone who has sat on several grant-awarding bodies will see other less desirable influences. The system provides no incentive to productivity. On the contrary, because many economically minded geologists work with industry in such a way that they do little research, the grants tend to be concentrated in those fields which are of the least economic interest. Another feature of the system seems to work against originality, the very aspect which the system was designed to encourage. Grant-giving bodies seem usually to be faced with a multitude of requests in a few fields and with none at all in others which seem to offer equal or greater opportunities. This has often come about because it happened that, a generation or more ago, one or two good teachers in one subject attracted many good students and thus started a tradition. These men all want to continue research in that field and the system offers no encouragement for anyone to change and enter new fields.

Three methods can be used to offset this. First, government bodies can and do encourage lagging fields by extra grants. Second, a major scientific discovery attracts good new workers. This happened when modern physics replaced classical physics as the center of interest and resulted in the preponderant emphasis on nuclear physics today. Third, it would help if industry stated their needs. If the mining and petroleum industries want more miners, geologists and geophysicists, they should say so more clearly. To many city students, the industry is only heard as a small and isolated voice from the distant and unattractive backwoods. Most Canadians have never seen a mine or an oil derrick. This is a very different situation from that in South Africa, where the largest city is also a mining camp. Alberta is an exception in that there are oil wells near cities.

The industry could get better results without even spending any more money. Existing scholarships and advertising could be more specifically directed to meeting needs. Of course, some companies already do a great deal through summer employment and scholarships. An outstanding example of this has been provided by Geophysical Service Inc., which, for 16 years in the United States and for 2 years in Canada, has conducted orientation sessions for their summer students. Dr. Cecil Green started these schemes with the altruistic view of aiding education, but the effect has been to encourage many top students to enter the earth sciences (Shrock 1966).

In spite of many excellent individual efforts, there is no industrywide program in Canada. The pulp and paper industry, on the other hand, supports a research institute, and in South Africa the Chamber of Mines supports an Economic Geology Research Unit in the University of Witwatersrand. Granted that conditions are not the same, but these efforts do deserve consideration.

Agriculture, which is to be seen everywhere, is no more important to the Canadian economy than mining, but it gets a great deal more help from universities, both through specialized agricultural colleges and through research in biology. The mining industry needs men and it needs basic research into the Earth to help prospecting. Both of these are best provided by universities, but, at a time when good science students have so many attractive alternatives, no industry is likely to interest the better students without saying it needs them and without presenting a modern attractive and clearly visible appearance. I believe it is fair to say that, in the opinion of most other scientists, neither geologists nor prospectors have appeared to understand properly the meaning of basic research. This is not because of any fault on the part of geologists. It is because the Earth is so complex that they have been limited to collecting data. The Earth is so much more complex than any of these matters studied in physics and chemistry that the science of geology has been correspondingly less developed. Suddenly, earth scientists appear to have been presented with two great opportunities.

The discovery of principles is always more exciting and more useful in science than the collection of data alone. It appears that a great new principle in Earth behavior may have just been discovered. This should be quickly and vigorously explored and exploited. It seems that we know now what is going on in the Earth. This could be as important to geology as Harvey's discovery of the circulation of the blood was to physiology or as evolution was to biology. This is the most exciting event in geology for a century and every effort in research should be bent toward it.

In the second place, it appears to follow that the Earth and its economic deposits are part of a system—an Earth system—all parts of which react on other parts. New methods for the first time give us the opportunity to explore the whole system—its interior and its ocean floors as well as its land surface. We are on the verge of exploring other nearby planetary systems. Orebodies and oil structures are parts of that system, not just accidents.

What an exciting challenge this is! What a chance for great discoveries! What an appeal to young men! We must realize that a great change has occurred. If we are to take advantage of it we must radically broaden our views and change our habits.

REFERENCES

Y. Belmonte, P. Hirtz & R. Wenger, *Salt basins around Africa,* v. 55-74, (Inst. Petrol., London, 1965).

P. M. S. Blackett, E. C. Bullard & S. K. Runcorn, editors, A Symposium on Continental Drift, *Phil. Trans. Roy. Soc.,* A258, (1965).

R. T. Chamberlin, *Theory of continental drift,* v. 87, (Amer. Assoc. Petrol. Geol., Tulsa, 1928).

R. R. Coats, 'Magma Type and Crustal Structure in the Aleutian Arc', Amer. Geophys. Union, Monog. 6, p. 92-109 (1962).

A. Cox, G. B. Dalrymple & R. R. Doell, 'Reversals in the Earth's Magnetic Field,' *Sci. Amer.,* v. 216, p. 44-54 (February, 1967).

Department of Manpower and Immigration, 'Career Outlook, University Graduates', Canada (1966-67).

R. S. Dietz, 'Passive Continents, Spreading Sea Floors and Collapsing Continental Rises', *Amer. J. Sci.,* v. 264, p. 177-193 (1966).

W. M. Elsasser, 'Thermal Structure of the Upper Mantle and Convection', in *Advances in earth science.* Editor, P. M. Hurley (M.I.T. Press, Cambridge, 1966).

W. Hamilton and W. B. Myers, 'Cenozoic Tectonics of the Western United States', *Rev. of Geophysics,* v. 4, p. 509-549 (1966).

B. C. Heezen & M. Ewing in M. N. Hill, *The sea,* v. 3, p. 388-410 (John Wiley, London, 1963).

J. R. Heirtzler, X. Le Pichon & J. J. Baron, 'Magnetic Anomalies over the Reykjanes Ridge', *Deep-sea research,* v. 13, p. 427-433 (1966).

H. H. Hess, 'History of Ocean Basins', in *Petrologic studies*, A. E. J. Engel et al., editors. (Geol. Soc. Amer., New York, 1962), p. 599-620.

A. Holmes, 'Radioactivity and Earth Movements', *Trans. Geol. Soc. Glasgow*, v. 18, p. 599-607 (1928-1929).

I. G. Gass & D. Masson-Smith, 'The Geology and Gravity Anomalies of the Troodos Massif, Cyprus', *Phil. Trans. Roy. Soc. London*, v. A255, p. 417-467 (1963).

B. Kummel, *History of the Earth*, (W. H. Freeman, San Francisco, 1961).

I. McDougall & F. H. Chamalaun, 'Geomagnetic Polarity Scale of Time', *Nature*, v. 212, p. 1,415-1,417 (1966).

B. E. McMahon & D. W. Strangeway, 'Kiaman Magnetic Interval in the Western United States', *Science*, v. 155, p. 1,012-1,013 (1967).

L. W. Morley & A. Larochelle, 'Paleomagnetism as a Means of Dating Geological Events', Roy. Soc. Canada, Sp. Pub. 9, 40-51 (1964).

A. C. Munyan, editor, 'Polar Wandering and Continental Drift', Soc. Econ. Paleontol. & Mineral., Sp. Pub. 10, (1964).

D. N. Ninkovitch, N. Opdyke, B. C. Heezen & J. H. Foster, 'Paleomagnetic Stratigraphy—deposition and tephrachronology in. North Pacific deep-sea sediments', *Earth Planet. Sci. Lets.*, v. 1, p. 476-492 (1966).

N. D. Opdyke, B. Glass, J. D. Hays & J. Foster, 'Paleomagnetic study of Antarctic Deep-Sea Cores', *Science*, v. 154, p. 349-357 (1966).

E. Orowan, 'Continental Drift and the Origin of Mountains', *Science*, v. 146, p. 1,003-1,010 (1961).

A. D. Raff & R. G. Mason, 'Magnetic Survey off the West Coast of North America, 40° N latitude to 52° N latitude', Geol. Soc. Amer. *Bull.*, v. 72, p. 1,267-1,270.

J. S. Royds, H. L. Thomson & J. W. Strickland. Amer. Assoc. Petrol. Geol. *Bull.*, v. 49, p. 2,269-2,287 (1965).

S. K. Runcorn, editor, *Continental drift* (Academic Press, New York, 1962).

R. R. Shrock, *A cooperative plan in geophysical education* (Geophysical Service Inc., Dallas, 1966, 143 p.).

L. R. Sykes, 'Mechanism of Earthquakes and Nature of Faulting on the Mid-Ocean Ridges', *J. Geophysics*, v. 72, p. 2,131 (1967).

D. C. Tozer, 'Heat Transfer and Convection Currents', *Phil. Trans. Roy. Soc. London*, v. A258, p. 252-271 (1965).

F. J. Vine, 'Spreading of the Ocean Floor: New Evidence', *Science*. v. 154, p. 1,405-1,415 (1966).

F. J. Vine & D. H. Matthews, 'Magnetic Anomalies over Oceanic Ridges', *Nature*, v. 199, p. 947-949 (1963).

F. J. Vine & J. T. Wilson, 'Magnetic Anomalies over a Young Oceanic Ridge off Vancouver Island', *Science,* v. 150, p. 485-489 (1965).

J. T. Wilson, 'Did the Atlantic Ocean Close and then Reopen?' *Nature,* v. 211, p. 676-681 (1966).

J. T. Wilson, 'A New Class of Faults and Their Bearing on Continental Drift', *Nature,* v. 207, p. 343-349 (1965).

An Open Letter to J. Tuzo Wilson

V. V. Beloussov
Chairman, Upper Mantle Committee
Moscow

Dear Professor Wilson,
It was very interesting to read your article *Revolution in the Earth sciences,* which you kindly sent me. You succeeded in making a striking case for the necessity of a general theory in the geosciences. It seems to me that our Upper Mantle Project is, in fact, intended either to work out such a theory, if possible, or, at least, to collect the necessary material.

You assert, however, that a theory already exists and that textbooks should be rewritten and education reorganized in accordance with it. An important advantage of the new theory you see is a chance of its captivating young people and diverting a certain number of them from physics and mathematics.

I would like to say a few words on this subject.

1. You are surely mistaken when you say that nothing important has happened in geoscience since the time of Lyell. I cannot understand your passing over the contraction hypothesis. It reigned in our science for almost a whole century and we owe to it great achievements in geology, especially in the second half of the last century.

I think that in geotectonics M. M. Tetyayev's ideas (1941) were also most revolutionary. He showed the regularity in relations between events previously regarded as disjointed, and established the principle of unity of geotectogenesis.

2. The example of the contraction hypothesis could serve as a warning to all those who are in a hurry to rewrite textbooks.

At the end of the last century and even at the beginning of this one, few people questioned the correctness of the contraction hypothesis. And all textbooks were written on the basis of that theory. The fundamental principles of the contraction hypothesis became so much a second nature to geologists that up to the present they have been strongly reflected in the majority of ordinary regional geological papers.

And yet the foundations of the contraction hypothesis collapsed. The difficulties with a geological theory nowadays are due mainly to the fact that, though the foundations collapsed, the store of particular ideas and concepts still remains.

The contraction hypothesis was undermined not only because a new physical phenomenon was discovered—radioactivity—that turned upside down all our ideas on the thermal regime of the Earth. It was undermined also because of its primitive nature; it schematized natural phenomena, reducing them to a state of complete distortion. It never could explain vertical (epeirogenetic) crustal movements. It failed when detailed investigations established that folding can be of various types and of an obviously different origin (for instance, what relation is there between horizontal compression and platform quaquaversal folding, so interesting to the oil man?). Beginning in 1906 with Ampferer (after whom Fourmarier wrote so wonderfully about it in 1947) it has been shown repeatedly that the structure of a geosyncline, in all its complexity and history, cannot be explained by a rough mechanism of crushing between two rigid blocks. The list of cases in which the contraction hypothesis proved to be too schematic could be much longer.

3. What is the situation now with the new theory that you are defending? There simply is no foundation to it: while, according to the then-accepted cosmogenic hypothesis of Laplace, it seemed most probable that the globe should grow cooler, nobody has shown that deep convection currents, which are necessary for your theory, really exist or even that they *can* exist. The computed schemes are even more primitive. For instance, they do not bother to take into consideration that the sources of heat are not outside the moving material, but in it. It is not clear at all where and at what depth these currents are flowing. In the oceans they should be at the surface. Well, and under the continents, too?

It is not so difficult to compile a scheme of currents under the Atlantic Ocean only, but nobody has yet succeeded in constructing a scheme of currents for the entire globe, even a speculative one, that is mechanically possible. You have made such an attempt (Wilson 1965). Consider it attentively: Africa with its rift and the adjacent Indian Ocean (also with a rift) present an insoluble problem. So, instead of drawing arrows that, according to the regularity that you consider to be fundamental, should point away from the rifts, you have drawn an arrow from the south northward under the entire African continent. Why? And why on the other side of the Indian mid-ocean ridge the arrows are also not normal to the rift, but parallel to it? And the northern part of the ridge: you have simply wiped it off the scheme. Why? Near the coast of California you show currents that meet at right angles. Is this possible according to elementary principles of hydrodynamics?

As to the schematization of natural phenomena, the situation is not at all an improvement on the contraction hypothesis. For the development of a geosyncline you actually accept the same mechanism of horizontal crushing already discredited. Epeirogenetic movements continuously and universally taking place (compare thickness and facies of deposits) are left unexplained. They are not even mentioned. All the structural development is reduced to rifts and geosynclines, which are two structural poles associated with each other. First, there are not only geosynclines on the continents. My country, for instance, is mostly beyond young geosynclines, and we simply cannot forget the laws governing the development of extra-geosynclinical areas. Second, geological data clearly indicate that rifts are very young: they originated much later than the beginning of the development of even the youngest (Alpine) geosynclines and, consequently, cannot be the cause of the origin of geosynclines. And what about the old geosynclines? The Hercynian, the Caledonian, the Precambrian? We have no data whatsoever about the existence of their corresponding rifts.

In the theory that you are so ardently advocating, the geological development of continents is much more schematized than was done in the contraction hypothesis. The geology of continents is simply and completely annihilated. Could it be that you really want to rewrite the textbooks, and throw overboard a great part of the outstanding achievements in the geology of continents?

This excessive schematization is caused by the fact that the oceans occupy first place in the new concept. Naturally, the successes in the study of oceans are tremendous and they provided quite a new color to our ideas about the Earth. But no matter how great these successes, we should not forget that, owing to obvious circumstances, our knowledge of the ocean areas remains not only very schematic, but to a great extent indirect, being founded on interpretation of indirect data. Suffice it to say that we still do not know of what material the consolidated ocean crust consists. No one has proved that it is serpentinic. On the other hand, we know a little about the crust of the continents.

In this way, the old, tested and rather precise geology of the continents is being sacrificed for the sake of as yet indefinite data on the structure of the oceans! You propose to use the process of repeated opening and closing of oceanic basins as the basis of historical geology. Would not it be better to wait until we get more direct evidence on the structure of the ocean floor? Would not such a theory have a reverse effect upon the young people to the one you expect, inasmuch as they are not going to be satisfied by an emotional effect, but will demand more exact justification?

4. Of course an advocate of any theory prefers data convenient for his purpose to the inconvenient ones. But when a person in doubt asks questions, they have to be answered. Otherwise an undesirable impression is created. I am enumerating here the questions that have been asked time and again, but for which no answers have been yet received:

(a) Since the spreading ocean floor predetermines that the crust in the ocean should be serpentinic, how can you explain such persistent thickness of the crust in the ocean and what is the Mohorovicic discontinuity in the ocean? (I think you will agree that the explanation suggested by Hess in 1962 is too artificial to be true.)

(b) What is the composition of the crust in such seas as the Sea of Okhotsk, Sea of Japan, the Caribbean Sea, the Black Sea? How was the crust formed there, as there are no rifts?

(c) If the currents involve the floor of the ocean up to the surface, how are they able to descend under the continents without any deformation of the ocean floor even at the edges of the continents?

(d) If two opposing currents flow under the continent and meet each other, the latter will stop, apparently at the place where the effect of the two currents is counterbalanced. The young folded zone of the Cordilleras and Andes is the edge of the continent.

Why does the current from one side pass under the entire continent, while the current from the other side stops at the edge? What happens to the currents under Africa?

(e) In connection with the aforesaid a question arises as to the depth of the currents. What are the reasons for the slow vertical movements of anteclises and syneclises on platforms? Where are the chambers of magma that is ejected in geosynclines, on platforms and in the oceans? In the mantle? At what depth? And, if the continents are displaced in respect to the mantle, how does it happen that anteclises and syneclises remain in the same place as the continent for hundreds of millions of years and the magmatic centers are just as stable during the tectonic cycles? How can a volcano that is displaced together with the floor of the ocean not lose its connection with its feeding chamber?

(f) What is the explanation for tectonic cycles with their regular and fundamentally uniform repetition of tectonic and magmatic phenomena? This question becomes especially acute because, according to the theory you are defending, at different times the continents were over different parts of the mantle.

(g) How can you explain the fact that the mean value of heat flows is the same on the continents as in the oceans?

(h) If a geosyncline originates at a descending current, that is, the coldest branch, how can you explain the signs of a substantial heating of the crust in the geosyncline (intrusions, regional metamorphism, granitization), confirmed by among other things measurements of old geothermal gradients by mineral geothermometers?

5. You believe that the way you indicate is the only one leading to generalization. The majority of Soviet geologists and geophysicists who study general problems of geology are of a different opinion. There is no doubt that there are also other ways. For instance, the ways I have tried to indicate in my papers. They seem to be more promising not only to me. Some people think that the old idea of the formation of continents by differentiation of the oceanic crust can still be revived. Your list of attempted generalizations published during recent decades would be much longer if you take into account the papers of Soviet researchers. From our point of view it is strange to pass over in silence the papers by A. D. Arkhangelsky (1941), V. A. Obruchev (1940), M. A. Ussov (1940), M. M. Tetyayev (1941), even without mentioning those still living. Our country occupies a sixth of the land on

the globe and among its population there are geologists who feel their great responsibility to science and the economy of their country.

6. It should be mentioned, of course, that modern interpretation of geomagnetic observations favors the theory you are defending. This is your strongest argument. But, strangely enough, it is the only one, because the others (like, for instance, the contours of the coasts, paleoclimatic data, and structural similarities) can always have another interpretation as well.

Naturally, geomagnetic data deserve close attention. But when they are placed against the enumerated impressive list of contradictions, a second thought is warrented on the significance of an extremely great scattering in the results of paleomagnetic data, on some very strange results obtained (indicating, for instance [Boer 1963], the movement of the Italian Alps from the Himalayas across Iran and Turkey, or the location of New Guinea, despite its primordial adherence to the circum-Pacific ring, during the Cretaceous in the middle of the Indian Ocean) and on those original assumptions on the constant structure of the magnetic field of the Earth, on which the interpretation is founded. Bearing in mind such a number of contradictions in other geological and geophysical data, should not this only positive and ponderous argument in favor of your theory be subjected to especially scrupulous and objective analysis?

Are not, for instance, these very confident references to magnetic anomalies parallel to ocean rifts premature if we take into account that after more detailed investigations the stretches of anomalies fall apart into rather irregularly scattered patches (Matthews et al. 1965; Loncarevic et al. 1966)? When making too-general comparisons it is always easy to become a victim of illusions: certain groups of patches could appear to look like stretches. Remember the canals on Mars! And, by the way, why are there no such stretches near the Red Sea or near the grabens of East Africa?

Thus I am entirely for a general theory, but for a theory based on deep and precise study of all the aspects of phenomena, taking place both in the oceans and on continents, and which would unite all these events. I am afraid that, for the time being, we have studied natural phenomena much too insufficiently to formulate any well-founded theory. Still, we now have at our disposal new and effective technical and methodological means that should help us to fill in the main blanks within a reasonable time. To do this we should call upon our colleagues, the young ones especially. And it is our duty to indicate, on the basis of our experience, which routes are best to follow in studying the interior of the Earth and the processes that are taking place there.

Admittedly, to choose a route one needs working hypotheses. Better yet if there are several. From this point of view both the theory of continental drift and the concept of spreading ocean floor could serve as such working hypotheses along with others (like the concept on the differentiation of the ocean crust or an oceanization of the continental crust). A great number of hypotheses guarantees a varied approach, and that is only beneficial for our science.

But we will be bitterly reproached (perhaps also ridiculed!) by the coming generations if we call one of such working hypotheses a final theory, if we assert that the truth is at our elbow and that we only have to stretch out our hand to pick the flower. We have dedicated our lives to a difficult science, which, unfortunately, is still assembling fundamental data. We have only just begun to penetrate the secrets of the very shallow interior. It would be most irresponsible of us to tempt the young people, saying that all the difficulties are behind us and, instead of leading them along a hard and strenuous path of search and menial labor, inevitable for a scientist, to lull them with delusive hopes and dreams.

REFERENCES

O. Ampferer, *Uber das Bewegungsbildung von Faltengebirge*, Austria, Geol. Reichsanst., Jahrb., Bd. 56, Wien (1906).

A. D. Arkhangelsky, *Geological structure and geological history of the USSR*, Gosudarst. Geol. Izd., Moscow (1941). (Russian)

J. de Boer, *Geology of the Vicentinian Alps (NE Italy)*, with special reference to their paleomagnetic history. Geologica Ultraiectina, n. 11 (1963).

P. Fourmarier, *Les forces en action dans la genese du relief tectonique*, Soc. Belge d'Etudes geog., *Bull.*, v. 17, n. 1 & 2, p. 20-57 (1947).

H. H. Hess, History of ocean basins, in *Petrologic studies*—a volume in honor of A. F. Buddington, Geol. Soc. Am., New York (1962).

B. D. Loncarevic, C. S. Mason & D. H. Matthews, *Mid-Atlantic ridge near 45° North,* 1, *The median valley,* Canadian Jour. Earth Sciences, v. 3 (1966).

D. H. Matthews, F. J. Vine, & J. R. Cann, *Geology of an area of the Carlsberg ridge, Indian Ocean,* Geol. Soc. Am., *Bull.,* v. 76, n. 6 (1965).

V. A. Obruchev, *A pulsation hypothesis in geotectonics,* Akad. Nauk SSSR Izv., Ser. Geol., n. 1 (1940).(Russian)

M. M. Tetyayev, *The Principles of geotectonics,* Gosudarst. Geol. Izd., Moscow (1941). (Russian)

M. A. Ussov, *A geotectonic hypothesis of self-development in the matter of the Earth,* Izv. Akad. Nauk SSSR, Ser. Geol. v. 1 (1940). (Russian)

J. T. Wilson, *Evidence from ocean islands suggesting movement in the Earth—a symposium on continental drift,* Phil. Trans. Roy. Soc. London, n. 1088 (1965).

A Reply to V. V. Beloussov

J. TUZO WILSON

Dear Dr. Beloussov:
It is kind of you to address an open letter to me. I agree with much of what you say and a few years ago I would have agreed with more. On the other hand the existence of important differences make me grateful to the editor of *Geotimes* for this opportunity to reply. May I assure you that these differences in no way diminish my regard for the great contributions you have made in promoting international cooperation as a member of the Special Committee for the International Geophysical Year, as president at the Berkeley meetings of the International Union of Geodesy & Geophysics in 1963 and as chairman of the Upper Mantle Committee.

I shall pass over some minor divergences in point of view in order to come to the fundamental argument between us. It is that you continue to believe that continents and ocean basins have always been fixed, whereas 9 years ago I was converted to the view that extensive horizontal movements have occurred. Indeed, I now hold that the lithosphere has repeatedly been broken into plates which have travelled extensively relative to one another and that this has gone on throughout much of geologic time and not only since the late Paleozoic as Wegener suggested.

In spite of this fundamental difference, I still accept much of what you say. I agree that most geological observations made on continents are accurate, important and not to be disregarded. I agree that most of the evidence for continental drift which you cite has been discussed for a long time with inconclusive results. In spite of the efforts of such men as Taylor (1910), Baker (1911), Wegener (1912), du Toit (1937) and more recently Carey (1958), Runcorn (1962), Blackett (1965), and Bullard (1965) you can justify your statement that 'the contours of the coasts, paleoclimatic data, structural similarities etc. can always have another interpretation' on the grounds that these arguments converted only a few to believe in drift. For many years after they had been advanced the great majority of geologists and geophysicists rejected them (R. T. Chamberlin 1928 and Jeffreys 1959).

It is also true that arguments based upon the postulated existence or nonexistence of mechanisms may be of doubtful value. Thus I agree that the arguments you mention would not by themselves have justified a radical revolution.

The chief difference between us is not about what you say, but arises from what you have omitted to say. Your arguments would have had some validity a few years ago, but it seems to me that they have now been superseded. It astonishes me that you almost completely ignore the tremendous recent discoveries. It is they, not the old topics you mention, which have changed so many opinions, which have inspired a revolution in the earth sciences, and which have produced the beginnings of an elegant and precise new explanation of the Earth's behavior (Morgan 1968; Le Pichon 1968, and Isacks, Oliver & Sykes 1968).

The new arguments which you dismiss or ignore, but which I consider conclusive, fall into three groups that may be classed under the headings of sea-floor spreading, paleomagnetism and seismology.

The notion that ocean basins have spread from mid-ocean ridges was advanced by Holmes (1931) and Hills (1947) and used in both editions of Holmes' textbook. It received impetus from the discovery by Ewing & Heezen (1957) of the continuity of mid-ocean ridge system. This led to the greatly improved views on spreading of Hess (1962) and Dietz (1961). Evidence supporting it has come from many directions.

It has been observed that islands and cores tend to increase in age from the crests of mid-ocean ridges towards the margins of ocean basins (Wilson 1963a, Funnell 1966, Burcke et al 1967). This supports the idea of spreading from the mid-ocean ridges as does Ewing & Ewing's (1967) observation that sediments increase in thickness in the same direction. The discovery and exploration of fracture zones by the U.S. Coast & Geodetic Survey, H. W. Menard (1964), Heezen et al. (1964) and others led to the hypothesis that growing oceans should be crossed by transform faults which should differ from transcurrent faults in three respects: (1) they should not enter the bounding continents, (2) their seismicity should be confined to the axial line and its offsets, (3) their directions of motion should be in the reverse directions to those for simple offsets. The demonstration that many fracture zones have these properties has gone far to prove that ocean basins are indeed spreading (Wilson 1965a, Sykes 1967). I should like to take this opportunity of acknowledging that my own views on ocean floors have been much helped by the splendid bathymetric charts published in the Soviet Union, notably by Udintsev (1964).

You dismiss the proposal by Vine & Matthews (1963) that the pattern of magnetic anomalies over mid-ocean ridges supports ocean-floor spreading. I consider that even the earliest evidence for regular patterns of magnetic anomalies on the ocean floor was clear (Raff & Mason 1961) and I cannot believe that it would be true now to dismiss the vast and regular patterns set forth by Vine (1966), Heirtzler et al. (1968) and Deminitskaya & Karisik (1966) as 'rather irregularly scattered patches'. You say nothing of the supporting evidence for paleomagnetic reversals by Cox, Dalrymple & Doell (1967), McDougall & Chamalaun (1966) and, in cores, by Opdyke et al. (1966).

The second group of modern discoveries that you also dismiss briefly are concerned with paleomagnetism. Some of the contradictions and difficulties you mention may be valid, but others are natural in the development of a new subject. What is more important is the much greater body of consistent observations. Those on young rocks agree with the present pattern of continents while those on older rocks suggest systematic changes in that pattern (Irving 1964; Krapotkin in press). No one has done more to show that the Earth's field has long been dipolar and hence to remove one source of doubt than your compatriots Khramov et al. (1966).

A third modern developement has been published only since your letter. Isacks, Oliver & Sykes (1968) say that modern views on drift can explain the distribution, mechanism, direction of motion of earthquakes with a detail that was quite lacking before.

I think that you should answer the grave objections to fixed continents which these three modern lines of evidence pose before you reject all the many possible hypotheses of moving continents. By the same token I am glad to try to reply to the eight specific questions your letter addressed to me, for I think it is possible to fit your observations into an appropriate version of continental drift. My answers:

(a) The explanation suggested by Hess (1962) seems excellent to me.

(b) I believe that except for thicker sediments the floor of the Sea of Okhotsk is the same as that of the adjacent Pacific because the Kuril arc is younger than the floors. The Caribbean is a tongue of the Pacific (Wilson 1966). The Black Sea, like the Mediterranean, is a remnant of the closing Tethys Ocean (Le Pichon 1968; Wilson in press). The Sea of Japan may be older and more complicated and I cannot yet explain it.

(c) The ocean floor is deformed into deep trenches at the edge of continents where plates of the lithosphere override one another (Isacks, Oliver & Sykes 1968).

(d) Your confusion arises from adhering to the outmoded views of Wegener. Continents are no longer regarded as rafts moving through a plastic ocean floor. The whole lithosphere, at least 50 km thick, is brittle and broken into plates of which continents form incidental parts. When viewed in this way your problems about the Americas do not arise. The question of how the mid-ocean ridges which encircle Africa (and Antarctica) can expand has puzzled many until they have realized that upward and outward flow beneath the ridges is possible if the ridges themselves are moving their positions outwards (Wilson 1965b).

(e) If flow occurs it is in the asthenosphere or low-velocity layer of seismology lying between depths of 50 and a few hundred kilometers (Anderson 1967). The other problems you raise cover some of the main problems of geology. Even if they have not yet been fully explained by the new ideas, they were certainly never fully comprehended by the old, and the problems are vast and would take too long to discuss here. Gorshkov and Eaton & Murata (1960) have discussed evidence for sources of magmas. I believe that in the oceans some volcanoes may become detached from their roots (Wilson 1963b), but this does not seem to happen under continents.

(f) Although major geological phenomena are recurrent, the problem of whether cycles are regular and universal is still debated. If drift has depended upon the fracture and relative movement of plates of the lithosphere, at any one place it would probably appear to be intermittent because it would depend upon the way in which the lithosphere had broken and moved. Ocean basins have, I believe, repeatedly opened, grown and closed again (Wilson in press). The life histories of all ocean basins might be generally similar and hence give the impression of tectonic cycles.

(g) Schuiling (1966) has given one explanation of why heat flow is much the same over ocean and continents. Knopoff, Von Herzen and your own distinguished colleague E. A. Lubimova have discussed this problem at length in a recent book (Gaskell 1967).

(h) This again raises major problems, but one explanation of why evidence of high temperatures can be found in places where currents have descended could be that at these places one plate of the lithosphere has been overridden by another, pressed down, heated, and then subsequently risen isostatically from great depths.

In conclusion, those of us who hold the new point of view do not wish to discredit or discard the old data. If we hold that it was insufficient, it is because we want to draw attention to data of the new kinds. Your lack of reference to the new shows how necessary this is, although at the recent International Geological Congress some of your colleagues assured me that 80 per cent of Soviet earth scientists have accepted continental drift (for example, Voronov 1968 and Krapotkin in press). What we want to change is not the data, but the frame of reference or the way of looking at data. (If two groups of scientists both studied whirlpools, and one group held that the water in them did not move they would never understand whirlpools no matter how much data they collected. The other group, which admitted that the water was moving, could understand the nature of whirlpools with little data.)

I hold that the trouble with the earth sciences (geology, geophysics and geochemistry alike) and the reason they have not progressed as they might has been not lack of or errors in data of traditional kinds, but a lack of the new kinds of information and an utterly wrong way of looking at the Earth. If indeed the Earth is, in its own slow way, a very dynamic body and we have regarded it as essentially static, we need to discard most of our old theories and books and start again with a new viewpoint and a new science.

I believe that what is happening in earth science is similar to what happened in chemistry about A.D. 1800, in biology when evolution was introduced, in physics when classical views were replaced by modern. It's not new data, but a change in outlook that marks a scientific revolution, as T. S. Kuhn (1964) has so elegantly pointed out. If a scientific revolution is in progress in the earth sciences it provides us all with an exciting opportunity, the prospect of a great revival, and I think we should embrace the change and expect the whole study of the Earth to move rapidly forward.

I believe this revolution will unite branches formerly fragmented and that the new unified science of a dynamic Earth deserves a new name—geonomy.

Thank you again for this opportunity to make my views clear. Because it was my ideas which you singled out for attack, they have received more attention in this reply than they would otherwise deserve.

REFERENCES

Recent publications:

P.M.S. Blackett, Edward C. Bullard & S.K. Runcorn, editors, *A symposium on continental drift:* Phil. Trans. Royal Soc. London, v. A258, p. 1-323 (1965).

Gerard Piel, editor, *Gondwanaland revisited: new evidence for continental drift,* Proceedings of the American Philosophical Society, v. 112, n. 5, (1968).

E. Takeuchi, S. Uyeda & H. Kanamori, *Debate about the Earth,* Freeman Cooper & Co., San Francisco, 253 p. (1967).

T.S. Kuhn, *The structure of scientific revolutions,* University of Chicago Press, 172 p. (1964).

In preparation or in press:
Marshall Kay, editor, *Symposium on the North Atlantic area,* Amer. Assoc. Petrol. Geol.

J. Tuzo Wilson, editor, *Symposium on continental drift in the South Atlantic region,* Unesco, Paris.

J.A. Jacobs, R.D. Russell & J. Turzo Wilson, *Physics and geology,* 2d edition, McGraw-Hill, New York.

3

GEOLOGY AND MANKIND

Erosion resulting after all the vegetation on this area of Shasta National Forest was killed by smelter fumes. U. S. Forest Service.

15
Polluting The Environment

LORD RITCHIE-CALDER

Modern technological society is based on oil, coal, uranium, iron, copper, and many other non-renewable geologic materials. Finding adequate supplies will require increasingly diligent and sophisticated uses of geology in the face of rising per-capita consumption and the exploding world population. These same factors compound yet another extremely critical problem — the disposal of industrial wastes. The next two essays relate to the inter-disciplinary field of environmental studies, in which geologic principles have an important place.

Lord Richie-Calder, Professor of International Relations at the University of Edinburgh and an outstanding science writer, discusses the general problem of pollution in no uncertain terms.

To hell with posterity! After all, what have the unborn ever done for us? Nothing. Did they, with sweat and misery, make the Industrial Revolution possible? Did they go down into the carboniferous forests of millions of years ago to bring up coal to make wealth and see nine-tenths of the carbon belched out as chimney soot? Did they drive the plows that broke the plains to release the dust that the buffalo had trampled and fertilized for centuries? Did they have to broil in steel plants to make the machines and see the pickling acids poured into the sweet waters of rivers and lakes? Did they have to labor to cut down the tall timbers to make homesteads and provide newsprint for the Sunday comics and the

From **The Center Magazine,** a publication of the Center for the Study of Democratic Institutions. Copyright 1969 by The Fund for the Republic, Inc. Reprinted by permission of **The Center Magazine.**

celluloid for Hollywood spectaculars, leaving the hills naked to the eroding rains and winds? Did they have the ingenuity to drill down into the Paleozoic seas to bring up the oil to feed the internal-combustion engines so that their exhausts could create smog? Did they have the guts to man rigs out at sea so that bore-holes could probe for oil in the offshore fissures of the San Andreas Fault? Did they endure the agony and the odium of the atom bomb and spray the biosphere with radioactive fallout? All that the people yet unborn have done is to wait and let us make the mistakes. To hell with posterity! That, too, can be arranged. As Shelley wrote: "Hell is a city much like London, a populous and smoky city."

At a conference held at Princeton, New Jersey, at the end of 1968, Professor Kingsley Davis, one of the greatest authorities on urban development, took the role of hell's realtor. The prospectus he offered from his latest survey of world cities was hair-raising. He showed that thirty-eight per cent of the world's population is already living in what are defined as "urban places." Over one-fifth of the world's population is living in cities of a hundred thousand or more. Over 375,000,000 people are living in cities of a million and over. On present trends it will take only fifteen years for half the world's population to be living in cities, and in fifty-five years everyone will be urbanized.

Davis foresaw that within the lifetime of a child born today, on present rates of population increase, there will be fifteen billion people to be fed and housed—over four times as many as now. The whole human species will be living in cities of a million and over and the biggest city will have 1,300,000,000 inhabitants. Yes, 1.3 billion. That is 186 times as many as there are in Greater London today.

In his forebodings of Dystopia (with a "y" as in dyspepsia, but it could just as properly be "Dis," after the ruler of the Under-world), Doxiades has warned about the disorderly growth of cities, oozing into each other like confluent ulcers. He has given us Ecu-menopolis—World City. The East Side of Ecumenopolis would have as its Main Street the Eurasian Highway, stretching from Glasgow to Bangkok, with the Channel tunnel as an underpass and a built-up area all the way. West Side, divided not by railroad tracks but by the Atlantic, is already emerging (or, rather, merging) in the United States. There is talk, and evidence, of "Boswash," the urban development of a built-up area from Boston to Washington. On the Pacific Coast, with Los Angeles already sprawling into the desert, the realtor's garden cities, briskly reenforced by industrial estates, are slurring into one another and presently will stretch all the way from San Diego to San Francisco. The Main Street of Sansan will be Route 101. This is insansanity. We do not need a crystal ball to foresee what Davis and Doxiades are predicting—we can see it through smog-colored spectacles; we can smell it seventy years away because it is in our nostrils today; a blind man can see what is coming.

Are these trends inevitable? They are unless we do something about them. I have given up predicting and have taken to prognosis. There is a very important difference. Prediction is based on the projection of trends. Experts plan for the trends and thus confirm them. They regard warnings as instructions. For example, while I was lecturing in that horror city of Calcutta, where three-quarters of the population live in shacks without running water or sewage disposal, and, in the monsoon season, wade through their own floating excrement, I warned that within twenty-five years there would be in India at least five cities, each with populations of over sixty million, ten times bigger than Calcutta. I was warning against the drift into the great conurbations now going on, which has been encouraged by ill-conceived policies of industrialization. I was warning against imitating the German Ruhr, the British Black Country, and America's Pittsburgh. I was arguing for "population dams," for decentralized development based on the villages, which make up the traditional cultural and social pattern of India. These "dams" would prevent the flash floods of population into overpopulated areas. I was *warning,* but they accepted the prediction and ignored the warning. Soon thereafter I learned that an American university had been given a contract to make a feasibility study for a city of sixty million people north of Bombay. When enthusiasts get busy on a feasibility study, they invariably find that it is feasible. When they get to their drawing boards they have a whale of a time. They design skyscrapers above ground and subterranean tenements below ground. They work out minimal requirements of air and hence how much breathing space a family can survive in. They design "living-units," hutches for battery-fed people who are stacked together like kindergarten blocks. They provide water and regulate the sewage on the now well-established cost-efficiency principles of factory-farming. And then they finish up convinced that this is the most economical way of housing people. I thought I had scotched the idea by making representations through influential Indian friends. I asked them, among other things, how many mental hospitals they were planning to take care of the millions who would surely go mad under such conditions. But I have heard rumors that the planners are so slide-rule happy they are planning a city for six hundred million.

Prognosis is something else again. An intelligent doctor, having diagnosed the symptoms and examined the patient's condition, does not say (except in soap operas): "You have six months to live." He says: "Frankly, your condition is serious. Unless you do so-and-so, and unless I do so-and-so, it is bound to deteriorate." The operative phrase is "do so-and-so." One does not have to plan *for* trends; if they are socially undesirable our duty is to plan *away* from them, and treat the symptoms before they become malignant.

A multiplying population multiplies the problems. The prospect of a world of fifteen billion people is intimidating. Three-quarters of the world's present population is inadequately fed — hundreds of millions are not getting the food necessary for well-being. So it is not just a question of quadrupling the present food supply; it means six to eight times that to take care of present deficiencies. It is not a matter of numbers, either; it is the *rate* of increase that mops up any improvements. Nor is it just a question of housing but of clothing and material satisfactions — automobiles, televisions, and the rest. That means greater inroads on natural resources, the steady destruction of amenities, and the conflict of interest between those who want oil and those who want oil-free beaches, or between those who want to get from here to there on wider and wider roads and those whose homes are going to collapse in mud slides because of the making of those roads. Lewis Mumford has suggested that civilization really began with the making of containers — cans, non-returnable bottles, cartons, plastic bags, none of which can be redigested by nature. Every sneeze accounts for a personal tissue. Multiply that by fifteen billion.

Environmental pollution is partly rapacity and partly a conflict of interest between the individual, multimillions of individuals, and the commonweal; but largely, in our generation, it is the exaggerated effects of specialization with no sense of ecology, i.e. the balance of nature. Claude Bernard, the French physiologist, admonished his colleagues over a century ago: "True science teaches us to doubt and in ingnorance to refrain." Ecologists feel their way with a detector through a minefield of doubts. Specialists, cocksure of their own facts, push ahead, regardless of others.

Behind the sky-high fences of military secrecy, the physicists produced the atomic bomb—just a bigger explosion—without taking into account the biological effects of radiation. Prime Minister Attlee, who consented to the dropping of the bomb on Hiroshima, later said that no one, not Churchill, nor members of the British Cabinet, nor he himself, know of the possible genetic effects of the blast. "If the scientists knew, they never told us." Twenty years before, Hermann Muller had shown the genetic effects of radiation and had been awarded the Nobel Prize, but he was a biologist and security treated this weapon as a physicist's bomb. In the peacetime bomb-testing, when everyone was alerted to the biological risks, we were told that the fallout of radioactive materials could be localized in the testing grounds. The radio active dust on The Lucky Dragon, which was fishing well beyond the proscribed area, disproved that. Nevertheless, when it was decided to explode the H-bomb the assurance about localization was blandly repeated. The H-bomb would punch a hole into the stratosphere and the radioactive gases would dissipate. One of those gases is radioactive krypton, which decays into radioactive strontium, a particulate. Somebody must have known that but no-body worried unduly because it would happen above the tropo-sphere, which might be described as the roof of the weather sys-tem. What was definitely overlooked was the fact that the tropo-sphere is not continuous. There is the equatorial troposphere and the polar troposphere and they overlap. The radioactive strontium came back through the transom and was spread all over the world by the climatic jet streams to be deposited as rain. The result is that there is radio-strontium (which did not exist in nature) in the bones of every young person who was growing up during the bomb-testing—every young person, everywhere in the world. It may be medically insignificant but it is the brandmark of the Atomic Age generation and a reminder of the mistakes of their elders.

When the mad professor of fiction blows up his laboratory and then himself, that's O.K., but when scientists and decision-makers act out of ignorance and pretend it is knowledge, they are using the biosphere, the living space, as an experimental laboratory. The whole world is put in hazard. And they do it even when they are told not to. During the International Geophysical Year, the Van Allen Belt was discovered. The Van Allen Belt is a region of magnetic phenomena. Immediately the bright boys decided to carry out an experiment and explode a hydrogen bomb in the Belt to see if they could produce an artificial aurora. The colorful draperies, the luminous skirts of the aurora, are caused by drawing cosmic particles magnetically through the rare gases of the upper atmosphere. It is called ionization and is like passing electrons through the vacuum tubes of our familiar neon lighting. It was called the Rainbow Bomb. Every responsible scientist in cosmology, radio-astronomy, and physics of the atmosphere protested against this tampering with a system we did not understand. They exploded their bomb. They got their pyrotechnics. We still do not know the price we may have to pay for this artificial magnetic disturbance.

We could blame the freakish weather on the Rainbow Bomb but, in our ignorance, we could not sustain the indictment. Anyway, there are so many other things happening that could be responsible. We can look with misgiving on the tracks in the sky—the white tails of the jet aircraft and the exhausts of space rockets. These are introducing into the climatic system new factors, the effects of which are immensurable. The triggering of rain clouds depends upon the water vapor having a toehold, a nucleus, on which to form. That is how artificial precipitation, so-called rainmaking, is produced. So the jets, crisscrossing the weather system, playing tic-tac-toe, can produce a man-made change of climate.

On the longer term, we can see even more drastic effects from the many activities of *Homo insapiens,* Unthinking Man. In 1963, at the United Nations Science and Technology Conference, we took stock of the several effects of industrialization on the total environment.

The atmosphere is not only the air which humans, animals, and plants breathe; it is the envelope which protects living things from harmful radiation from the sun and outer space. It is also the medium of climate, the winds and the rain. These are inseparable from the hydrosphere, including the oceans, which cover seven-tenths of the earth's surface with their currents and evaporation;

and from the biosphere, with the vegetation and its transpiration and photosynthesis; and from the lithosphere, with its minerals, extracted for man's increasing needs. Millions of years ago the sun encouraged the growth of the primeval forests, which became our coal, and the life-growth in the Paleozoic seas, which became our oil. Those fossil-fuels, locked in the vaults through eons of time, are brought out by modern man and put back into the atmosphere from the chimney stacks and exhaust pipes of modern engineering.

This is an overplus on the natural carbon. About six billion tons of primeval carbon are mixed with the atmosphere every year. During the past century, in the process of industrialization, with its burning of fossil-fuels, more than four hundred billion tons of carbon have been artificially introduced into the atmosphere. The concentration in the air we breathe has been increased by approximately ten per cent; if all the known reserves of coal and oil were burned the concentration would be ten times greater.

This is something more than a public-health problem, more than a question of what goes into the lungs of the individual, more than a question of smog. The carbon cycle in nature is a self-adjusting mechanism. One school of scientific thought stresses that carbon monoxide can reduce solar radiation. Another school points out that an increase in carbon dioxide raises the temperature at the earth's surface. They are both right. Carbon dioxide, of course, is indispensable for plants and hence for the food cycle of creatures, including humans. It is the source of life. But a balance is maintained by excess carbon being absorbed by the seas. The excess is now taxing this absorption, and the effect on the heat balance of the earth can be significant because of what is known as "the greenhouse effect." A greenhouse lets in the sun's rays and retains the heat. Similarly, carbon dioxide, as a transparent diffusion, does likewise; it admits the radiant heat and keeps the convection heat close to the surface. It has been estimated that at the present rate of increase (those six billion tons a year) the mean annual temperature all over the world might increase by 5.8 F. in the next forty to fifty years.

Experts may argue about the time factor or about the effects, but certain things are observable not only in the industrialized Northern Hemisphere but also in the Southern Hemisphere. The ice of the north polar seas is thinning and shrinking. The seas, with their blanket of carbon dioxide, are changing their temperatures with the result that marine life is increasing and transpiring more carbon dioxide. With this combination, fish are migrating, even changing their latitudes. On land, glaciers are melting and the snow line is retreating. In Scandinavia, land which was perennially under snow and ice is thawing. Arrowheads of a thousand years ago, when the black earth was last exposed and when Eric the Red's Greenland was probably still green, have been found there. In the North American sub-Arctic a similar process is observable. Black earth has been exposed and retains the summer heat longer so that each year the effect moves farther north. The melting of the sea ice will not affect the sea level because the volume of floating ice is the same as the water it displaces, but the melting of the land's ice caps and glaciers, in which water is locked up, will introduce additional water to the oceans and raise the sea level. Rivers originating in glaciers and permanent snowfields (in the Himalayas, for instance) will increase their flow, and if the ice dams break the effects could be catastrophic. In this process, the patterns of rainfall will change, with increased precipitation in areas now arid and aridity in places now fertile. I am advising all my friends not to take ninety-nine-year leases on properties at present sea level.

The pollution of sweet-water lakes and rivers has increased so during the past twenty-five years that a Freedom from Thirst campaign is becoming as necessary as a Freedom from Hunger campaign. Again it is a conflict of motives and a conspiracy of ignorance. We can look at the obvious—the unprocessed urban sewage and the influx of industrial effluents. No one could possibly have believed that the Great Lakes in their immensity could ever be overwhelmed, or that Niagara Falls could lose its pristine clearness and fume like brown smoke, or that Lake Erie could become a cesspool. It did its best to oxidize the wastes from the steel plants by giving up its free oxygen until at last it surrendered and the anaerobic microorganisms took over. Of course, one can say that the mortuary smells of Lake Erie are not due to the pickling acids but to the dead fish.

The conflict of interests amounts to a dilemma. To insure that people shall be fed we apply our ingenuity in the form of artificial fertilizers, herbicides, pesticides, and insecticides. The runoff from the lands gets into the streams and rivers and distant oceans. DDT from the rivers of the United States has been found in the fauna of the Antarctic, where no DDT has ever been allowed. The dilemma becomes agonizing in places like India, with its hungry millions. It is now believed that the new strains of Mexican grain and I.R.C. (International Rice Center in the Philippines) rice, with their high yields, will provide enough food for them, belly-filling if not nutritionally balanced. These strains, however, need plenty of water, constant irrigation, plenty of fertilizers to sustain the yields, and tons of pesticides because standardized pedigree plants are highly vulnerable to disease. This means that the production will be concentrated in the river systems, like the Gangeatic Plains, and the chemicals will drain into the rivers.

The glib answer to this sort of thing is "atomic energy." If there is enough energy and it is cheap enough, you can afford to turn rivers into sewers and lakes into cesspools. You can desalinate the seas. But, for the foreseeable future, that energy will come from atomic fission, from the breaking down of the nucleus. The alternative, promised but undelivered, is thermonuclear energy—putting the H-bomb into dungarees by controlling the fusion of hydrogen. Fusion does not produce waste products, fission does. And the more peaceful atomic reactors there are, the more radioactive waste there will be to dispose of. The really dangerous material has to be buried. The biggest disposal area in the world is at Hanford, Washington. It encloses a stretch of the Columbia River and a tract of country covering 650 square miles. There, a twentieth-century Giza, it has cost much more to bury live atoms than it cost to entomb all the mummies of all the Pyramid Kings of Egypt.

At Hanford, the live atoms are kept in tanks constructed of carbon steel, resting in a steel saucer to catch any leakage. These are enclosed in a reenforced concrete structure and the whole construction is buried in the ground with only the vents showing. In the steel sepulchers, each with a million-gallon capacity, the atoms are very much alive. Their radioactivity keeps the acids in the witches' brew boiling. In the bottom of the tanks the temperature is well above the boiling point of water. There has to be a cooling system, therefore, and it must be continuously maintained. In addition, the vapors generated in the tanks have to be condensed and scrubbed, otherwise a radioactive miasma would escape from the vents. Some of the elements in those high-level wastes will remain radioactive for at least 250,000 years. It is most unlikely that the tanks will endure as long as the Egyptian pyramids.

Radioactive wastes from atomic processing stations have to be transported to such burial grounds. By the year 2000, if the present practices continue, the number of six-ton tankers in transit at any given time would be well over three thousand and the amount of radioactive products in them would be 980,000,000 curies—that is a mighty number of curies to be roaming around in a populated country.

There are other ways of disposing of radioactive waste and there are safeguards against the hazards, but those safeguards have to be enforced and constant vigilance maintained. There are already those who say that the safety precautions in the atomic industry are excessive.

Polluting the environment has been sufficiently dramatized by events in recent years to show the price we have to pay for our recklessness. It is not just the destruction of natural beauty or the sacrifice of recreational amenities, which are crimes in them-selves, but interference with the whole ecology—with the balance of nature on which persistence of life on this planet depends. We are so fascinated by the gimmicks and gadgetry of science and technology and are in such a hurry to exploit them that we do not count the consequences.

We have plenty of scientific knowledge but knowledge is not wisdom: wisdom is knowledge tempered by judgment. At the moment, the scientists, technologists, and industrialists are the judge and jury in their own assize. Statesmen, politicians, and administrators are ill-equipped to make judgments about the true values of discoveries or developments. On the contrary, they tend to encourage the crash programs to get quick answers—like the Manhattan Project, which turned the laboratory discovery of uranium fission into a cataclysmic bomb in six years; the Computer/Automation Revolution; the Space Program; and now the Bio-engineering Revolution, with its possibilities not only of spare-organ plumbing but of changing the nature of living things by gene manipulation. They blunder into a minefield of undetected ignorance, masquerading as science.

The present younger generation has an unhappy awareness of such matters. They were born into the Atomic Age, programmed into the Computer Age, rocketed into the Space Age, and are poised on the threshold of the Bio-engineering Age. They take all these marvels for granted, but they are also aware that the ad-vances have reduced the world to a neighborhood and that we are all involved one with another in the risks as well as the opportunities. They see the mistakes writ large. They see their elders mucking about with *their* world and *their* future. That accounts for their profound unease, whatever forms their complaints may take. They are the spokesmen for posterity and are justified in their protest. But they do not have the explicit answers, either.

Somehow science and technology must conform to some kind of social responsibility. Together, they form the social and economic dynamic of our times. They are the pacesetters for politics and it is in the political frame of reference that answers must be found. There can never be any question of restraining or repressing natural curiosity, which is true science, but there is ample justification for evaluating and judging developmental science. The common good requires nothing less.

16

The Perils Of Polluting The Ground

DAVID M. EVANS

David Evans is a geological engineer at the Mineral Resources Institute of the Colorado School of Mines. He was the first to suggest that recent earthquakes in Denver, Colorado, resulted from the disposal of waste at the Rocky Mountain Arsenal by pumping it down a deep well. Much has been written about pollution of the air, rivers, and lakes, but Evans adds still another dimension—fouling of the ground itself by disposing of wastes in ponds and shallow wells.

An aroused public demands today that we clean up our smog-filled air and polluted water. Looking for another cheap place to dump poisonous waste, some government agencies and industries are turning to underground dumping.

To see what happens when industrial poisons get loose underground, come with me to the farm owned by Jesse and Gladys Powers northeast of Denver.

In 1954 disaster struck the farm. Mrs. Powers recalls what happened: "First it was our meadow grass and sugar beets. They just puckered up and wilted away—the more we'd water them, the shorter they'd get."

The grass in the meadows died. The beet crop died. The cows aborted and lost their calves. Many of the young pigs and some of the registered heifer calves died.

What had gone wrong?

With the help of the Colorado State agricultural agent, Jesse Powers finally discovered that his three shallow water-wells were pumping water—and the weed killer, 2,4-D!

Seven years and two Government investigations later, Jesse Powers was able to prove that the pollution in his wells was coming from his neighbor—the U.S. Army's Rocky Mountain Arsenal. Since 1943, the Army had been dumping wastes from the manufacture of poison gas into holding ponds scooped out of the sandy soil. From the ponds, the poisons seeped into the ground, poisoning the water beneath 4,200 acres.

Particularly ominous was the fact that although 2,4-D was in the well-water, none was dumped into the Army's holding ponds. Finally, experts concluded that the 2,4-D had been *formed spontaneously* from other substances discharged from the Arsenal. When she described this incident in her book *Silent Spring,* Rachel Carson pointed out that quite without the intervention of scientists the holding ponds had become laboratories for the production of a chemical fatally damaging to much of the plant life it touched.

The Army stopped dumping wastes into the holding ponds in 1957. The other day I asked Powers if his water supply was still poisoned. "I had it checked a couple of months ago," he said. "It's as bad as ever."

In 1961, Powers was paid $52,000 in damages, which he figures was a reasonable amount for loss suffered to that date. However, he has not been compensated for losses since 1961 as a result of being unable to use the wells.

Recently, the editor of a chemical trade magazine told me that a dumping expert had said, "There is as much space (for dumping) underground as exists in all of this nation's bodies of water." The editor asked me what I thought of dumping this much poison underground.

I told him it would be sweeping our pollution problem under another rug.

When you realize that "all of this nation's bodies of water" includes not only polluted Lake Erie, and the polluted Potomac, Connecticut, Hudson and Ohio Rivers but all the rest of the Great Lakes, the Mississippi and Columbia Rivers, not to mention countless other rivers, lakes and streams—then you get some idea of what might be in store. Underground disposal dumps will leak like the holding ponds at the Rocky Mountain Arsenal. When they do, the problems that plagued Jesse and Gladys Powers will hit many other American families across the nation.

When these families discover poisons in their basements or drinking water, it will be too late to stop the trouble—the harm will have been done, and the monetary compensation will mean little. Even so, before they can collect damages they will have to figure out what the poison is and where it is coming from. Then, they must hire lawyers to prove their claims. What if, like the Powers family, they discover that the poison is different from any being dumped? They might find it impossible to prove who is to blame.

If waste chemicals reacted to form new poisons in the holding ponds of the Rocky Mountain Arsenal, might not other chemicals dumped underground react similarly—perhaps producing fantastically lethal compounds? As Rachel Carson said, "... the story of the Colorado farms transcends it local importance. What other parallels might there be, not only in Colorado but wherever chemical pollution finds its way into public waters?"

Officials claim that some wastes cannot be cleaned up—they must be dumped. If this is so, then (as is done with high-level radioactive poisons) they must be stored where they can be guarded and kept out of man's environment. If chemists and physicists are smart enough to make poisonous wastes and are not smart enough to un-make them, they should pay for keeping them out of harm's way.

I agree with the Task Force on Environmental Health that told the Secretary of Health, Education and Welfare: "It is obvious that one way to halt the contamination of the environment is to prohibit automobiles, stop the generation of electricity, and shut down industry—but it is just as obvious that this way is impossible ... What is possible is to find ways to eliminate the contamination at its source. Or, next best, to capture a pollutant and use it in a non-harmful way; or, finally, to bring the level of pollution down to a point compatible with the requirements of human health and welfare."

The enthusiasts for underground dumping say that unlike dumping wastes into holding ponds, they will dig "deep" wells to underground porus reservoirs of sandstone or limestone. Watertight casing will be run into the wells and cemented all the way to the top. After the wastes are treated and filtered (if necessary), they will be injected into the "disposal zone." A disposal zone 50 feet thick, and with 20 per cent porosity (holes between grains of rock) will hold 2 billion gallons of waste per square mile, engineers say. They point out that for years the oil industry has been disposing of billions of gallons of salt water (produced with oil and natural gas) by pumping it underground.

The disposal well people don't tell about disposal projects in Texas that have erupted like the Norris Geyser basin in Yellowstone National Park. Or about salt water injected into the ground in Canada that came out of the ground in Michigan. Or about the failure of a recently completed $2 million disposal well project on the shores of Lake Erie. The "watertight" casing in one well parted and blew 30 feet into the air. The "blow-out" spewed 150,000 gallons of industrial waste into the lake every day for three weeks. At last report, they were drilling two more disposal wells.

For years, Frank B. Conselman, president of the American Association of Petroleum Geologists, has pointed to the seriousness of the salt water disposal problem in Texas. A single (42-gallon) barrel of oil field brine with 100,000 parts per million of chloride from salt can ruin 20,000 gallons of drinking water.

Dr. Conselman shows dozens of pictures of faulty injection programs: creek banks white with salt seeping to the surface; salt water springs spouting through covered-over and long forgotten wells; and windmills that have pumped sweet water for 50 years suddenly producing salt water. He says "When formations receiving the brine fill up and pressures accumulate, something gives. Usually, the result is rupture toward the surface."

In one case near Abilene, Tex., salt water pumped underground ruptured a limestone formation under a sewer line and burst the sewer line as well. Conselman says, "It created, literally quite a stench."

What if these were chemical or radioactive poisons bursting through to the surface? How poisonous are these wastes liable to be? Testifying before the Colorado Water Pollution Control Commission, Dr. William A. Colburn, president of the Atomic Storage Corporation, said, "... (radioactive waste) is deadly, (for) ... a few hundred years minimum (to) they say two thousand years. But, when you think of the chemical waste that never dies down, it has infinite danger ..."

If the injected poisons are going to be dangerous forever, will the watertight casing in the disposal wells hold forever? Or will they be leaking some oil wells that are 20 or 30 years old?

Last November, a family in Los Angeles discovered that about 80 gallons of heavy black oil a day was oozing into their basement. Their home had been built over an oil well that had been abandoned 30 years earlier. The oil company was not held responsible for the damage since it had complied with the laws of California when the well was abandoned. The oozing disaster that struck the California family was considered "an act of God."

On the banks of the Pecos River in west Texas an oil spring is yielding a hundred dollars worth of oil a day to its owner. The oil is seeping to the surface through the corroded casing of active oil wells in a nearby field.

One reason why steel casing corrodes in the ground is that it upsets the electrochemical balance of the rocks and water. When nature tries to restore this balance the casing is corroded. Recently I heard of an instance where the casing of an industrial disposal well corroded within a few years. The old well was plugged and a new well was drilled at great expense. This time, stainless-steel casing was welded to the regular casing through the troublesome salt water zone. The combination of two metals and a weld in a salt-water solution made such a good battery that this time the casing corroded out in three months!

A second way in which casing can be corroded is from the inside by the highly corrosive liquids—hydrochloric acid, sulfuric acid, chromic acid and carbolic acid, to name a few—that are injected. True, the pipes carrying these acids are plastic-lined. But will the plastic lining last forever—or will it last about as long as the "perpetual" glass lining in my hot-water heater?

My friend Travis Parker, professor of geology at Texas A&M College, recently mentioned another disposal-well problem: Pressure. "What the disposal well people forget," he said, "is that these so-called disposal zones are already full of salt water before they start injecting into them. The result of pumping wastes into these zones is to raise the pressure higher and higher."

This high pressure will leak out through cracks already in the rocks, or it will rupture the rocks and form new cracks. Frank Conselman says, "Water injected underground, whether for flooding (to increase oil recovery in oil fields) or disposal, is a prime source of trouble. It is difficult to control, and may be diverted by any paths of least resistance known or unknown, natural or artificial. These escape routes multiply with nearness to the surface of the ground." Also, escape routes multiply as pressures are raised.

A good case in point is the offshore oil well near Santa Barbara which recently leaked 20,000 gallons of oil a day from an underground reservoir into the ocean. What caused this leak? Normal pressure from a deep oil and gas zone escaped up the hole being drilled and out into a shallow oil zone. Although the pressure was normal for the deep zone, it was much above normal pressure for the shallow one. Result: Rupture along a geologic crack (fault) to the surface. Once pressure opened the crack, oil in the shallow zone escaped. This oil smeared California beaches for miles.

As this is being written, the area around Marshall in southern Michigan was still having its troubles—caused by pressure getting loose underground. The toll thus far: two drilling rigs collapsed into craters, a park closed, cratering around various wells, and eruptions of gas, oil and water over a 12-mile area.

It all started in late 1968 when a drilling rig hit a high-pressure zone at a depth of 4,550 feet. When gas started blowing out of the hole, the driller plugged the well at the top of the hole. Within half an hour an abandoned water-well half a mile to the west turned into a 130-foot water spout. Nearby farm wells began to spout and also blow highly explosive natural gas. A number of families were asked to move out for their own safety. Next, two drilling rigs in a nearby oil field hit high pressures at shallow depths. Rigs were toppled and gas blew to the surface around steel casing.

Thanksgiving Day a tremendous explosion, followed by violent gas and water spouts and caving sink holes, shook the outskirts of the town of Marshall, 12 miles from any previous trouble. Christmas Eve, a water-well at a roadside park 3 miles north of Marshall suddenly erupted.

There is no proof yet that any of these events are related. At last account, state officials and industry were cooperating to try to pin down causes and prevent a recurrence.

Pumping fluids underground under high pressure can cause earthquakes, and we need go no farther than Denver for proof. When it was shown that the holding ponds at Denver's Rocky Mountain Arsenal were the source of pollution in the local well water, the Army drilled a 12,040-foot-deep disposal well near one of the ponds. Waste injection into the well began March 1962, and the following month Denver had an earthquake—the first in 80 years.

Pumping into the well and the earthquakes continued. None of the earthquakes were particularly damaging, but they were strong enough to get the undivided attention of everyone in the Denver area.

In November 1965, I pointed out that the 710 Denver earthquakes recorded by that time correlated with the volumes of fluid pumped into the Arsenal well. Periods of high injection resulted in many quakes, low pumping rates resulted in fewer earthquakes. When injection pressures were doubled, the number of earthquakes doubled. All earthquake epicenters (where they originated) were near the well.

How can fluid pressure trigger earthquakes? By forcing open cracks in the rocks far enough to allow the rocks to slide past each other. When the rocks slide seismographs record the vibration as an earthquake.

When it was determined that injection into the Arsenal well was causing the Denver earthquakes, all disposal pumping into the well was stopped. Since then, the disposal program at the Rocky Mountain Arsenal has been completely revamped, with the contaminated water being recycled for reuse. Now, no disposal ponds or wells are needed. If the authorities had decided to eliminate the contamination at its source to begin with, U.S. taxpayers would have been saved millions of dollars for damages to crops and real estate by pollution, another $1.5 million for the cost of the disposal well—and Denver would still be an earthquake-free city.

Disposal wells are usually drilled right at the factory, which means that most disposal wells are near cities. Is there some way of spotting trouble so that widespread contamination can be prevented?

Here are two "checks" offered by an expert with the Federal Water Pollution Control Administration: First, carefully monitor the fresh-water wells in the area! The expert doesn't say what to do next. If a few billion gallons of carbolic acid or cyanide compounds showed up in the city's water supply, my inclination would be to leave town quickly.

Second: If the amount of waste injected into the well at a given pressure suddenly increases (for example, if the well should suddenly start taking 500 gallons a minute at 100 pounds pressure when it had been taking 100 gallons) this is an indication that either the disposal zone, or the casing in the well, has ruptured. Once again, when this happens, it may be too late to prevent eventual widespread contamination. Like atomic power plants and rocket trips to the moon, the margin for error in disposal wells is zero.

Disposal engineers are careful to point out that poisonous wastes will only be injected into disposal zones containing salt water. What do they mean by that?

Geologist William C. Finch of Houston, Texas, who knows more about salt water than anyone I know, explains the folly of this course of action. "To begin with," Bill told me, "there are three kinds of water—fresh water that has less than 1,000 parts per million of dissolved salts; brackish water that contains less salt than sea water and more than fresh water. Sea water has 35,000 parts per million dissolved salts. When water contains more salt than sea water it is called brine."

The so-called salt water in most potential disposal zones is in reality brackish water. There is hundreds of times more brackish water underground than fresh water. Brackish water can be desalted at about one-fourth of the cost of desalting sea water. Brackish water is likely to be our next big source of fresh water as human and industrial demands arise. It is a priceless natural resource and Bill Finch contends it would be the wildest kind of folly to contaminate it with industrial waste.

"We can desalt brackish water with 2,500 parts per million salts for 25 cents a thousand gallons, against 98 cents a thousand gallons for sea water—and desalting costs are going down all the time," Finch says.

"Brackish water is located where it will soon be needed—beneath the deserts of Arizona and Nevada and the plains of west Texas and eastern Colorado, for example."

Bill Finch is not alone in opposing the disposal well idea. Many other geologists warn that we just don't know enough about underground conditions to predict what will happen when both pressure and poison are turned loose. Who expected well disposal to cause a thousand earthquakes in Denver? Who expected tens of thousands of gallons of oil to erupt from the ocean floor at Santa Barbara?

Testifying before the Daddario subcommittee on Environmental Quality in 1968, Dr. William T. Pecora, director of the U.S. Geological Survey, made three statements that show the conclusions of this great earth science organization:

"There is a constant interaction of the water locked up in the porous and permeable rocks of the earth, a constant interaction of this water supply with the streams."

". . . no one is properly prepared to recharge water or liquid wastes into complex underground systems with certainty of the results that such an operation may yield."

"Hopefully the time is approaching when large quantities of polluted material cannot be dumped into the ground without considerable thought being given to the consequences."

Nevertheless, there is great pressure in some states, including Colorado, to encourage industry by providing a cheap place to dump wastes underground. There are industrial disposal wells in 17 states. Tom Ten Eyck, chairman of the Colorado Water Pollution Control Commission, said recently of industrial disposal wells: "It's a coming thing and we must be prepared to deal with it."

In a report issued February 4, 1969, the Colorado Water Pollution Control Commissioners said, "It (the Commission) recognizes the need for such systems if Colorado industry is to continue to expand and the health and economic Welfare of its citizens are to be advances."

But the experience of many states has proved the disposal well is not the solution to the industrial waste problem. The answer is to reuse the water when possible, clean it up so that disposal is unnecessary, or store it safely where it will be available when needed. Many companies have learned this lesson.

Two years ago, I was in Houston. A rotten odor filled the air. "Smell that?" a friend asked. "That's the smell of dead fish in the Houston Ship Channel. The pollution is so bad it kills fish by the million."

On the same trip I visited with officials of a major oil company who did not believe in wholesale dumping into the oceans and rivers. They were using 12 disposal wells for underground dumping instead.

A few weeks ago I visited the same company again. Officials told me they have plugged all but four of their disposal wells. They are getting rid of the remaining four as fast as they can. Why? They have decided that in the long run it will be cheaper and less troublesome to clean up pollution at the plants.

Disposal wells, they told me, are unpredictable, and expensive. The risk of expensive lawsuits resulting from pollution breaking out at unknown times and places is too great. "We finally decided that we were turning snakes loose in the grass that might come back and bite us," said one official.

I visited a new $4 million plant that cleans every drop of water from the company's oil refining and petrochemical complex on the Houston Ship Channel. After being cleaned, the water is discharged into an artificial lake. The plant superintendent proudly pointed to healthy fish living in this water that had just come through a maze of pipes and chemical reactions. "See those fish?" he said, "They are inspected every day. If one of them gets sick, there's hell to pay in the plant until we find out what's bugging him. Then we clean up the water."

17

The Race To The Bottom Of The Sea

WOLFGANG FRIEDMANN

The last two selections show how geological developments can complicate the lives of men, or vice versa.

An exciting new geological frontier has opened with the exploration of the nearly three-quarters of the earth's surface that lies beneath the oceans. Incredibly, the whole region is already beset by legal problems. They are discussed by Wolfgang Friedmann, Professor of International Law at Columbia University.

Until very recently, the ocean bottoms, inaccessible and remote from the affairs of mankind, were the subject only of dreams and legends, of sunken cities or sea monsters. But the exploration, exploitation, and possible occupation of the ocean bed—a subject of science fiction little more than a generation ago—has, since the end of the last world war, and particularly during the last few years, become an increasingly practical possibility.

This exploration is a development hardly less dramatic than the journeys into outer space. But whereas moon landings and space satellites will, for the foreseeable future, have mainly scientific and perhaps strategic importance, the opening up of the ocean floor may, within a decade, decisively alter the political, economic, and military face of the earth. The crucial question before mankind is whether this conquest of the ocean bed will

Reprinted from the COLUMBIA FORUM, Spring, 1969, Vol. XII, No. 1. Copyright 1969 by The Trustees of Columbia University in the City of New York.

proceed by international cooperation or by an increasingly frantic competitive race among nations.

Historically, it has been the surface of the seas that mattered, an area covering nearly three quarters of the surface of the earth. The most important achievement of international law relating to the seas, as it has been fashioned since the beginning of the seventeenth century, has been the freedom of all to sail on and fish in them. Without this freedom—which has been maintained through peace and war in succeeding centuries—the exploration and the colonization of the newly discovered continents and the worldwide spread of international commerce and communications would have been all but impossible. But the freedom of the seas, solemnly reaffirmed by the Geneva Convention on the High Seas in 1958, is being steadily whittled away and may be little more than a fiction within less than a decade.

In an historic speech delivered at the United Nations on November 1, 1967, the Maltese Ambassador, Mr. Arvid Pardo—as a preliminary to certain proposals for the international control of the seabed resources—surveyed the principal developments that have occurred within the last 25 years, including the accelerating rate of technological progress and its momentous strategic, economic, and legal consequences. A draft treaty submitted in March 1968 by Senator Claibrone Pell on the same subject, and the numerous official, semiofficial, and private studies now in progress, indicate that a subject that might well have remained the preserve of oceanographers, naval strategists, and oil economists is now arousing worldwide concern. As well it should, for the manner and form of the control of the ocean beds may radically change the map of the world.

What can revolutionize the political and economic topography of the earth is the rapidly growing ability of man to dwell for prolonged periods at the bottom of the oceans. at depths of many hundreds of feet, and the corresponding capacity to exploit the untapped mineral resources of the ocean beds at depths far exceeding the relatively shallow explorations that have hitherto been made. Not long ago an expert predicted that before 1975 there would be colonies of aquanauts living and working at depths of the order of 500 meters. The United States Navy, which maintains a number of sea laboratories, announced that it has been possible for aquanauts to move freely for limited periods at depths of well over 250 meters.

The technique used by Jacques Yves Cousteau, the navy divers, and others, is the creation of an artificial atmosphere, composed generally of oxygen, nitrogen, and helium, adjusted to the divers' operating depths and equalized to the pressure of the gases dissolved in the fluids and tissues of the divers' bodies. Teams of aquanauts have spent periods of several weeks in enclosed submersibles at the bottom of the ocean. There is little doubt that within less than a decade the establishment of settlements at the bottom of the ocean, as well as transportation along the ocean beds, will be practical. Pipelines will be laid to transport oil and natural gas along the bottom of the seas. Experts estimate that the cost of installations and transportation along the ocean bed may well be much lower than the cost of operating vehicles at lesser depths.

The enormous significance of such technological progress has to be seen in both economic and military terms. Recent studies, such as John L. Mero's 1964 book *The Mineral Resources of the Sea,* have estimated that the total ocean bed reserves of such vital commodities as oil, manganese, aluminum, copper, nickel, cobalt, and molybdenum exceed by many hundreds of times the known terrestrial reserves. A token of the possible developments is the already very active exploration of the natural gas resources of the North Sea. In 1967, in the *Oil and Gas Journal,* the reserves of the Dutch North Sea concession alone were estimated at 40 trillion cubic feet, or roughly one third of the latest estimates of natural gas reserves in the United States.

The strategic implications of deep sea settlements and stations are no less momentous. Submarine or nuclear missile bases established at the bottom of the seas could push military outposts close to potential enemy shores, and the present relative openness of mutual intelligence operations—which may be beneficial rather than injurious to the maintenance of peace—would be replaced by the secrecy of bases spread beneath the oceans. But how, it will be asked, is all this compatible with the freedom of the seas? The fact is that over the last twenty years the legal and actual freedom of the seas has been steadily eroded and is in danger of becoming little more than a pious phrase.

It was the so-called Truman Proclamation of September 28, 1945, that started the now-universal extension of national jurisdiction over the continental shelf. It is proclaimed that "the exercise of jurisdiction over the natural resources of the ¡subsoil and sea-bed of the continental shelf by the contiguous nation is reasonable and just . . . since the continental shelf may be regarded as an extension of the land mass of the coastal nation and thus naturally appurtenant to it, . . . and since self-protection compels the coastal nation to keep close watch over activities off its shores, . . . " the United States "regards the natural resources of the subsoil and seabed of the continental shelf beneath the high seas but contiguous to the coasts of the United States as appertaining to the United States, subject to its jurisdiction and control."

The Truman Proclamation spread like wildfire. During the subsequent decade, one nation after another claimed national jurisdiction over its continental shelf in terms similar to, or exceeding those of, the Truman Proclamation. So rapid was the general acceptance of the new doctrine that little more than a decade later, in Geneva in 1958, a multilateral convention on the continental shelf, ratified by nearly 40 states and in force since June 1964, turned the continental shelf doctrine into a general principle of international law.

Although this has been hailed by many as a triumph of international cooperation, it is essentially a triumph of nationalism over the international sharing of the resources of the seas, which would flow from the legal principle of the freedom of the seas. Moreover, the Geneva Convention definition of the continental shelf went beyond that of the Truman Proclamation by not strictly limiting it to a maximum depth of 200 meters. The extension of the depth "beyond that limit to where the depth of the superjacent waters admits of the exploitation of the natural resources of the said areas . . . " might have been regarded as relatively insignificant even a decade ago. With the explosive progress of the technological possibilities of probing into ever greater depths, this clause may make some claims to continental shelves potentially unlimited. Indeed, at the Conference of the International Law Association last August in Buenos Aires, the American Branch proposed that national claims should extend to a depth of 2,500 meters.

Some international lawyers have maintained that there is nothing to prevent a state from claiming exclusive jurisdiction over a continental shelf to an unlimited length and depth, or until it met with the competing claim of another state, in which case, under the terms of the Convention, the boundary between the two would be the median line. At the present time, continental shelf explorations, mainly oil drillings, take place close to the national shores, within 20 miles from the coast, and at much lesser depths than 200 meters. But then, the present state of ocean technology would have been regarded as fantasy a few decades ago. Furthermore, the continental shelves are very unequal in depth, length, and exploitability. In width, they may range from one to 800 miles; and the width and importance of the continental shelf bear no relation to the size of the coastal state. Among the new "ministates" are small islands that dispose of an enormous continental shelf area. Moreover, the dwindling colonial empires own islands thousands of miles from the mother country. If, as has been suggested, the continental shelf area is measured from the island coast, and not from that of the mother state, remarkable consequences could follow. Thus, the French-owned islands of St. Pierre et Miguelon in the Gulf of St. Lawrence could give France virtual control of the Gulf. The continental shelf claims of such tiny islands as Clipperton off Mexico (French), or St. Helena—of Napoleonic fame—in the Atlantic (British), could extend national rights over seabeds of thousands of miles.

Nor could the proliferation and extension of continental shelf exploitation by the various coastal states remain without effect on the freedom of navigation and fisheries. Although the Geneva Convention expressly safeguards these and other aspects of the freedom of the seas against the continental shelf claims and provides for safety installations as a means of protection, the growing exploitation of the mineral seabed resources is likely to curtail the exercise of such freedoms.

If extended far from the shores, the various stations and artificial islands may come to constitute increasing hazards to navigation, while fish life and other forms of sea vegetation may be seriously affected by pollution through oil, radioactive fallout, and other by-products of modern industrial progress. It seems that man reaches for new resources with one hand, while with the other he destroys or pollutes the very bases of his life.

The lack of a clear territorial limitation of the continental shelf, and the now-concrete possibility of continental shelf explorations reaching deep into the oceans, would be serious enough to demand international action. But the dangers to the freedom of the seas do not stop there. There is the possibility that portions of the open ocean beds, outside of any continental shelf, may be appropriated as parts of national territories. Unfortunately, international law does not forbid this. Most authorities—writing in times when this was mainly a theoretical question—hold that the freedom of the seas is limited to the waters and does not prohibit the appropriation of some sectors of the seabed.

The prospects are truly alarming. There are many plateaus in the seas, which are relatively easy to occupy. Not only may major maritime nations wish to appropriate these plateaus for strategic or economic purposes, but also many landlocked states may seize this opportunity to turn themselves into maritime nations by claiming sovereignty over tiny islands erected somewhere in the seas, with all the appurtenant territorial waters, continental shelves, and other aspects of national sovereignty. And since most of the land-locked states will hardly be in a position to give effective protection to their newly claimed islands, they are likely to depend on the support of one or another of the major naval powers, thus enlarging the worldwide strategic, political, and ideological struggle. At the same time, fierce competition among the states, or among private enterprises in a growing number of states, may push explorations and mining of the continental shelves further and further into the deep seabeds. What can be done?

A position by the United Nations on the problem is the first and obvious step. Following his speech to the General Assembly in the fall of 1967, the Maltese Ambassador introduced a draft resolution stipulating that "the seabed and the ocean floor underlying the seas beyond the limits of national jurisdiction as defined in the treaty, are not subject to national appropriation in any manner whatsoever." In response to this proposal, the General Assembly appointed a special ad hoc committee which prepared a report recognizing the existence of an area of the seabed and the ocean floor underlying the high seas that is beyond the limits of national jurisdiction. Acting on this report the General Assembly, in December 1968, established a permanent committee "to study the elaboration of the legal principles and norms which would promote international cooperation" in the field and requested the Secretary-General to study the question of establishing "appropriate international machinery" to promote exploitation of seabed resources beyond the limits of national jurisdiction in the interest of mankind "irrespective of the geographical location of States, and taking into special consideration the interests and needs of developing countries." Nevertheless, "due respect for national jurisdiction" is to be given in any such long-term program. The permanent committee will report to the 1969 session of the General Assembly when it meets in the fall.

Such a General Assembly resolution, while helpful in that it recognizes the existence of the seabed and ocean floor beyond the limits of national jurisdiction, loses much of its usefulness because these limits have not been defined.

In the present situation of indefinite claims on continental shelves, it would be unrealistic to expect that any state would denounce the exclusive rights to the continental shelf exploration conceded by the Geneva Convention. But it is perhaps not too much to hope that states would agree to a definition of limitation, in terms of either maximum depth (200 meters) or maximum distance from the shores. Otherwise any affirmation of the freedom of the ocean bottoms might be made meaningless by over-expanding claims to continental shelves, which would in effect divide up the ocean floor.

The next step, also proposed in the Maltese Resolution—and, more recently, in Senator Pell's draft treaty—might be the establishment of an international agency to control the various national activities undertaken in the deep seas and on the ocean floor. The best way of doing this—proposed in the Pell draft treaty—would be to constitute a specialized United Nations agency as the licensing authority over deep sea operations beyond the redefined continental shelf jurisdictions. Licenses could be issued to governments, which would either direct the entrepreneurs or sublicense to public or private corporations. In either case the license fee would be paid to the UN agency. The revenue—which would initially be small but in later years might become very large—could be used to bolster the budget of the United Nations, or for some other agreed purpose. At this point the proposals of the Maltese Resolution and the Pell draft diverge. The former would use at least part of the revenues to aid the poorer nations. This is an attractive idea as development aid has been lagging in recent years and falling far behind the minimum needs of the developing countries. The Pell draft refrains from such an ambitious plan, perhaps because the Senator has little hope that an increasingly aid-hostile Congress would even consider it.

A new agency would probably best be constituted as a semi-autonomous body, somewhat on the pattern of the World Bank or one of the other functional agencies. Disputes on such matters as the extent of license rights and undue interference by any licensee with navigation, fishing or other legitimate activities, could be submitted to an international judicial body, perhaps on review from a fact-finding panel. The obvious judicial organ would be the International Court of Justice.

These details could be worked out, but the big question is whether the nations will in fact agree on any form of international control. A report of a subcommittee of the House of Representatives issued in 1967 is hardly encouraging. The report says that it would "be precipitate, unwise, and possibly injurious to the objectives that both the United States and the United Nations have in common, to reach a decision at this time regarding a matter that vitally affects the welfare of future generations." It specifically opposes any United States support for "the vesting of title to the seabed, the ocean floor, or ocean resources, in any existing or new internation organization." It concludes that "hasty action in this field could create more problems than it will solve or avert."

The fallacy of this reasoning is that in the vacuum created by inaction, extravagant continental shelf claims and seabed appropriations may be established that will be difficult to reverse. And in less than a decade this might happen.

The challenge posed by the accessibility of the ocean bed is symbolic of a wider issue: whether the pace of military, political, and economic competition between nations will be further accelerated, or whether the overwhelming common problems of mankind—the population explosion, the threatening exhaustion of resources by ruthless and uncontrolled competition, soil erosion, extinction of whales and fisheries, water and air pollution, and the extension of the nuclear weapons race to outer space and the deep seas—will bring about the necessary minimum of international cooperation for survival.

18

The Great Landslide Case

SAMUEL L. CLEMENS

The concluding essay, is by a man who needs no introduction.

The mountains are very high and steep about Carson, Eagle, and
Washoe Valleys—very high and very steep, and so when the snow
gets to melting off fast in the spring and the warm surface-earth
begins to moisten and soften, the disastrous landslides commence.
The reader cannot know what a landslide is, unless he has lived
in that country and seen the whole side of a mountain taken off
some fine morning and deposited down in the valley, leaving a
vast, treeless, unsightly scar upon the mountain's front to keep
the circumstance fresh in his memory all the years that he may go
on living within seventy miles of that place.

General Buncombe was shipped out to Nevada in the invoice
of territorial officers, to be United States Attorney. He considered
himself a lawyer of parts, and he very much wanted an opportunity to manifest it—partly for the pure gratification of it and
partly because his salary was territorially meager (which is a strong
expression). Now the older citizens of a new territory look down
upon the rest of the world with a calm, benevolent compassion,
as long as it keeps out of the way—when it gets in the way they
snub it. Sometimes this latter takes the shape of a practical joke.

One morning Dick Hyde rode furiously up to General Buncombe's door in Carson City and rushed into his presence with-

Chapter 34, "The Great Landslide Case," from **Roughing It,** Vol. I, by
Samuel L. Clemens. Reprinted by permission of Harper & Row, Publishers, Incorporated.

out stopping to tie his horse. He seemed much excited. He told the General that he wanted him to conduct a suit for him and would pay him five hundred dollars if he achieved a victory. And then, with violent gestures and a world of profanity, he poured out his griefs. He said it was pretty well known that for some years he had been farming (or ranching, as the more customary term is) in Washoe District, and making a successful thing of it, and furthermore it was known that his ranch was situated just in the edge of the valley, and that Tom Morgan owned a ranch immediately above it on the mountainside. And now the trouble was, that one of those hated and dreaded landslides had come and slid Morgan's ranch, fences, cabins, cattle, barns, and everything down on top of *his* ranch and exactly covered up every single vestige of his property, to a depth of about thirty-eight feet. Morgan was in possession and refused to vacate the premises—said he was occupying his own cabin and not interfering with anybody else's—and said the cabin was standing on the same dirt and same ranch it had always stood on, and he would like to see anybody make him vacate.

"And when I reminded him," said Hyde, weeping, "that it was on top of my ranch and that he was trespassing, he had the infernal meanness to ask me why didn't I *stay* on my ranch and hold possession when I see him a-coming! Why didn't I *stay* on it, the blathering lunatic—by George, when I heard that racket and looked up that hill it was just like the whole world was a-ripping and a-tearing down that mountainside—splinters and cordwood, thunder and lightning, hail and snow, odds and ends of haystacks, and awful clouds of dust!—trees going end over end in the air, rocks as big as a house jumping 'bout a thousand feet high and busting into ten million pieces, cattle turned inside out and a-coming head on with their tails hanging out between their teeth!—and in the midst of all that wrack and destruction sot that cussed Morgan on his gatepost, a-wondering why I didn't *stay and hold possession!* Laws bless me, I just took one glimpse, General, and lit out'n the county in three jumps exactly.

"But what grinds me is that the Morgan hangs on there and won't move off'n that ranch—says it's his'n and he's going to keep it—likes it better'n he did when it was higher up the hill. Mad! Well, I've been so mad for two days I couldn't find my way to town—been wandering around in the brush in a starving condition—got anything here to drink, General? But I'm here *now,* and I'm a-going to law. You hear *me!*"

Never in all the world, perhaps, were a man's feelings so outraged as were the General's. He said he had never heard of such high-handed conduct in all his life as this Morgan's. And he said there was no use in going to law—Morgan had no shadow of right to remain where he was—nobody in the wide world would uphold him in it, and no lawyer would take his case and no judge listen to it. Hyde said that right there was where he was mistaken —everybody in town sustained Morgan; Hal Brayton, a very smart lawyer, had taken his case; the courts being in vacation, it was to be tried before a referee, and ex-Governor Roop had already been appointed to that office, and would open his court in a large public hall near the hotel at two that afternoon.

The General was amazed. He said he had suspected before that the people of that territory were fools, and now he knew it. But he said rest easy, rest easy and collect the witnesses, for the victory was just as certain as if the conflict were already over. Hyde wiped away his tears and left.

At two in the afternoon referee Roop's Court opened, and Roop appeared throned among his sheriffs, the witnesses, and spectator, and wearing upon his face a solemnity so awe-inspiring that some of his fellow-conspirators had misgivings that maybe he had not comprehended, after all, that this was merely a joke. An unearthly stillness prevailed, for at the slightest noise the judge uttered sternly the command:

"Order in the Court!"

And the sheriffs promptly echoed it. Presently the General elbowed his way through the crowd of spectators, with his arms full of law-books, and on his ears fell an order from the judge which was the first respectful recognition of his high official dignity that ever saluted them, and it trickled pleasantly through his whole system:

"Way for the United States Attorney!"

The witnesses were called—legislators, high government officers, ranchmen, miners, Indians, Chinamen, negroes. Three-fourths of them called by the defendant Morgan, but no matter, their testimony invariably went in favor of the plaintiff Hyde. Each new witness only added new testimony to the absurdity of a man's claiming to own another man's property because his farm had slid down on top of it. Then the Morgan lawyers made their speeches, and seemed to make singularly weak ones—they did really nothing to help the Morgan cause. And now the General, with exultation in his face, got up and made an impassioned effort; he pounded the table, he banged the law-books, he shouted, and roared, and howled, he quoted from everything and everybody, poetry, sarcasm, statistics, history, pathos, bathos, blasphemy, and wound up with a grand war-whoop for free speech, freedom of the press, free schools, the Glorious Bird of America and the principles of eternal justice! [Applause.]

When the General sat down, he did it with the conviction that if there was anything in good strong testimony, a great speech and believing and admiring countenances all around, Mr. Morgan's case was killed. Ex-Governor Roop leaned his head upon his hand for some minutes, thinking, and the still audience waited for his decision. And then he got up and stood erect, with bended head, and thought again. Then he walked the floor with long, deliberate strides, his chin in his hand, and still the audience waited. At last he returned to his throne, seated himself, and began, impressively:

"Gentlemen, I feel the great responsibility that rests upon me this day. This is no ordinary case. On the contrary, it is plain that it is the most solemn and awful that ever man was called upon to decide. Gentlemen, I have listened attentively to the evidence, and have perceived that the weight of it, the overwhelming weight of it, is in favor of the plaintiff Hyde. I have listened also to the remark of counsel, with high interest—and especially will I commend the masterly and irrefutable logic of the distinguished gentleman who represents the plaintiff. But, gentlemen, let us beware how we allow mere human testimony, human ingenuity in argument and human ideas of equity, to influence us at a moment so solemn as this. Gentlemen, it ill becomes us, worms as we are, to meddle with the decrees of Heaven. It is plain to me that Heaven, in its inscrutable wisdom, has seen fit to move this defendant's ranch for a purpose. We are but creatures, and we must submit. If Heaven has chosen to favor the defendant Morgan in this marked and wonderful manner; and if Heaven, dissatisfied with the position of the Morgan ranch upon the mountainside, has chosen to remove it to a position more eligible and more advantageous for its owner, it ill becomes us, insects as we are, to question the legality of the act or inquire into the reasons that prompted it. No—Heaven created the ranches, and it is Heaven's prerogative to rearrange them, to experiment with them, to shift them around at its pleasure. It is for us to submit, without repining. I warn you that this thing which has happened is a thing with which the sacrilegious hands and brains and tongues of men must not meddle. Gentlemen, it is the verdict of this court that the plaintiff, Richard Hyde, has been deprived of his ranch by the visitation of God! And from this decision there is no appeal."

Buncombe seized his cargo of law-books and plunged out of the court-room frantic with indignation. He pronounced Roop to be a miraculous fool, an inspired idot. In all good faith he returned at night and remonstrated with Roop upon his extravagant decision, and implored him to walk the floor and think for half an hour, and see if he could not figure out some sort of modification of the verdict. Roop yielded at last and got up to walk. He walked two hours and a half, and at last his face lit up happily and he told Buncombe it had occurred to him that the ranch underneath the new Morgan ranch still belonged to Hyde, that his title to the ground was just as good as it had ever been, and therefore he was of opionion that Hyde had a right to dig it out from under there and—

The General never waited to hear the end of it. He was always an impatient and irascible man, that way. At the end of two months the fact that he had been played upon with a joke had managed to bore itself, like another Hoosac Tunnel, through the solid adamant of his understanding.

Technology and Society: ³⁶⁸

A CANADIAN PERSPECTIVE

Technology and Society:

A CANADIAN PERSPECTIVE

John Goyder

broadview press

Canadian Cataloguing in Publication Data

Goyder, John, 1946- .
Technology and society : a Canadian perspective

Includes index.
ISBN 1-55-111-048-2
1. Technology - Social aspects - Canada. I. Title.

T14.5.G69 1997   303.48'3'0971   C97-930452-0

Broadview Press
Post Office Box 1243, Peterborough, Ontario, Canada K9J 7H5

in the United States of America:
3576 California Road, Orchard Park, NY 14127

in the United Kingdom:
B.R.A.D. Book Representation & Distribution Ltd.,
244A, London Road, Hadleigh, Essex SS7 2DE

Broadview Press gratefully acknowledges the support of the Canada Council, the Ontario Arts Council, and the Ministry of Canadian Heritage.

PRINTED IN CANADA

To Jo-Anne

Contents

Preface

Job losses due to automation in the 1990-94 recession made the impact of technology on Canadian society a more pressing issue than ever before. The implications of technology for their future lives became *the* salient issue for many of today's undergraduates. The same students enjoyed more gadgets than their parents ever dreamed of in their youths, for it was a time of wondrous technological inventiveness as well. When Canadians living at the end of the twentieth century were not playing with their new computers or pondering the jobs lost due to those same miracle machines, they were being reminded time after time on the nightly news hour about the environmental degradation that seems to be another feature of technological society. Issues such as the depletion of fishing stocks, ozone layer thinning, global warming, and air pollution reminded Canadians that, for all its magnificence, technology has come to dominate and perhaps threaten nature. For all that they can in theory serve equally the causes of good and ill, the inventions of humankind seem to many a tempting but dangerous siren.

Technology is one of the most global of phenomena, and perhaps that explains why most Canadian publisher's lists carry many interesting items by British and US authors related to the sociology of technology, history of technology, or ethical issues in technology. Much more rare is a book which, without becoming parochial, takes Canadian society as the reference point. The motive for undertaking

the present volume was this sense that Canadian college and university students should be using a book that takes their own country as the baseline, and it was the start of a very gratifying writing project when the editors at Broadview Press agreed.

There were in print several fine Canadian books on specific technology issues such as the workplace or communications as well as highly distinguished more general contributions from a generation ago, especially in the works of Harold Innis, Marshall McLuhan, and George Grant. These three prompted Kroker's contention that "Canada's principal contribution to North American thought consists of a highly original, comprehensive, and eloquent discourse on technology." I saw, however, the need for an up-to-date work that could simultaneously reacquaint the present cohort of readers with the rich Canadian tradition in research on technology and society, deal with technology over a broad selection of issues, and be explicitly oriented to this country while not losing sight of the global scale of technology.

In attempting to address these objectives, I found myself, although a sociologist by training, reading anthropological approaches to the evolution of human societies. That is how I made sense of George Grant's insistence that technology is "a fate," the theme which opens the first chapter and concludes the last. The project also required considerable reading in history and economics. While the backbone of this book is sociological, it is consistent with the best of the Canadian tradition to draw from whichever disciplines lead us to insights into the connection between technology and society.

The first three chapters address the social and historical antecedents of the creation and transmission of technology. Technology here is the dependent variable; something to be explained. Chapter 1 defines the phenomenon and suggests a way of classifying the vast array of tools and techniques along three dimensions. A general conceptual scheme of interrelationships between society and technology appears in Chapter 2, while technological diffusion is addressed in the third chapter. This amount of attention on diffusion acknowledges that the development of up-to-date technology is not as taken for granted in our country as it is, for example, in the powerful land to the south. It is a legitimate concern of Canadians that we tend to be mere importers of technology, with possible adverse consequences for our economic competitiveness. Beginning with Chapter 4, the focus shifts more to the social con-

sequences of technology — technology as independent variable — perhaps the more traditional concern of sociologists. A Canadian audience, it is felt, needs to consider carefully the linkage of technology with economic development. On the issue of the technocratic value system, examined in Chapter 5, there is a strong Canadian tradition ranging from the past work of George Grant to the present writings of figures such as Ursula Franklin. We have national issues too with technology and work (Chapter 6), for automation has particular importance for a country possessing what some term a "branch plant" economy. Do Canadians suffer twice over in the automated work force, first losing traditional jobs and then being excluded from new jobs in the technology-creation, "R&D," sector? Technology and communications are addressed in Chapters 7 and 8, beginning with the indigenous tradition dating back to the work of Innis in the 1950s, then moving to some ideas of Marshall McLuhan. The new leading technologies of artificial intelligence and biotechnology are addressed in Chapter 9, while a view of how the public in Canada and abroad assess issues in technology is presented in Chapter 10. The final chapter offers a short re-statement of the main themes.

Bibliographically, I have tried to cite and suggest enough sources to give undergraduate students a reasonable feel for the society and technology literature. It is a vast and rapidly growing body of knowledge, and the present book of relatively small size yet ambitious range could never pretend to provide a definitive bibliography. Sources are cited in three forms. Newspaper and magazine articles generally are referenced within the text, to differentiate them from more conventional and perhaps permanent academic sources, and also to keep the number of footnotes from becoming intimidating. Sources cited "for the record," material which does not seem to me a high priority for students to actually read, receive mention within the numbered footnotes. Finally, sources that students would more likely profit from consulting in the original appear within the Further Readings sections at the end of each chapter. Most of these further readings are discussed within the text, but occasionally I include items without commentary as a way to plug gaps within this short volume. The "scientific" style of referencing in my opinion does not succeed for undergraduate textbooks; it becomes too tempting to overwhelm the reader with citations, few of which will be worth the student's consulting. Only the scholarly pretensions of the

author are satisfied thereby.

On the perennial stylistic issue of personal and impersonal pronouns, a reviewer noted seeming inconsistency and so the following clarification is offered: by and large I try to avoid writing "I," but when this self-reference does appear it is a signal that a personal opinion is being offered rather than some truth taken for granted throughout the technology and society literature. "We" refers when employed to seminar-like moments in the book when the writer is inviting the reader to join him in some thought experiment, turning of attention to a new issue, or other such collective intellectual endeavour.

Technology and Society: A Canadian Perspective presupposes no prior grounding in sociology nor in any other discipline at the post-secondary level. With that audience in mind, various social science terms which may be unfamiliar appear in *"italics within quotation marks"* and are defined in a glossary.

As I complete work on this book I am mindful of the many friends and colleagues who in various ways assisted my endeavours. Grace and Harry Logan (University of Waterloo) coached me on issues in linguistics. My wife Jo-Anne did some good copy-editing and allowed me to talk about the book any time I paid for a dinner out. I learned more about the ideology behind technocracy in discussions with Peter Eglin, Department of Sociology and Anthropology at Wilfrid Laurier University. Keith Warriner from my own department helped me with the environmental literature. Thanks to my good colleague Jim Curtis for running tabulations for me from his copy of the *World Values Study* data. Two University of Waterloo Sociology graduate students, David Cameron and Margie Lambert-Sen, gave me feedback on some of the earlier chapters during a reading course on technology and society. Cathy Newell (Arts Computing Office, UW) was immensely helpful with assistance in preparing diagrams in Word Perfect. Three science colleagues at University of Waterloo, Gerry Toogood, Lew Brubacher, and John Honek, checked over Chapter 9 for me, and warned me against my most obvious layperson errors. I enlisted Alicja Muszynski to be my political correctness advisor. One of the reasons I enjoyed writing this book so much was the opportunity to collaborate once again with Michael Harrison at Broadview Press. Thanks also to the copy editor, Betsy Struthers, and to Broadview's Barbara Conolly for careful work on the manuscript.

For permission to reproduce previously published material, I gratefully acknowledge Elsevier Science Limited, The Conference Board of Canada, Universal Press Syndicate, and The Orion Publishing Group Ltd. Kathryn McMullen was kind enough to confirm figures from the draft version of her report on work skill, cited in Chapter 6.

All photographs were taken by the author excepting the Avro Arrow photo which Les Wilkinson very generously lent me from his extensive collection. Thanks to the National Science Museum, London, England, for permission to take photographs and print them herein, and to the staff of the National Museum of Science and Technology in Ottawa for permisssion to go beyond the normal public access areas for purposes of shooting photos.

I want to acknowledge the valuable services to academics provided by Statistics Canada. The 1994 edition of their *General Social Survey* was an invaluable resource for the preparation of this volume. For permission to use on-line survey results about Internet use, grateful acknowledgement to Mike Perry.

Finally, one of the happy reciprocities of academic life is the custom of scholars unselfishly performing manuscript reviews for publishers. It is a time-consuming task, minimally remunerated, and I keenly feel my indebtedness to the two anonymous reviewers engaged by Broadview. They examined the manuscript with great care, allowed me to write the book I wanted, but offered many helpful suggestions for improving and clarifying the presentation. The third reviewer of the manuscript was not an established professor, but rather a most gifted undergraduate student, Tim Huh, who generated many pages of valuable, mature commentary on the manuscript. We like to talk in academe of professors learning from students, and in this instance I really did.

Waterloo, Ontario

Part I.

HUMAN BEINGS AND TECHNOLOGY: BASIC RELATIONSHIPS

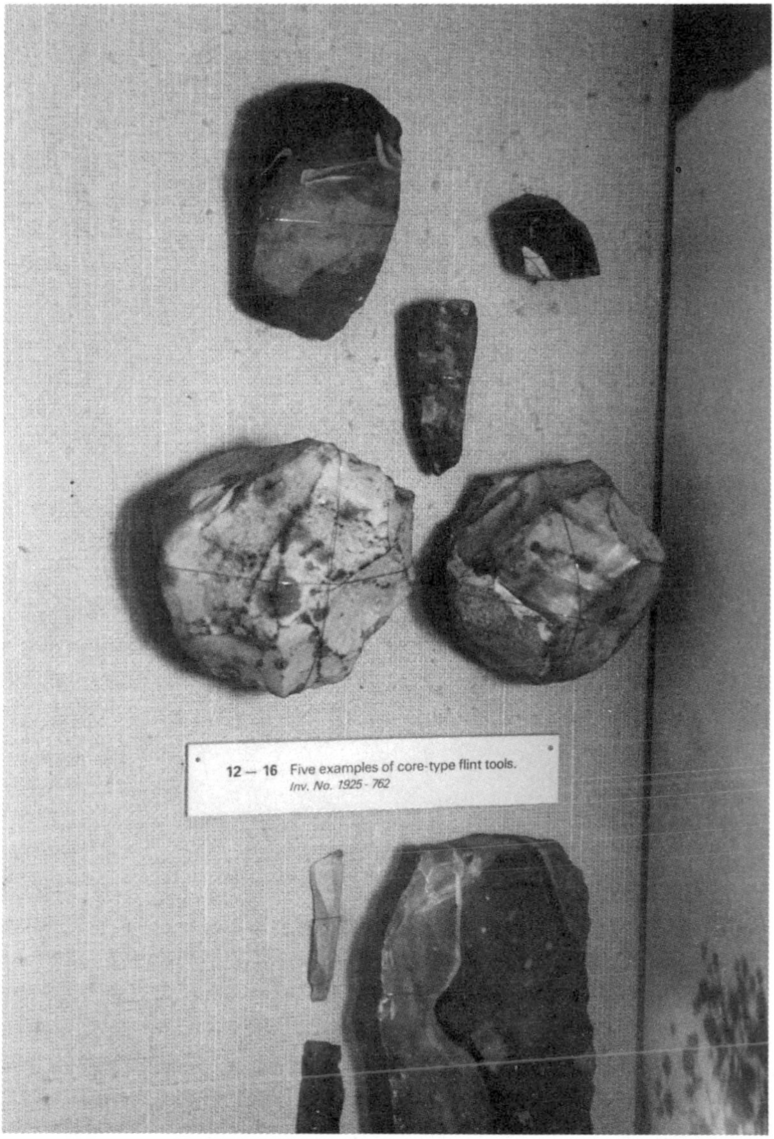
12 — 16 Five examples of core-type flint tools.
Inv. No. 1925 - 762

The First Technology: Stone Age tools. *The Science Museum,* London, UK. Photo taken with permission

Chapter 1: Technology as Fate

Technology as Evolutionary Fate

George Grant

Technology, the Canadian philosopher George Grant once wrote, is "a fate."[1] In this chapter, we shall examine that view, and in so doing arrive at both a definition and a *"typology"* of technology. Grant is one of the most colourful figures in the past half century of Canadian academe. Born in 1918, he would, if still alive today, be nearer the age of the grandfathers than the fathers of most readers of this book. He was a passionate, complex man, the sort of professor remembered years after a course has ended. Students in his classes would become mesmerized by the ever-lengthening tip on the cigarette he perennially smoked during lectures, waiting for the inevitable cascade of ash down the front of his jacket. Yet these entertainingly sloppy habits only accentuated by contrast Grant's clear and far-reaching mind. A former student recalled for CBC Radio:

> I still remember his first entrance into class. He came walking in with a rather dishevelled outfit, but then when he turned to the class, he spoke with tremendous clarity, dignity, almost defiance.

I sort of thought of him as a member of an aristocracy, most of whose members had been recently guillotined. He was the remnant, as it were.[2]

Grant's strong religious faith explains much of his thinking. He was a pacifist who, moving to England to take up a Rhodes Scholarship in the fall of 1939 just as World War Two was breaking out, served first in an ambulance unit and then as an air raid warden in London. In heeding his conscience, Grant accepted many of the risks of combat service with none of its glamour. After the war, he took a job teaching philosophy at Dalhousie University, but it was unusual within university philosophy departments to encounter Grant's blending of philosophy with a Christian commitment that extended far beyond Sunday church attendance. He made enemies and became frustrated in seeking to pursue his own rather unique form of humanistic-Christian philosophy in the university system of the late 1950s. Even faculties of arts were becoming increasingly research oriented in what he saw as a busy but un-profound "publish or perish" style. In 1961 he joined the newly founded religion department at McMaster University, which provided a congenial academic home for much of his career.

When Grant said at the outset of his 1986 book *Technology and Justice* that "technology is fate," he seemed to be making a spiritual point about the nature of human beings. We are by nature complex mixes of reason and impulse who cannot resist experimenting with technology and therefore should not be surprised to encounter both desired and undesired consequences. To view technology as a fate is a liberating viewpoint even though it does not initially sound so, for the issue of ultimate causation gets set to rest: the problem of whether the growth of technology has determined the evolution of societies around the world, or alternatively whether the evolution of society caused the great increase in technology. Both are true, but within a process extending back through vast stretches of time. It is not that causality in the shorter time scale is without interest — repeatedly in this book such issues shall be considered — but the counsel taken from Grant is that the creation and use of technology is deeply embedded into the very definition of humanity. Our conception will be neither that of technology inexorably "driving" the forms of society, *"technological determinism"* as that is termed, nor one of technology as a mere epiphe-

nomenon of society. Instead, the orientation is one of thinking about reciprocal causation between society and technology.

Stone Age Tools and Hominid Evolution

We are not going to take the technology as fate theme in a quite so literally spiritual sense as Grant portrayed it. Instead in the next few paragraphs let us reflect back on what research by *"paleoanthropologists,"* those who study the ancient prehuman era using mainly the methods of *"archaeology,"* has revealed about the place of primitive technology in the evolution of humankind. Suppose that Grant's thought about technology as fate refers not to just one generation and moment in history, but to an evolutionary process occurring over a vast stretch of time.

To trace back to the very earliest examples of technology is to speak in millions of years. It is hard to imagine even a unit like a million dollars. What, in contrast, is a million years? It is forty thousand generations, as rough approximation, and perhaps twenty thousand lifetimes. Over such intervals, the forces of evolutionary change get to work, and are detectable thanks to the men and women dedicated to discovering the origins of humanity by spending the prime years of their lives digging by hand under the hot equatorial sun. These archaeologists search for the fossilized remains and artifacts of hominids, the near human ancestors who were more than apes but less than the modern species which emerged in the past few thousand years. Some of these remains are skeletal, others are only imprints of a hand or a foot.

Archaeologists understand ancient technology to mean "tools" in a quite everyday sense of the term. Time and again descriptions of excavation sites refer to the "tool kit" of a hominid species, referring mainly to stone versions of the hammer, the chisel, the knife, the axe, etc. Such tools are so primitive that often artifacts must be evaluated on a statistical basis to be distinguished from stone fragments formed by natural processes such as erosion.[3] The photograph at the opening of this chapter illustrates such ancient tools.

The earliest tools discovered so far are stone implements from some 2.5 million years ago. It is a rather moving irony that the forward racing technology in our own age is helping to push prehistorical time backwards with respect to dating and classification of hominid remains. "Mitochondrial DNA" analysis, for example, an important

genetics-based approach to classification, was introduced in 1987 and reoriented paleoanthropological thinking about the evolutionary tree for *Homo sapiens sapiens*, the technical name for modern humanity. Thus the study of human origins is a vibrant field not just because of discoveries in fresh dig sites, but due also to increasingly more precise analysis of remains such as the "Taung baby" discovered as long ago as 1924 and analyzed by Raymond Dart in South Africa.[4]

In this ancient primal world, what was the place of tools in the evolution of hominids? It is true that other species are tool-using. Chimpanzees, for example, will push a stick into a termite nest in the ground, extract the stick, then lick off the bugs. Such non-hominid examples are trivial, however, in comparison to the extent of tool-making and tool-using among the direct ancestors of modern human beings. Tools, most anthropologists seem to agree, are so fundamental and ancient that they are part of the evolution of humanity, part of several developments which began to separate humans from other primates. Richard Leakey, one of the legendary names in archaeological research on human origins, visualizes a scheme reproduced in Figure 1.1.

Figure 1.1: A Model of Human Evolution and Toolmaking

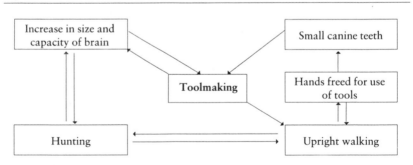

Source: Richard E. Leakey and Roger Lewin, *Origins: What New Discoveries Reveal about the Emergence of our Species and its Possible Future* (New York: Dutton, 1977), 73.

Six factors — increasing size & capacity of the brain, hunting, upright walking, freeing of hands, small canine teeth, and tool-making — functioned together as an evolutionary trajectory. Interacting with one another, these features were the ingredients in the natural selection process leading to humanity. Paleoanthropologists speak of the mutual reinforcement as "feedback" loops, the same term used to describe a sound system that amplifies its own hum. The "canine teeth" noted in the diagram are the "fangs" of animals such as tigers. The increasing use of weapons as hunting tools gradually meant that the teeth of hominids did not have to be so long. Expanding brain capacity was both a response to the mental requirements for using existing tools and the agent for enhancing the mental powers needed to make and use new ones. For a modern analogy, an anthropologist might say that the notorious complexity of the programmable VCR set is placing new demands on the contemporary human brain. But to recall the spans of time involved in evolutionary change, it would hastily be added that hundreds of generations will be required before brain chemistry shifts toward any detectable augmented skill in handling VCRs, even assuming that is an adaptation enhancing the survival capacity of the species!

Hunting is skilled work, as cat owners who have watched their pet roaming the backyard looking for prey know. That is why the Leakey scheme sees predatory activity as tied up with brain capacity. It was in part because hominids were evolving as upright walking, "bipedal" beings that their hunting had to rely on tools and brains rather than the instincts, speed, and agility of other species. The other primates, such as monkeys and even the more genetically hominid-like chimpanzees and gorillas, were not evolving into full bipedalism. Anthropologists place much stress on the growth of the open grassy "savanna" in Africa ten million or more years ago. The species that led to our own were the most adventuresome in coming out of the trees and into the open, bipedal, with "hands freed for use of tools," as Leakey remarked (73).

George Grant might not entirely have approved of transforming his notion of technology as fate into an argument based on Darwinian evolution. The contention in this chapter, however, is that the two emphases do not have to be in contradiction, and the anthropological approach helps accommodate Grant's somewhat mystical notions with a more conventional perspective.

Technology as Sociological Fate

Grant sensed what was a key point for sociologists: that the world which every person is born into and raised within is another type of "fate." This is a fate more reversible, more accommodating to the free will of the individual, than the sociobiological fate shaped by evolution. Nevertheless this "fate" too is a powerful force joining technology and human beings. Due to our long history as tool users, we and our ancestors back for countless generations have all been born into a technological environment. Grant recognized this legacy when writing in *Technology and Justice* of technology and the scientific mind-set as "civilizational destiny," meaning "the fundamental presuppositions that the majority of human beings inherit in a civilisation, and which are so taken for granted as the way things are that they are given an almost absolute status" (22).

I do not refer to fate as though to imply that nothing is left to be studied within the field of technology and society, for that is not so. The theme of fate is merely to set a beginning stage or context. This fate does not forbid humans to exercise choices and make decisions about technology. It does, however, mean that those choices and decisions have ultimate boundaries. From our evolved nature, it is impossible to imagine some wholesale renunciation of technology by humankind. Further, a recurring theme throughout our deliberations in the chapters to follow is the constraint imposed by existing social arrangements and conventions, especially those involving multiple facets of economic differentiation between people within nations and between nations themselves.

Tools, Technology, and Technique

It is casting a broad net to include stone age tools in the same category as the latest computer used in Canadian businesses or homes. Yet the two have essential features in common. Looking back to the simplest examples of technology can clarify thoughts for the analysis of the most complex forms. The archaeologists with first hand knowledge of stone age implements had no doubt they were discovering forms of technology, and we shall share that position. Indeed the opening words in Leakey and Lewin's *Origins*, the book already noted above, read: "

... a primitive human picked up a water-smoothed stone, and with a few skilful strikes transformed it into an implement. What was once an accident of nature was now a piece of deliberate technology ..."(8).

The reference to action being "deliberate" is for most students of technology a crucial definitional point. Technology, as the term will be used in this book, is something that is used with deliberation, for a purpose. These purposes are *"cultural,"* meaning that they are part of a body of knowledge passed from person to person in a society, in a chain stretching from one generation to the next. To understand what is cultural, consider what is non-cultural. If an insane person talks back to a radio, believing it to be a living thing, this troubled person is not participating in the culture of present society; in an argument with the plastic box filled with electronic circuits and components, the radio is not being used as technology. So strong is the consensus about the meaning of most technology that any serious disagreement about such meaning could in fact be a criterion for judging sanity.

Technology is, in addition, based on knowledge of some form. The stone age hominids may have been highly superstitious and mystical beings, full of what we might regard as non-scientific knowledge. Tool use, however, must have been part of the most rational portion of the psyche of these beings. That can be claimed because using tools is so firmly embedded in the multimillion year process of evolution, the succession of trials and errors in which successful adaptations to the environment advance into the next generation. It was the brute rationality of evolution that caused such outcry in Darwin's time (see Chapters 9 and 10 in *Charles Darwin*, Further Readings). Admittedly, it would be naive to suppose that hominids were experimenting with lifestyles deliberately calculated to propel them toward full humanity. Surely they were simply looking for the food and shelter to remain alive. Tools and techniques for living must have been selected for all kinds of reasons including faith and chance. It was nature, not the hominid, which was so highly rational, for nature rewarded hominids when they did stumble in the right direction.

Technological knowledge may at times be "wrong" or based on inaccurate or misunderstood principles, but increasingly over history it has been derived, justified, and advanced by experimentation. Measurement of results occurs in some form, if even impressionistically, and ideas within science are applied to all sorts of problems, some-

times quite unrelated. Time and again, as Chapter 2 shall illustrate, a technological advance comes about from realizing how a scientific principle or an engineering achievement in one domain has implications for another.

Here is a summary definition of *"technology"* proposed by the noted American sociologist Wilbert Moore and to be adopted herein:

Technology is best understood as the application of knowledge to the achievement of particular goals or to the solution of particular problems.[5]

This quite abstract definition of technology does not mention hardware, machines, and the material realm in general. Some students of technology and society would prefer a definition making specific reference to technology as necessarily including machines, meaning that if there is no hardware, it's not technology.[6] The difference perhaps is disciplinary in part: sociologists likely favour a definition like Moore's (above), while historians and those with a more engineering-oriented background are impressed with the physical traces of technology. The sociological definition makes it easier and more logical to conceive of technology as a generic phenomenon related to all other facets of society. It includes computer software as technology, for example, whereas the more restricted definition might not.

"Technique" shall be understood as referring to some of the particular pieces of knowledge that often form part of a technology. One takes ski classes to learn a technique for parallel turns. The technology of Alpine skiing includes not just the technique for turning, but kinetic knowledge about weight transfer, the mechanics of cambered edges, and, in this particular example, much expensive hardware such as boots, skis, poles, and clothes. In some writing, however, especially in European work such as Jacques Ellul's famous book *The Technological Society* (see Further Readings), technique is synonymous with technology.

It is consistent with a view of technology as a defining feature of humanity to encounter reference to how such tools and techniques are a deeply embedded aspect of all human cultures. "Technology is not simply the ensemble of things and machines, it is the character of modern life," as one contemporary Canadian author has said.[7] George

Grant, in *Technology and Justice*, wrote of technology as a "co-penetration of the arts and sciences" (12). He had investigated the term's origin or etymology, and was impressed that technology derives from the Greek words for art (*techne*) and systematic study (*logos*). In referring to "art," Grant meant not just "high art" such as painting or sculpture, but rather the "making" of things in general. The co-penetration of arts and sciences equated bringing together knowing and making and, as Grant put it, "in each lived moment of our waking and sleeping, we are technological civilization." The primeval "fate" referred to earlier has become perhaps the dominant trait of humanity today.

The Heterogeneity of Technology

Technology has developed in so many directions over the past century that it is almost bewilderingly varied. A look at what "technology" means to the reporters of current events is one way of taking inventory of modern technology, and in Table 1.1 below appears a summary of technologies referred to in Canada's *The Globe and Mail* newspaper during the winter of 1995. Reference to "technology" was found 133 times over the period mid-January to the end of March. Technology in these sentences referred 28% of the time to specific domains such as "police technologies" (17 March, A21), "computer technology" (28 February, B1), or "reproductive" technology (16 February, A22). Nearly 50% of the sentences coded made reference to technology in a general sense such as "the new generation of technology" (25 March, D1) or "technology's blandishments" (21 January, D8), while the remaining 26% pertained to clusters such as "information technology." In 34% of the passages, technology was mentioned in the same sentence with adjectives such as "high," "new," "advanced," or "latest."

One of the difficulties in the analysis of technology and society is whether the myriad specific technologies have any common denominator beyond the most basic definition. If not, each technology should be analyzed separately, with great caution about drawing general conclusions. In the Canadian tradition, George Grant was skilled at viewing technology as a unified entity, while figures such as Harold Innis and Marshall McLuhan were more conscious of the need to tailor each analysis to the technology at hand.

How, then, can technologies be classified into a manageably small,

Table 1.1: References to "Technology" in the *Globe and Mail,* Winter 1995

	Count	Percentage
Types of Technology		
References to technology in general	61	46%
References to clusters of technology	35	26%
References to specific technologies	37	28%
Adjectives used in referring to technology		
"new"	18	16%
"high-tech" or "high-technology"	13	11%
"advanced"	7	6%
"latest"	3	3%
"modern"	2	2%
At least one of the above adjectives	39	34%

"Types" percentages based on the 133 appearances of the term "technology." "Adjectives" percentages based on modifiers used within the 114 sentences referring once or more to "technology."

logically distinct, number of categories? The construction of "typologies" starting from some key dimensions is a basic analytical tool throughout the social sciences, and is now undertaken below.

Dimensions of Technology: Material/Power/Knowledge

A typology of technology can be built by taking several of the criteria already in use within the literature on technology and society and combining them into a single scheme. Let us begin, then, with examination of these criteria separately, within sub-headings.

Material

Pre-historical times commonly carry the designation "Stone Age," and the importance of stone tools as the inaugural technology for human beings has already been stressed. It follows that the material a tool is fabricated from should be seen as a dimension of technology. A distinction between stone, bronze, and iron is standard in any historical reference book. To add the ages of steel and plastic is to represent two key materials, although not the only ones, used within the technology

of the later nineteenth and twentieth centuries.[8] Alongside stone and the various metals, forget not humble wood. Being less durable than stone, there are no wooden artifacts comparable to the stone tools of over one million years of age, yet some anthropologists think that the use of wood goes back just as far as stone tools.[9] It was wooden farm implements (the "digging stick" for seeding and the plough for preparing land for planting) that "some time before 4000 BC," according to James Burke, appeared along the banks of the Egyptian Nile and in Mesopotamia. These wooden tools held pride of place as the opening examples cited by Burke in *Connections,* which was made into a television series in the 1970s (details in Further Readings).

Stone and wood were being used as materials for technology up to 2.5 million years ago, followed much later by bronze circa 3,500 BC, and the Iron Age around 900 BC. Beginning in the mid-nineteenth century, the development of the steel industry proceeded rapidly. Primitive steel-making was achieved by 1847, but a viable steel industry relied on several key advances, three separate ones in 1856 alone (by Bessemer, Mushet, and Siemens). By 1867, Confederation Year in Canada, steel rails were in commercial production in Pennsylvania. Plastic was in commercial production by the beginning of the twentieth century, although several of the key patents for various types of plastic material date well back into the nineteenth century. An inventor named Alexander Parkes, for example, had exhibited moulded decorations made of "parkesine" in 1862.[10]

The displacement of one material for technology by another can occur with dramatic abruptness. In naval warfare, the age of wood passed with the appearance of the Confederate "ironclad" warship *Merrimac* during the American Civil War. In the Battle of Hampton Roads, the stretch of water leading to Norfolk, Virginia, *Merrimac* had on a March day in 1862 the rare military luxury of invulnerability while it attacked a helpless wooden-hulled enemy fleet. Next day, the Federal navy's own ironclad, named *Monitor,* arrived and fought the *Merrimac* to a draw. Naval battle had been revolutionized within two days, and by the end of the nineteenth century steel had become the chosen material for all shipbuilding. This, incidentally, torpedoed shipbuilding in Nova Scotia, which had stood fourth in the world during the days of wood, but lost much ground to other countries once the transition to iron and steel began.

The materials used in technology set limitations and potentialities for technological development. It was, for example, the beginning of metal working, initially with bronze, that made possible tools for shaping the first wooden ploughs. These were preconditions, a technological "trigger effect" as James Burke would phrase it, for the substitution of agriculture in place of *"hunting and gathering."* This progression in the economic organization of societies was a pivotal step for humankind, opening the way to the accumulation of wealth which, even at an early stage of history, has been a prerequisite for further technological innovation.

In this first encounter with the relationship between technology and the *"culture"* and *"social structure"* of societies, to use sociological nomenclature, we need to remember no claim is being made that the invention of the plough single-handedly caused agricultural societies. Although perhaps a *necessary* cause, the technology was not the *complete* cause. The motivation to adopt agriculture had to be present too, and that same motivation would be a factor in devising the first farm tools. The related notion that the adoption of agricultural society indirectly enabled further technological innovation is another theme that shall reoccur in later pages.

Source of Power

Power is so tied up with technology, particularly in the domain of economic production, that source of power belongs in the typology. R. A. Buchanan, a leading British authority on the history of technology, used a diagram in his book *The Power of the Machine* (see Further Readings) in which "sources of power" form the nucleus in a series of concentric circles representing "elements of technological revolution" (5). In biblical times, human power in the form of slave labour was an important source of power. We have conclusive evidence from Hollywood movies of the "spectacular, cast of thousands" 1950s genre, and also, less facetiously, from the preserved records of the Egyptian and Roman Empires. Human power may seem to bear little implication for technology, but the noted technology writer Lewis Mumford felt that the coordination of the power of massed labourers was an important step toward mechanization by machine later in history.[11] By the early Bronze Age, animals such as oxen were being harnessed as a

source of energy, initially for ploughing, according to the medieval historian Lynn White (see Further Readings). By the late 1600s, the period just before the steam engine was developed, it has been estimated that nearly 60% of all power in Europe came from animals, mainly some 14 million horses and 24 million oxen. About one-quarter of available energy took the form of combustion of wood. Water-wheels supplied an estimated 10% of energy at that time, with human power accounting for about 5% and wind on sails some 1%.

Steam had a long run, lasting to the beginning of the twentieth century, as a premier source of power. By then electricity was beginning to take over. In the United States in 1900, for example, only 5% of mechanical drive capacity was electrical. That figure rose to 25% in 1909 and then to a clearly dominant 78% by 1929, just prior to the Great Depression of the 1930s.[12]

Knowledge-Base

Even in the Stone Age it was not just the physical forms of the tools themselves that were so remarkable, for, as noted, they were in the most ancient times scarcely distinguishable from natural stone shapes. Rather, it was the realization that conscious beings who thought much as we do made these tools. Similarly, in the present day what is striking is not so much the plastic covers of objects displayed for retail in places such as Future Shop or Radio Shack nor even the sight of the electronic components inside, but the complexity of the accumulated scientific thought behind the gadgets so quickly taken for granted. The centrality of knowledge for technology is articulated in the definition by Wilbert Moore, cited above.

Such knowledge-bases are a crucial link between technology and society, for even highly "technical" knowledge is embedded within broader bodies of thought. It is only in the comparatively recent past that the knowledge-bases for technology have been partitioned off as esoteric scientific specialities. As the historian of technology George Basalla points out in a theme reminiscent of George Grant, in earlier times art and craft skills were the main knowledge-base for technology.[13] Even modern science requires knowledge such as "good lab technique" which at least in part simply means the standards of cleanliness expected of any well run restaurant. Another example of the wider

knowledge-base for modern technological research is office management skills in handling grant funds and the coordination of personnel. Scientific research is, indeed, often most vulnerable where the broader knowledge-base has been ignored. For example, the Berkeley team who developed the Mitochondrial DNA technique were criticized, justifiably social scientists would feel, for a casual and improvised sampling plan for obtaining placenta samples for different races (see Brown, 98-104, in Further Reading). They were tracing genetic commonalities between the peoples of each continent, yet sampled from the San Francisco area because this convenient home base is ethnically diverse!

In classifying types of knowledge-base for technology, a watershed occurred between mechanical technologies founded on the Newtonian physics of the seventeenth century, and electronic technologies rooted in the twentieth century physics of Albert Einstein. The typical components of mechanical technologies are wheels, gears, valves, and cylinders. The water-wheel is an example still impressive, its antiquity notwithstanding. The forces at work on a water wheel command our attention from the moment the rushing water of the "mill race" pushes the giant wheel into motion, causing the axle to rotate. The photo facing Chapter 2 illustrates this. It is unmistakable that the rotation is harnessed and redirected with gears and drive belts. Electronic technology, like its mechanical predecessors, obeys laws of physics, and indeed social scientists describe any social theory modelled closely along the natural science laws of orderly forces as "mechanistic." But the forces behind electronic technology are hidden, miniaturized, and instantaneous in comparison with mechanical technologies. One observer after another has felt that somehow these distinctions have a cumulative and diffuse impact on a society. As far back as 1934, for example, Lewis Mumford had surveyed the history of technology and found that "in the seventeenth century the world was conceived as a series of independent systems" while "today ... the world has conceptually become a single system."[14]

Some of the boldest, most original, some would say far out, ideas on the consequences of the mechanical versus the electronic knowledge-base for technology come from Marshall McLuhan. He is no less famous a Canadian name in the analysis of technology and society than George Grant. McLuhan, indeed, was undoubtedly the better known of the two outside this country, hobnobbing with prominent

people in the US and elsewhere. For a period in 1967, for example, he exchanged personal letters with the Vice-President of the United States.[15]

A prolific writer, McLuhan produced some eighteen books and scores of articles over a career spanning 1936 to 1980. Although trained to be a professor of English in the traditional style of looking reverently back at the old classics, as a student at Cambridge he became interested in modern literature. Later, working as a lecturer in a midwestern American college, his attention turned to popular culture as represented in media such as the comics, an interest which evolved into a lifetime study of the impact on societies of different forms of communications technology. McLuhan's name remains in the public eye even more than 15 years following his death, due largely to his sonorous aphorism "the medium is the message." The phrase still appears regularly in newspapers and magazines (e.g., *Macleans*, 13 March 1995, 68).

McLuhan's 1964 volume *Understanding Media* gives the clearest statement of this lively thinker's assessment of mechanical versus electronic technologies. We encounter the University of Toronto scholar in this book nearing the summit of his fame and intellectual powers.[16] "After three thousand years of explosion, by means of fragmentary and mechanical technologies, the Western world is imploding," read the opening words of *Understanding Media*. McLuhan must have had in mind that, up to the middle of the twentieth century, technology was constantly being deployed for expanding the frontiers of land and ocean. Consider, for a very literal example, the use of the technology of railroad transportation for the opening of the Canadian West. Possibly another aspect of McLuhan's "explosion" was the growing bulk, complexity, specialization, and speed of most mechanical technology. This sense of growth was fully realized late in the nineteenth century, becoming a point of great pride. In his book on Niagara Falls (see Further Readings), Pierre Berton reports that the construction of the first electric generator plant by Canada was described as "a triumph of human enterprise" by a journalist writing in 1892 (217).

Mechanical technologies were "fragmentary" in the sense that actions and reactions were slow enough to be amenable to analysis in sequence. "In the mechanical age now receding," wrote McLuhan, "many actions could be taken without too much concern. Slow movement insured that the reactions were delayed for considerable periods

of time" (20). There is more to it than that, however. McLuhan's ideas can be difficult due to his habit of packing several only partially related thoughts into one cryptic phrase such as "fragmentary." With this term, he also meant sense of detachment, citing as example the surgeon who performs operations without becoming "humanly involved" with the patient. By putting these last two thoughts together — mechanical technology as imparting a fragmentary quality to societies due both to time delay between action and reaction and in the sense of emotional fragmentation — we see just how abstract McLuhan's thought was. Abstract, wonderfully original, with great powers of synthesis, he was also less than totally logical at times, and seldom backed his statements with much evidence.

In all this discussion of mechanical technology, McLuhan was mindful of a contrast with the "electronic age" (viii). By the 1960s, the western world had moved well into that new "implosion" phase, which had begun to accelerate with a series of electronic knowledge-base inventions from the nineteenth century. The territorial frontiers were occupied and the outward "explosion-like" flow of population had slowed if not ceased. This certainly was so with respect to migration in Canada. Similarly, in a symbolic sense, machines were imploding, as reflected by the miniaturization of all manner of electronics from calculators to computers. Continuing with the implications of McLuhan's theory, if mechanical technology was "fragmented" then electronic technology must have been "defragmented" although that latter adjective is my wording, not McLuhan's. In contrast to mechanical technology's delays in action-reaction, with electronic technology, events occur nearly instantaneously. Consider examples in the technology of diplomatic relations; the transformation wrought by the hot line, the spy satellite, the fax machine, etc.; and how radio, television, and opinion polls by telephone can crystallize and reflect public opinion so quickly.

The more ethereal dimension of defragmentation is encountered early in *Understanding Media* with what has become a famous McLuhanism: "As electrically contracted, the globe is no more than a village" (20). The notion of the global village had already appeared in his earlier books (for example, *Gutenberg Galaxy*, 21) and referred, so McLuhan believed in perhaps a rather utopian way, to a heightened sense of mutual involvement among human beings. A raised "human awareness of responsibility to an intense degree" indeed may occur

sporadically, as when a widely publicized disaster such as famine in a Third World country precipitates mass relief donations from the public. But a generalized, sustained, enhanced sense of mutual global human involvement today compared to 100 years ago seems unlikely.

We may not agree, then, with every detail of McLuhan's analysis, but he is far from alone in stressing the distinction between technologies founded in a mechanical versus an electronic knowledge basis. A similar notion appears, for example, in Bolter's notion of each age having a "defining technology" which acts "as a metaphor, example, model, or symbol."[17]

Electricity was not the only emerging knowledge-base at the onset of the twentieth century. "Parallel to the advances that took place in electricity and metallurgy from 1870 onward were the advances that took place in chemistry," Lewis Mumford noted in his classic text.[18] By mid-twentieth century, the advanced knowledge-bases were entangling in complex ways. The early 1950s work on DNA to be discussed in Chapter 9, for example, mixed knowledge about physics, mathematics, chemistry, and biology. The "biotechnology" resulting from the DNA research has, along with applications in such varied fields as agriculture, medicine, and criminology, had implications for the most basic theoretical presuppositions throughout social science. Such research stimulated the field known as "sociobiology," a fusion of sociology and biology spearheaded by figures such as Edward Wilson (see *The Diversity of Life*, Further Readings). Looking toward the new century, it was said that "we stand on the threshold of the Biological Century" (see *The World of 2044*, 100, in Further Readings).

Alongside tool-based technologies, a social science knowledge-base with origins in ancient philosophy has accumulated over a long period. From psychology, economics, sociology, and political science, the ideas of figures such as Max Weber, Sigmund Freud, John Meynard Keynes, Jean Piaget, B.F. Skinner, and Talcott Parsons (for a very brief list of many possible names) provide a knowledge-base behind non-tool technologies for steering human thoughts and actions in such fields as marketing, education, office management, psychoanalysis, and counselling. These various theories about humanity become important, for example, when in later chapters ideas such as Taylorism in relation to work and technology are encountered.

Main Directions in Technological Evolution

Technology has an accretionary, cumulative quality that we are reminded of by thinking in terms of materials, power sources, and knowledge-bases. As the repertoire of materials for technology has expanded along with the scientific knowledge-base, ever more complex mixes of materials within technologies and increasingly exotic composite materials can be found. Within the sport of sailing, for example, old timers reared on cotton sails and wooden spars now examine specifications on carbon fibre masts and discuss Spectra versus Kevlar cloths with their sailmaker. Power sources too have accumulated and interacted with other dimensions. Figure 1.2 itemizes some of the historical progression in terms of the three dimensions of technology.

To demonstrate how typologies can lead from description into analysis, a moment's reflection should be spared for whether the "weighting" of each dimension in the material/power/knowledge scheme has shifted over the centuries. Surely it has! Up to the time of Jesus Christ, technology was mainly about materials. This shifted gradually toward increasing emphasis on power source, culminating perhaps with the perfection of the steam engine. Looking back at the twentieth century, it is not for a moment said that materials and power sources have been unimportant, but scores of authors — Daniel Bell just to name one — seem to say that the knowledge-base has become the truly essential dimension.[19] Figure 2.1, to look ahead to Chapter 2, is a time chart of technology that may illustrate movement toward the primacy of knowledge-bases.

Emphasizing material/power/knowledge as three dimensions of technology admittedly is an arbitrary choice.[20] The *purpose* of technologies, for example, is not suggested as a dimension within the typology presented. As Moore's definition (see above) rightly contends, human technology is purposeful, but examining these purposes and deciding how closely they are realized has to do with the extended analysis of technology rather than its mere classification. It is a topic for later chapters, especially given the sociological truism that actions can have both intended and unintended consequences. Indeed some have claimed that differentiating between the two essentially marks what the discipline is all about. Throughout Part II (Chapters 4 to 9),

Figure 1.2 Some Steps in Material, Power Source and Knowledge-Base for Technology

Approximate Dating	Material	Power Source	Knowledge-Base
Up to 2.5 million BC to 4,000 BC	Stone, wood	Human	Arts
3,500 BC	Bronze	Animal	
900 BC	Iron		Mechanical
By AD 1600		Wind, Water, Thermal	
late 1800s	Steel	Electrical	Electrical
early twentieth century	Plastic		Chemical
mid twentieth century		Atomic	
late twentieth century	Many new composites		Biological

examples of that ambiguity shall be encountered, ranging from the printing press to the computer.

To end this discussion of dimensionality, if purpose were the fourth dimension of technology, ethics would be the fifth, but that too is treated as an issue later (see Part III). The approach in those final two chapters will be to analyze and account for the normative or ethical positions taken by members of the public rather than to prescribe. That is work for the reader him or herself, working from the best information available.

Conclusion

Technology, we learn from Grant, is a fate, interpreted within the present pages to refer to an evolutionary path of humanity. This is consistent with anthropological research on hominid societies, and provides a stance on causal issues. Technology and society exist in a dynamic,

reciprocal relationship; they act on, and react to, one another. It is therefore not the argument of this book that liberal *"ideology"* single-handedly caused technology to evolve as it did, nor that technology alone has caused social arrangements such as capitalist society. Technology involves both knowledge and purpose: "The application of knowledge to the achievement of particular goals or to the solution of particular problems" is our chosen definition.

Notes

1. George Grant, *Technology and Justice* (Toronto: Anansi, 1986), 9. This biographic section draws from William Christian's *George Grant: A Biography* (Toronto: University of Toronto Press, 1994).
2. Louis Greenspan, on "Ideas" program of January 27, 1986. From transcript published by the CBC.
3. John E. Pfeiffer, *The Emergence of Man*, 2nd. ed. (New York: Harper & Row, 1972), 90.
4. Described in Adam Kuper, *The Chosen Primate: Human Nature and Cultural Diversity* (Cambridge, Mass.: Harvard University Press, 1994), 36.
5. Wilbert E. Moore, ed., *Technology and Social Change* (Chicago: Quadrangle Books, 1972), 5.
6. Rudi Volti, for example, in his well-respected textbook sees technology as: "a system based on the application of knowledge, manifested in physical objects and organizational forms, for the attainment of specific goals" (*Society and Technological Change*, 3rd. ed. [New York: St. Martin's Press, 1995], 6). An example of an historian's "hardware-oriented" approach to defining technology is George Basalla's *The Evolution of Technology* (Cambridge: Cambridge University Press, 1988), 30. A yet more focussed definition is Donald Cardwell's, in *The Fontana History of Technology* (London: Fontana, 1994). "Technics," which are tools based on art and craft knowledge, here are distinguished from the technologies based on "systematic or scientific knowledge" (4). Cardwell also understands technologies to be "innovations" in working production rather than pre-technological "inventions."
7. Eric Stowe Higgs, "The landscape evolution model: A case for a paradigmatic view of technology," *Technology in Society* 12 (1990), 498.

8. For example, William L. Langer, ed., *An Encyclopedia of World History*, 4th. ed. (Boston: Houghton Mifflin, 1968), 14-15. A Canadian document from 1918 reads "the present century is surely an age of steel" (see B. Sinclair, N.R. Ball, and J.O. Petersen, *Let Us Be Honest and Modest: Technology and Society in Canadian History* [Toronto: Oxford University Press, 1974], 156). Robert J. Weber, who insists that "civilizations are defined by the materials they use," terms ours "the synthetic age" (*Forks, Phonographs, and Hot Air Balloons: A Field Guide to Inventive Thinking* (New York: Oxford University Press, 1992), 167.

9. See for example, John E. Pfeiffer, *The Emergence of Humankind*, 4th. ed. (New York: Harper and Row, 1985), 74 and 116-17.

10. Mary Maxwell's *Human Evolution: A Philosophical Anthropology* (London: Croom Helm, 1984) has a useful "timetable of evolutionary events" in Appendix I. Dating of the ages of steel and plastic appears in Colin Chant, ed., *Science, Technology and Everyday Life 1870-1950* (London: Routledge, 1989), and Kenneth Chew and Anthony Wilson, eds., *Victorian Science and Engineering Portrayed in* The Illustrated London News (London: The Science Museum, 1993), 17. Also, Richard Shelton Kirby, S. Withington, A.B. Darling, and F.G. Kilgour, *Engineering in History* (New York: McGraw-Hill, 1956), Chapter 10. Nathan H. Mager, *The Kondratieff Waves* (New York: Praeger, 1987), 70, refers to and dates the age of electricity, the age of chemicals, the age of plastics, etc. The importance of electricity is underlined in Zbigniew Brzezinski's phrase "technetronic" (technology and electricity), which is part of the title of his book *Between Two Ages: America's Role in the Technetronic Era* (New York: Viking, 1970).

11. Lewis Mumford, *Technics and Civilization* (New York: Harcourt, Brace & World, 1963 [1934]), 41.

12. Pre-steam figures from Fernand Braudel, *Civilization and Capitalism. Volume 1, The Structures of Everyday Life: The Limits of the Possible* (London: Collins Fontana, 1981), Chapter 5. Electrical estimates from Alan I. Marcus and Howard P. Segal, *Technology in America: A Brief History* (San Diego: Harcourt Brace Jovanovich, 1989), 228-29.

13. Basalla, 27.

14. Mumford, 369.

15. Matie Molinaro, Corinne McLuhan, and William Toye, eds., *The*

Letters of Marshall McLuhan (Toronto: Oxford University Press, 1987), 342, 349.

16. Marshall McLuhan, *Understanding Media: The Extensions of Man* (New York: Signet, Mentor, 1964). In Philip Marchand's biography *Marshall McLuhan: The Medium and the Messenger* (New York: Ticknor and Fields, 1989), the chapter covering the years beginning in 1964, when *Understanding Media* was published, is entitled "Canada's Intellectual Comet." For further background on McLuhan, see Further Readings for chapter 7.

17. J. David Bolter, *Turing's Man: Western Culture in the Computer Age* (Chapel Hill: University of North Carolina Press, 1984), 11.

18. Mumford, 228.

19. Daniel Bell, *The Coming of Post-Industrial Society: A Venture in Social Forecasting* (New York: Basic Books, 1973) is the classic source on the ascent to primacy of theoretical knowledge.

20. Henceforth, these terms shall be italicized when I want to remind the reader that these are three dimensions of technology adopted for special attention in this book.

Further Readings

For connections between evolution and tool-making:

Aitken, A.J *Science-Based Dating in Archaeology.* London: Longman, 1990.

Bowler, Peter J. *Charles Darwin: The Man and his Influence.* Oxford: Basil Blackwell, 1990.

Brown, Michael H. *The Search for Eve.* New York: Harper and Row, 1990.

Washburn, Samuel and Ruth Moore. *Ape Into Man: A Study of Human Evolution.* Boston: Little, Brown, and Company, 1974. See especially the chapter entitled "Tools Makyth Man."

Wills, Christopher. *The Runaway Brain: The Evolution of Human Uniqueness.* New York: Basic Books, 1993.

For alternative definitions of technology:

Ellul, Jacques. *The Technological Society.* New York: Vintage Books, 1967. He says: "The term technique, as I use it, does not mean machines, technology,

or this or that procedure for attaining an end. In our technological society, technique is the totality of methods rationally arrived at and having absolute efficiency ... in every field of human activity" (xxv).

McGinn, Robert E. "What Is Technology?" in Larry A. Hickman, ed., *Technology as a Human Affair.* New York: McGraw-Hill, 1990, 10-25.

Ridington, Robin. "Technology, world view, and adaptive strategy in a northern hunting society." *Canadian Review of Sociology and Anthropology* 19 (November, 1982), 469-81.

Teich, Albert H. *Technology and the Future.* 6th.ed. New York: St. Martin's Press, 1993. Compare, for example, passages on pp. 76, 198 and 267.

For an historical approach to technology:

Berton, Pierre. *Niagara: A History of the Falls.* Toronto: McClelland and Stewart, 1993., Chapter 7.

Buchanan, R.A. *The Power of the Machine: The Impact of Technology from 1700 to the Present.* London: Penguin [Viking], 1992.

Burke, James. *Connections.* London: MacMillan, 1978.

White, Lynn. *Medieval Technology and Social Change.* London: Oxford University Press, 1962.

For future directions and sociobiology implications:

Sheffield, Charles, Marcelo Alonso and Morton A. Kaplan, eds. *The World of 2044: Technological Development and the Future of Society.* St. Paul, Minnesota: Paragon, 1994.

Wilson, Edward. *The Diversity of Life.* Cambridge, Mass.: Belknap Press of Harvard University, 1992.

Working example of a water wheel, originally dating from circa 1820. *Museum of Science and Industry*, Manchester, UK. Note resemblance to early steam engine in photo facing chapter 4.

Chapter 2: Creating Technology

Three Environments for Technology

Overview

Let us think about three environments within which technology develops: the geographical, the sociobiological, and the social. These are outlined diagrammatically in Figure 2.1 below. Viewed on the vast time scale of human evolution, the geography of the earth has changed as startlingly as technology itself. At one time, some 225 million years ago, the world was a universal continent. The subsequent splitting of the land masses had implications for human development, including migration from Siberia to Alaska and down the Canadian Pacific coast. Anthropologists think that most of the North American native peoples arrived from a land or ice connection across the Bering Strait around 11,800 years ago, with the possibility of an earlier influx 30,000 years ago. Between those dates, ice conditions made passage down from Northwestern Canada impossible. The cycle of ice ages approximately every 100,000 years is schoolbook knowledge, but the "spike" in average temperatures over much shorter intervals is less frequently realized. Europe had a "little Ice Age" in the late thirteenth and fourteenth cen-

turies, a time when winters turned measurably longer and colder, food was scarce, and the need for heating fuels became acute.[1]

Sociobiology is included in Figure 2.1 to denote how the use of technology, "tool using" in paleoanthropology, is part of a defining statement about humanity that differentiates us from other living creatures. It is a way of writing into the diagram the "technology as a fate" theme from Chapter 1.

Social environment, in contrast, refers to human arrangements that are not necessarily biologically inevitable. This lower portion of Figure 2.1 amounts to a crash course in Introductory Sociology. "*Institutional domain*" makes reference to the arrangements made by a society to meet various types of collective need. Every society, for example, has an economy of some sort to meet the basic and not so basic material needs of a people. The Canadian society moving into the twenty-first century would be described as a postindustrial or information economy, in contrast to the manufacturing-centred industrial economy of earlier in the twentieth century. The institutional domains cross-cut one another; the family, for example, performs the function within the economy of providing for the food, shelter, and development of children, adolescents, and even young adults who may, for example, attend university while being financially supported by parents. Sociologists would point out, however, that a successful family is more than merely a banking arrangement; it also answers important emotional needs. Analysis of institutions should also acknowledge that the linkage from need to institutions is complex. The assumptions and procedures in place within a given institutional domain may serve particular interests, but not the general interest. A simple example from the mid 1990s might be the high profit-taking that the society tolerates from the chartered banks. Institutional human arrangements in addition often have considerable inertia, remaining in place with little change even if the originating "need" has shifted.

Sociology distinguishes between social structure and culture, terms already encountered in Chapter 1. These words have a somewhat complex meaning about which social scientists sometimes disagree among themselves. For present purposes, social structure refers to the ways people and groups of people are organized and connected within a society. A homogeneously middle class suburb, for example, implies a different social structure than a downtown area where the very rich may

Figure 2.1: The Environments of Technology

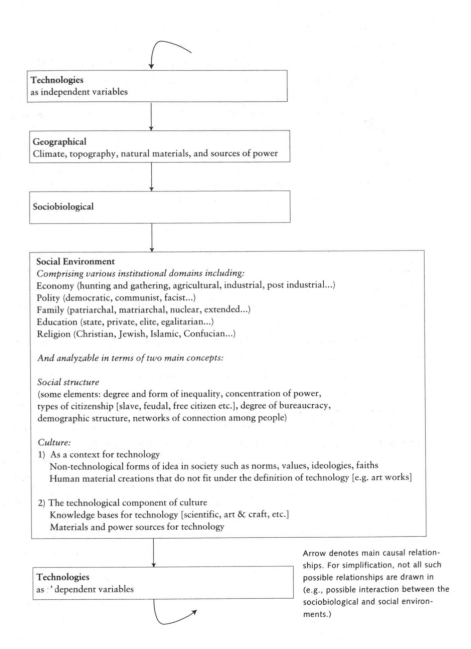

| Technologies |
| as independent variables |

| Geographical |
| Climate, topography, natural materials, and sources of power |

| Sociobiological |

Social Environment
Comprising various institutional domains including:
Economy (hunting and gathering, agricultural, industrial, post industrial...)
Polity (democratic, communist, facist...)
Family (patriarchal, matriarchal, nuclear, extended...)
Education (state, private, elite, egalitarian...)
Religion (Christian, Jewish, Islamic, Confucian...)

And analyzable in terms of two main concepts:

Social structure
(some elements: degree and form of inequality, concentration of power,
types of citizenship [slave, feudal, free citizen etc.], degree of bureaucracy,
demographic structure, networks of connection among people)

Culture:
1) As a context for technology
   Non-technological forms of idea in society such as norms, values, ideologies, faiths
   Human material creations that do not fit under the definition of technology [e.g. art works]

2) The technological component of culture
   Knowledge bases for technology [scientific, art & craft, etc.]
   Materials and power sources for technology

| Technologies |
| as ' dependent variables |

Arrow denotes main causal relationships. For simplification, not all such
possible relationships are drawn in
(e.g., possible interaction between the
sociobiological and social environments.)

rub shoulders with the poor. *"Bureaucracy"* is a form of social structure in which people relate to each other in a clearly defined, formal manner. It is, in a theme pursued in Chapter 8, an innovation in social structure when people become connected with distant relatives due to having e-mail addresses or home pages on the Internet.

Culture consists of both material things and ideas. The "things" refer to all recurring human material creations, whether it be a stone club or an aircraft carrier, a mud hut or the CN Tower. Many forms of human thought are also considered cultural, some examples of which are offered in Figure 2.1.

The main focus of this chapter will be the reciprocal impacts of social environment and technology, but first let us consider more explicitly the logic for sequencing the environments in the order of geographical, sociobiological, and social.

From Geography to Sociobiology to the Social Environment

There are some good reasons why scientists suspect a link between geography and sociobiology. William Calvin (see Further Readings), an American neurobiologist, makes an especially forceful case for the connection between temperature fluctuations over the past several hundred thousand years and the evolution of human intelligence. "I suppose," Calvin says, "that we could have quadrupled our brains without the virtues of winter, but I'll bet that it would have taken forever. So much brain enlargement in a mere 2.5 million years is awfully quick by the standards of evolutionary biology. Yet winter once a year, an abrupt climate change every few millennia, and an ice age every hundred thousand years will speed up things ever so nicely" (xvi). Calvin sees the geography of climatic change as the ultimate explanation driving all the other components of sociobiology, that is, such aspects of human evolution as bipedal walking, increasing brain size, and tool use. As the quotation suggests, geography was an accelerator that put the evolution in brain size onto "a fast track" (53). A seasonal alternation of summer and winter places increased adaptive demands on a species. While the simple gathering of fruits and berries might suffice for summertime survival, hunting skills would probably be needed to survive a winter. The longer climatic cycles had the same effect, according to Calvin, but on an "amplified" basis. For example, he believes that in

the warmer periods between ice ages a "juvenilization" tended to appear in the "gene repertoire," meaning earlier puberty and age of adulthood (128, 133). Since growth slows after puberty, an evolutionary movement toward such modern human features as small canine teeth would occur. It is not our task here to assess the accuracy of such bold scientific speculations, but only to become conscious of how varied the effects of geography on human development might be.

Awareness of the impact of geography on the social environment is part of Canada's historical legacy from the last century. Not much more than 100 years ago, the formative influence was "survival" on the frontier, the term adopted by Margaret Atwood in her famous commentary on Canadian literature (see Further Readings). By the end of the twentieth century this country had become predominately urban and postindustrial in economic form. Canadians had "lost some of their geographical consciousness," as the author of a noted geography text once wrote.[2]

The intention in Figure 2.1 is not to present the social environment as a simple mathematical derivative of geography or sociobiology. The fallaciousness of such a view is the lasting lesson of cultural anthropology. It was in the 1920s and 1930s that anthropologists such as Margaret Mead and Ruth Benedict began to show how greatly cultures differed even among primitive peoples still living very close to the geography of land forms, vegetation, and climate. In addition, common themes occur in the evolution of societies situated in very differing geographical environments.

For example, as a civilization develops, institutional domains become increasingly differentiated and complex, sociology being the discipline that particularly emphasizes that transition. The pioneering French sociologist Emile Durkheim made what is probably the best known statement about social complexity, basing his analysis on the growing "*division of labour.*" It "was only at the end of the eighteenth century," Durkheim felt, "that social cognizance was taken of the principle." Publishing his famous book on the social consequences of the division of labour in 1893, Durkheim had witnessed what was then a comparatively recent development. His study worked out how societies retain a sense of common purpose and nationhood in the face of ever-increasing complexity both in the economic division of labour and "in the most varied fields of society."[3]

Technology and the Social Environment

Technologies are, as noted repeatedly now, both part of and products of the social environment. That is a hard point to represent diagrammatically, but an attempt is made in Figure 2.1. Technologies are "part of" the social environment in the sense that the technological repertoire forms part of a society's culture, represented in Figure 2.1 by the notation "the technological component of culture." If culture indeed includes material artifacts having shared meaning within a population, then that must logically include technological hardware of all sorts.[4] And "non-machine" technology such as computer software is equally cultural since shared ideas also fit into this basic sociological concept.

Technologies as variables "dependent" on social environment (Figure 2.1) refers to the way in which the institutional domains and their related social structural and cultural context shape technology. Although unfamiliar words may be appearing here, the ideas are not difficult. Indeed, quite a simple notion is being phrased in the most general possible way. It was social environment shaping technology, for example, when the US mounted the Apollo moon landing program in the 1960s. The 1969 landing was made not only because it had become technically feasible, but because it suited the foreign policy needs of the US during the Cold War, and the vested interests of many large firms, not to mention the ego needs of the American population. Exploration of the planet Mars today is as much an issue of political will as technological capability.

The following quotation from William Leiss expresses another way in which social environment is linked to technology: "(T)echniques are almost always combined with class, status, and role determinations that specify who can perform the operations associated with techniques and under what culturally legitimated conditions. This combination is what I call a technology ... " (see Further Readings, 31). Although not mentioned in this quotation, the gender relations within a society are also a key aspect of social structure bearing on technological creation. The odds of becoming a James Watt, an Alexander Bell, or part of a modern research team working for William Gates at Microsoft are not evenly spread across the population. All the names in the preceding sentence refer to males, for the reason that the social structure of western societies has historically been biased that way. The onset of mass

entry of women into science and technology is indeed one of the notable social structural changes of the 1990s.

Reference to "technologies" in Figure 2.1, rather than to "technology," is deliberate. An emphasis on the heterogeneity of technology will be recalled from Chapter 1. The "technologies" in the present diagram mix various *materials, power sources,* and *knowledge-bases,* with the recipe constantly evolving over time. A time chart for a sample of key inventions over history appears in Figure 2.2. The list will serve as a reference chart for many of the technologies discussed in both this and later chapters. The earlier items listed all receive extended discussion in any history of technology. From 1900 on I exercised selectivity to avoid too massive a list, placing emphasis on defining and exemplary technologies for the twentieth century.

That it is far easier to decipher the causes and consequences of individual technologies than of "technology" in general is, reiterating a theme from Chapter 1, part of the stance toward technology being adopted in this book. It is not to deny the reality of "defining technologies" of an age, nor the usefulness of technology as an adjective, such as in *The Technological Society,* the English language title of the influential book by Jacques Ellul (see Chapter 1). Sociologists will always wish to reach conclusions about technology as a general phenomenon, but such attempts most succeed when cross-fertilized with the approach, more characteristic of historians, involving case studies of individual technologies. A work such as Alvin Toffler's *The Third Wave* (Further Readings) is vulnerable, despite the merits of its sweeping vision, to the charge of stereotyping technology.

To conclude this orientation to Figure 2.1, were it not for the risk of damage to the handsome volume resting in the reader's hands, it would be suggested that the page containing Figure 2.1 be folded top over bottom into a circle so that the entries for "technologies as dependent variables" and "technologies as independent variables" overlapped. Doing so would restore the proper symmetry to a discussion of technology and society, putting the necessary conceptual warp into physical expression. Technologies are so much a part of, so embedded into, societies that they constantly are both affecting the society while being affected by it. Any two dimensional scheme on paper has difficulty showing that.

While retaining in mind this duality of technology as effect and as

Figure 2.2: Time Chart of Major Events in the History of Technology

Name of Technology (inventor if known)	Date	Commentary	Source
Plough	8000 to 4000 BC	Basic tool for agriculture (wooden later metal-tipped)	(i)
Stirrup	2nd c. BC (India)	Implications for warfare, therefore for social structure	(ii)
Gunpowder	10th c. (China) 1318 (west)	Step toward thermal power	(iii)
Horse collar	pre-AD 1000	Increased "horse power"	(iv)
Clock, weight-driven	1286	New conception of time	(v)
Printing press (Gutenberg)	1440s	Democratized access to knowledge Various "McLuhanesque" effects	(vi)
Newcomen steam engine (Thomas Newcomen)	1712	Important step in power sources for Industrial Revolution	(vii)
Flying shuttle (John Kay)	1733	Mechanized weaving, early example of factory production, created need for mechanical spinning machine.	(viii)
Watt's steam engine (James Watt)	1769-patent	Main power source for Industrial Revolution, one of the defining mechanical technologies	(ix)
Spinning Jenny (Boulton, from work by Arkwright)	1770	Mechanical Spinning	(x)
Cotton gin (Eli Whitney)	1793	Mechanized separation of cotton fibres from seed, reducing the demand for slave labour in the American South	(xi)
Electric Telegraph (Samuel Morse)	1847	First long distance instant communication	(xii)
Transatlantic cable	1866	First step toward the electronic "global village."	(xiii)
The telephone (A.G. Bell)	1876	A new mode of social network, key electric knowledge-base technology	(xiv)
Steam turbine (Parsons)	1884	Integrated steam and electric power	(xv)
Gasoline internal combustion engine (Daimler)	1883	Became power source for the automobile and the airplane	(xvi)

Safety Bicycle (Starley)	1885	Democratized travel, especially for women	(xvii)
Radio (Marconi)	1896-patent	Continued communications revolution	(xviii)
Airplane (Wright brothers)	1903	Flight in a machine heavier than air	(xix)
Penicillin (Alexander Fleming)	1928	Key medical discovery	(xx)
Project Manhattan	1943-45	First atomic bomb	(xxi)
ENIAC	1946	Early mainframe computer	(xxii)
Transistor (Bell laboratories)	1947	"Birth of the age of electronics"	(xxiii)
Structure of DNA (Watson and Crick)	1953	Opened the way to biotechnology	(xxiv)
Sputnik	1957	First satellite in space, USSR	(xxv)
The genetic code	1966	prerequisite for genetic engineering of DNA	(xxvi)
Microprocessor* (Noyce/Kilby/Hoff)	1971	Complex circuits printed on to one silicone chip — thus miniaturization of electronics, basis for "smart cards"	(xxvii)
Mitocondrial DNA analysis	1987	Reinterpreting the human family tree	(xxviii)
"Convergence of modes"*	1990s	E.g., computer messages sent via telephone lines	(xxix)
"Dolly" the cloned lamb	1997	Highly publicized genetic engineering experiment	(xxx)

(i) James Burke (reference in Chapter 1), 9; (ii)White (Ch. 1), 14; (iii)Cardwell (Ch. 2), 11; Basalla (Ch. 1), 171; (iv)White, 61-64; (v) Cardwell, 13; (vi) Mumford, 439; (vii) Basalla, 95; (viii) Cardwell, 76; (ix) Cardwell, 85-5; (x) Cardwell, 95; (xi) Richard C. Dorf, *Technology, Society and Man* (San Francisco: Boyd and Fraser, 1974), 47; (xii) Kirby et al., (Ch. 1) 340; (xiii) J.J. Brown (Ch. 3), 115; (xiv) Pier L. Bargellini, "An engineer's review of Antonio Meucci's pioneer work in the invention of the telephone." *Technology in Society* 15 (1993), 410; (xv) Cardwell, 171; (xvi) Frederich Klemm, *A History of Western Technology* (London: Allen and Unwin, 1959), 342; (xvii) Cardwell, 199; (xviii) Cardwell, 186; (xix) Dorf, 47; (xx) Brian Bunch and Alexander Hellemans, *The Timetables of Technology* (New York: Simon and Schuster Touchstone, 1993), 367; (xxi) Marcus (Ch. 1), 304-09; (xxii) Marcus, 298; (xxiii) Diebold (Ch. 2); (xxiv) James Watson (Ch. 9); (xxv) Marcus, 323; (xxvi) Francis Crick, *What Mad Persuit: A Personal View of Scientific Discovery* (New York: Basic, 1988), 3; (xxvii) Marcus, 332; (xxviii) Michael Brown (Ch.1); (xxix) Bunch and Hellemans, 450; (xxx) *Globe and Mail*, 1 March 1997, A8

*See figure 8.1 for a more detailed timeline for computers.

cause, the focus in the pages below is with the creation of technology, beginning with the part played by economic surplus. Chapter 3 then examines the spread of technologies, to complete the "basic relationships" part of this text, in which technologies are being approached as the "dependent variables" denoted in Figure 2.1

Economic Surplus as Precondition for Technological Development

It seemed at first an incongruous choice when James Burke began *Connections* with the lowly plough (Further Readings, Chapter 1). A technology fabricated initially just from wood, using human or animal power and taking the same knowledge base as a Grade Six science project on growing plants from seeds, seems unremarkable. Yet as Burke argued, "(t)his simple implement may arguably be called the most fundamental invention in the history of man [and woman], and the innovation that brought civilization into being, because it was the instrument of surplus" (9).

Economic surplus, the condition of producing more than what is required merely to subsist, allows some people the time and provisions to think, to experiment, and to build, and also permits more than the minimal population densities possible in a hunting and gathering economy. In an economy based on hunting, surplus is hard to accumulate in part due to the short "shelf life" of meat; refrigeration, apart from by natural ice, did not exist until 1834.[5] "Gathering," referring both to the scavenging of kills by animal hunters and to the collection of naturally growing foods such as fruits and berries, succeeds only with the lowest of population densities. It is one of the key sociological lessons about technology formation that innovation is a social, and relatively high density, activity, meaning that progress comes from one person reacting to another's ideas.

In simpler times technologies could sometimes come about as part of casual trial and error in the workplace, but by the twentieth century developments increasingly necessitated investment of surplus capital, risked over long spans of time from conception to production. Machines were becoming larger, requiring high costs in materials and in plant space for assembly. Technology, like much of the rest of life, was becoming bureaucraticized. Invention began to move from the domain of the lone, often impecunious, visionary into the salaried per-

manent career conducted within university, government or corporate organizations with rules, offices, and hierarchies. One marker of bureaucratization was the development of mechanical and engineering associations. "Mechanics' Institutes" began appearing in British cities by the 1820s. In the US, specialized engineering associations formed, beginning with the American Institute of Mining Engineers (1871), followed by the American Society of Mechanical Engineers (1880), the American Institute of Electrical Engineers (1884), the American Institute of Chemical Engineers (1908), and the Institute of Radio Engineers (1912). The Canadian Society of Civil Engineers, founded in 1887, was concerned with such issues as "increased status, more self-control, and a clearer public identity." Engineering had begun to enter the post-secondary education curriculum, initially in Canada at King's College (now the University of New Brunswick) in 1854 and by the 1870s at institutions such as McGill, l'École polytechnique de Montréal, and the School of Practical Sciences (later part of University of Toronto).[6]

Research on the transistor illustrates the bureaucratization of invention. This key step toward reliable and miniaturized electronics began at the Bell Laboratories through an accidental discovery in 1939 about the properties of silicone. By 1948, after a million dollars worth of research, the transistor had been patented. It took another four years of equally intense development work before the transistor began to appear in telephones. The Bell scientists were salaried employees rather than independent entrepreneurs. Their employment contracts required them to transfer to Bell, at nominal or nil charge, all patent rights to discoveries. Any chance finding was, as standard procedure, passed up the chain of command for consideration at the top. It is not that Bell Labs were inappropriately authoritarian or stifling of creativity. The Lab was simply a bureaucracy, obviously a productive one which "encouraged individual initiative within a team framework," as Diebold relates in his account of the transistor (Further Readings).

The economist John Kenneth Galbraith, an Ontario farmboy who became one of our country's most notable academic exports to the United States, once illustrated the growth of investment-based technology using the following comparison. In 1903, the Ford Motor Company was formed on June 16, with the first car appearing on the market by October. The company had issued stock for $100,000. Sixty

years later, it took three-and-one-half years to get from design to production, with costs of $50 million to produce a new model.[7] Even over a much briefer historical period the same intensification of investment has taken place. The first fabrication plant for the Intel microprocessor, for example, cost $1 million in 1971. Only 25 years later the cost of designing a new chip was being set at $2 billion (*Globe and Mail*, 18 November 1996, A8).

Capitalism: The Technology Accelerator

Capitalism refers to an arrangement whereby individuals have a legal right to private ownership of land, buildings, equipment, and economic resources in any form including stocks, bonds, investment certificates, and cash itself. The primary connection with technology is, then, that capitalist activity is all about the accumulation of surplus by private citizens. Lewis Mumford, however, suggested another point in *Technics and Civilization*. Capitalism, he argued, "brought the new habits of abstraction and calculation ..." People under capitalism became "abstracted" in the sense that they were valued not as "personalities" or "souls," as they would in at least the most perfect type of precapitalist society, but now simply as commodities useful as labourers or as consumers. The valued attribute under capitalism was profit-making potential. This abstraction left an ethical vacuum: it has become morally unproblematic to deploy machines in replacement of workers or to displace workers into less skilled and lower paid positions. Another thread of Mumford's argument about capitalism was that "the abstractions of capitalism preceded the abstractions of modern science...."[8] This aspect of the innate nature of capitalism, he was arguing, stimulated the scientific thinking which built up the *knowledge-base* supporting technology, thereby fostering the development of technology itself.

Capitalist societies use patent laws to create an environment which makes invention financially attractive. Countries such as Canada, the US, and Great Britain enacted such laws years ago. British patent law extends back to the sixteenth century, with an important modernization in 1852. Such legislation grants inventors a monopoly for a period of years in the marketing of their ideas. During that time, others who wish to use the idea must negotiate a licensing arrangement or a pur-

chase of the patent outright from the holder. Canadian patents now last 20 years, after which patented innovations become open for anyone to use. Patent laws thus provide a protected interval allowing reasonable prospects of recovery of costs invested in a new technology. Economic historians investigate the effectiveness of patent law for national productivity (see MacLeod in Further Readings for an example). Too strong a patent law, with a watertight protection over a long period, may ironically be counterproductive because the dissemination of new ideas is thereby retarded. An unenforceable patent law, in contrast, or one protecting over a very short period, becomes a disincentive to invention. Many of the important patents in history, Watt's steam engine and Bell's telephone to name just two, had to be defended in costly courtroom battles.

In the case of the Bell patent, some convincing historical evidence shows that the essential principles of the telephone had been discovered in 1849 by an Italian named Antonio Meucci in Havana Cuba (see notes to Figure 2.2). Meucci, poor and non-English speaking, had no access to the kind of financial backing amassed by Bell. The Italian's patent filed in 1871 was only in "caveat" form (a statement of intention to patent) because he could not afford the higher fees for a full patent. By 1885, when Meucci's claim was judged against Bell's in an American court, the American Bell Telephone Company was a powerful institution which could afford to use every strategy, bribery included it is alleged, to prevail.

Patented ideas are a form of private property, capitalism's bedrock. Admittedly there are alternative forms of social structure by which the economic surplus and investment necessary for technological advance can be achieved. Collective ownership could in theory be equally profitable, and such communal endeavours may work especially well in agriculture. A systematic form of communal ownership, communism in other words, contrasts with capitalism by permitting nothing but state ownership of economic resources while sharing with or exceeding capitalism in the bureaucratic form of organization. Profound shock seized the western world when news arrived of "Sputnik," the first satellite in space, launched by the Soviet Union in 1957. The capitalist nations rushed to bolster scientific training in the schools, and all through the early 1960s there was missile envy because the Russians had more powerful rockets than the Americans. For certain big science

endeavours involving team research and concentration of state resources, there was nothing untechnological about communism as it was practiced in the Soviet Union.

So to identify capitalism as the technology accelerator is primarily to make an historical argument about social change out of the medieval era. At least at that stage of history, to be discussed more specifically below, the individualistic ethos and financial vigour of capitalism seemed to supply exactly the ingredients needed to promote technology. In the present day there is little left to argue about, with respect to the competing claims of capitalism versus communism. The former "Soviet bloc" disintegrated by the beginning of the 1990s, and most of the economies in contemporary developed societies use a blended form of capitalism within the context of some state ownership and considerable state regulation.

Technology as a Value

It is not just material and organizational conditions like economic surplus and bureaucratic social organization that determine the amount of technological creation. Collective priorities are important too, and so we would say that a favourable *"value system"* within societies aids technological development (see Figure 2.1, culture). Little is known or can be known about the values of Stone Age peoples. Perhaps the only clues are the cave wall paintings discovered in places such as the famous Lascaux cave in France, dating from as long ago as 30,000 BC.[9] The priority of hunting in the lives of those people is arrestingly portrayed in these ancient pictures.

To arrive at the modern view of technology it is well established, however, that a sharp reorientation away from the values of the medieval period from about AD 300 to 1500 had to occur. It is hard if one is not a professional historian whose mind "lives" in those medieval times of the Middle Ages to imagine a life so different from our own. This was the era of knights and jousting in Europe, of a feudal way of life with not just one class of citizenship but several, such as serf, villain, freeman, lord. It was a time when physics was organized around a conception of four basic elements — earth, air, fire, and water — whose properties accounted in some way for almost every event. If objects fell when dropped, for example, the explanation proffered was

that their main element was earth, which by nature seeks to reach the centre of the earth.

A marker of the end of this medieval life was a change in values so profound as to be known in history books as "The Scientific Revolution." While one cannot in a brief synopsis give a full account of such a massive transition, there is agreement that by the 1500s the social environment of the medieval period was losing its equilibrium. The medieval historian Norman Cantor writes that "in the fourteenth and fifteenth centuries ... European societies suddenly stopped expanding."[10] There was a geographical dislocation in the form of the pocket ice age, and an onset of economic depression just as devastating as those of our own time. Bubonic plague hit in 1348-49 and killed off population on the demoralizing scale of 25% to 40%. It was a time of rebellion by both urban workers and peasants, the breakdown in civil order being symbolized by roaming bands of mercenary soldiers.

Some of the principles of scientific method had been articulated as far back as the thirteenth and fourteenth century, and Cantor devotes several pages to these predecessors to The Scientific Revolution. The ideas did not break through, however, until the value system — "the social background in which these men worked" as Cantor terms it — had shifted. "The men who pursued these new studies did so on their own time; they had no social encouragement."[11] Somehow out of the pressures of geographic, biological, economic, and social dislocations this new set of ideas supporting a scientific ethos of scepticism and experimentation was able to emerge, and by the 1500s and 1600s become associated with a cluster of familiar names such as Copernicus, Bacon, Galileo, and Descartes. Newton followed only shortly afterward (1642-1727). The Protestant Reformation was under way by 1517 (the date of Luther's Ninety-five Theses), the connections between this and the growth of values conducive to accelerated capitalism being the topic of Max Weber's famous sociological study (see Further Readings). Some historians, however, feel that even the more traditional Catholic Church helped create a social environment for more technological innovation than occurred in the "classical" period of Greek and Roman times.[12]

Consistent with the scheme in Figure 2.1, we can identify at least one technology acting as independent variable during the transition from medievalism. The Gutenberg printing press (see Figure 2.2) arrived

early in The Scientific Revolution, and "was one important factor, for it allowed dissemination of the new ideas not as ill-copied manuscripts with multiplying errors but as consistent treatises...."[13] Chapter 7 will describe Marshall McLuhan's near obsession with the consequences of printing.

Technologies Causing Technologies

George Basalla, whose work was mentioned in Chapter 1, would maintain that it is primarily technology that causes technology. Artifacts "only arise from antecedent artifacts," as he says.[14] But in emphasizing the way that one technology leads to another, Basalla was not denying that technology is created within a larger context like the one mapped out in the present chapter. Indeed, he specified at the outset that "if technology is to evolve, then novelty must appear in the midst of the continuous." He surveyed "the varied sources of novelty — the human imagination, socioeconomic and cultural forces, the diffusion of technology, the advancement of science," and knew that "selection [of technology] is made in accordance with the values and perceived needs of society" (viii).

The explanation for technology which Basalla wants to put aside is the *"teleological"* one of necessity. It would be tempting, for example, to conceive of a technology as simple as the wheel as a response to the need for transportation in Bronze Age times. Basalla's historical research shows in fact that the first known application of the wheel, around 4,000 BC, was not for heavy transport but rather for "ritualistic and ceremonial purposes" such as burial ceremonies (8). Later the more varied applications of the wheel, such as for military and agricultural transportation, began to be realized. Only once the wheel existed did people begin to attribute a "need" to be met by the new technology, and by around 2400 BC wagons were known to be in use on farms. Basalla's point is to refute the simple thesis that farmers needed better transportation than just the backs of horse and ox, so therefore the wheel was invented. In "Mesoamerica" (present day Mexico and Central America), miniature wheeled objects dating from the fourth to fifteenth centuries have been discovered, and yet wheeled transportation did not begin until introduced by the Spanish following their conquests of the sixteenth century. In another contradiction to an argu-

ment of necessity, the wheel was abandoned in the Near East and North Africa between AD 300 and 700 because the terrain made transportation by camel more efficient.

When Basalla concludes that the wheel was "a culture-bound invention whose meaning and impact have been exaggerated in the West" (11), his reasoning is consistent with my scheme in Figure 2.1. "Need," after all, is largely perceived. A conception that has withstood the test of time is the "hierarchy of needs" spelled out by the psychologist Abraham Maslow in 1948.[15] Needs, for Maslow, are hierarchial in the sense that once the "lower" needs such as food have been met the satisfaction of that requirement becomes taken for granted, and the need people feel most acutely moves up the hierarchy to material possessions or needs such as self-esteem and satisfaction with life.

If Maslow is correct, and few would deny that he was articulating a deeply intuitive piece of "folk sociology," then Basalla's thesis about unnecessary invention should become increasingly persuasive as he moves his discussion from primitive to advanced technology. By and large it seems to be true, although perhaps better documentation exists on the creation of technology in modern times. One could look up old newspapers, for example, to verify Basalla's contention that "the automobile was not developed in response to some grave international horse crisis or horse shortage. National leaders, influential thinkers, and editorial writers were not calling for the replacement of the horse..." (6). As he says, the car was a rich person's toy at the beginning of the twentieth century.

The personal computer circa 1980 fits that interpretation equally well. Sherry Turkle's *The Second Self: Computers and the Human Spirit* (see Further Readings for Chapter 10) is a rich source on the early years of the PC. She conducted her research over the years 1978-1981 when interest in computers exploded following the 1975 introduction of the Altair, the first home computer, assembled from a kit. Turkle recounts how, although the manufacturer stressed "instrumental" uses for the Altair (financial planning, etc.), "from the beginning it was clear that this utilitarian rhetoric was not the source of the real excitement." Most people "described a point at which their sense of engagement with the computer had shifted to the non-instrumental. They spoke about 'cognitive play' and 'puzzle solving.'"... Once people actually had a computer in their home, the most interesting thing about it became

Table 2.1: Technology Creating Its Own Need: Perceptions of the Usefulness of Computers, Gallup Data for Canada

"All of us, in some way or another, have been affected by the introduction of, and the increased use of computers. As far as you are concerned, do you think this increased use of computers will, on the whole, make life easier or more complicated for you?"

	Easier	More complicated	No difference	Unsure	(Total %)
Canada-wide					
1992	65%	20	11	4	(100)
1988	56%	26	15	3	(100)
1983	39%	37	18	6	(100)
Age Breakdown, 1992 data					
18-29	80%	15	4	1	(100)
30-39	75%	17	6	2	(100)
40-49	67%	22	8	3	(100)
50-64	55%	24	16	5	(100)
65 and over)	28%	30	32	10	(100)

Source: Gallup Organization, *World Opinion Update*, February 1993, 23.

the computer itself, not for what it might do, but for how it made them feel" (167-68). The sense that computers are a necessity, meeting crucial needs, came later although by the 1990s that sentiment was well established, especially among the generation of Canadians who grew up with computers.

Gallup poll data reproduced in Table 2.1 show the evolving perception of the necessity of computers. In 1983, the era of the original IBM-PC, opinion was evenly divided as to whether computers made life easier or more complicated. By 1988, and even more clearly by 1992, a clear majority perceived the computer as making life easier. The perception of the computer as meeting needs was even more impressive by 1992 among the younger generation reared in the era of the home PC. The 29-year-olds in Table 2.1, the top end of the youngest age band, were born in the year of John Kennedy's assassination. By 1975, the year of the Altair, these 1963 newborns were already 12. They were the front-edge generation growing up with home computers, and show so

clearly that it is when technology exists that we are most likely to perceive a need for it.

Basalla devotes much attention to the evolution-like linkage between technologies. The steam engine, often named as the "defining technology" of the industrial era, is an especially clear example. It is simple enough in conception for the lay person to understand, but sufficiently complex that its technological and engineering antecedents involve some of the central insights of The Scientific Revolution noted earlier. James Watt's famous steam engine of 1775 improved on the Newcomen engine from the period 1705-12. (See photo at beginning of Chapter 4).

Newcomen's engine was "atmospheric," meaning that the source of power was a vacuum created in a cylinder. To understand this machine we have to know that it was designed for pumping flood water from mine pits. It was thus a slow-acting engine (12-14 strokes per minute), and the piston was attached to a heavy shaft connecting through a pivoting beam to the pump mechanism. The piston moved up and away from the bottom of the cylinder due essentially to the dead weight of the pump. Steam was introduced into the cylinder, and then cooled by injection of cold water. The condensing steam created a vacuum in the cylinder, driving down the piston by atmospheric force. The key *knowledge-base* for this technology was symbolized by the barometer, the instrument for measuring atmospheric pressure, an invention of 1643.[16] Steam as a means of creating vacuum had been adopted around 1690 by the French scientist Denis Papin after experiments with inducing low pressure in a cylinder by means of gunpowder.

Watt's contribution was to increase greatly the efficiency of steam engines by using the expansion of water boiled into steam to power the up-stroke and by more efficiently engineering the contraction into the vacuum stage of the piston's cycle. The steam engine is a leading example of the growing importance of the knowledge-base within technology. The materials used for the engines were not crucial since copper for boilers had been in existence for centuries, as had thermal power in its various forms. The relatively complex knowledge-base was the innovating feature. The fifteenth century printing press, which could claim social effects of equivalent importance to the steam engine, was also mainly a conceptual innovation in engineering (see Chapter 7 for details).

One invention within a production process often creates a bottle-neck requiring remedy through another invention. Need would appear to be present here, but Basalla would insist that whether or not a bot-tleneck is responded to reflects the social definition of need. An impor-tant example of bottleneck need is the flying shuttle in the textile indus-try. This device for mechanized weaving multiplied the consumption of yarn. The resulting shortage in yarn production was addressed by Arkwright's mechanical spinning device. See Figure 2.2 for dates.

To conclude these comments on technology as cause of technology: Basalla is not denying that people feel needs and believe that technolo-gies are essential for meeting them. Rather, he posits that at any point in history a far more diverse range of technological possibilities exists than people realize. Such options arise from the curious, inventive nature of humankind, from scientific developments, and from the cumulation of one technology opening the door to another. Technologies come to be regarded as "necessary," as meeting some "need," through a process analogous to evolutionary selection in nature. Instead of biological selection based on survival of a species, however, the technological process selects according to needs defined, rather arbitrarily Basalla would say, within the social environment. Often, as seen with the personal computer, people do not realize that they "need" a technology until it has been marketed.

Let us now apply this general framework developed in the preceding pages to the more specific issue of the Canadian environment for tech-nological innovation.

Creating Technology in Canada

Canada has the stereotype of being more an importer than an origina-tor of technology. "The driving force of contemporary Canada has not been indigenous," as the noted sociologist John Porter remarked once.[17] It is also true, however, that history has generally not been slanted to promote Canada as a technologically innovative society, so that the legacy of Canadian invention has been more neglected than need be in our national identity. All in all, the Canadian case is a revealing socio-logical laboratory within which to examine the importance of social environment for the creation of technology.

It is part of the general concept from Figure 2.1 that geography sets a background for technological development. Geography in interaction with social environment is especially evident in the earlier technology of this country. The only truly distant history we have, that of the aboriginal peoples, is the story of simple societies in close contact with nature. That can be a source of moral orientation when modern Canadians consider ethical issues in technology. It is not to claim that geography is irrelevant to European history, but these connections can become obscured by the very complexity of the European societies even by medieval times. European political history tends to be the most clearly remembered for the tales of mighty battles, titanic power struggles, and momentous clashes between competing idea-systems. So our schoolbook conception of Europe tends to dwell on the Battle of Hastings, or the rule of Charlemagne, or the Protestant Reformation, not so much on how ordinary people organized their lives while working the land. (For an example of the "social history" of Europe, see Braudel in Further Readings). In contrast, the study of Canadian history before settlement by Europeans is largely a history of everyday life.

The original inhabitants, the North American First Nations, had a highly successful technological culture flourishing in a climate that people find formidable even in the present day of heated and cooled cars and houses. The familiar birch-bark canoe, for example, was ideally suited to the transportation needs of its makers. The Iroquois nations were active in farming and supported large villages. The west coast Indians, especially, had considerable skills in carpentry. A good general review of the aboriginal peoples of Canada appears in *Canada's First Nations* (Further Readings).

After European settlement, too, Canadian technology was often a reaction to geography. J.J. Brown's *Ideas in Exile* (again, source in Further Readings), a history of technology in Canada, notes the foghorn for coastal navigation and the snowmobile for arctic travel as just two of the truly indigenous technologies that became important worldwide. The harshness and vastness of Canadian geography at one time kept this country at the forefront of communications technology. An early example is the "disposable ship." The shipment of timber to Britain was an important trade for Lower Canada in the 1820s. When the British government decided to tax oak and squared pine, Canadians built disposable ships, which could be sailed to England and then easi-

ly broken up and sold as timber. Two such vessels were constructed and delivered before the law was amended to remove the import tax.

To note other examples of Canadian innovation in communications technology, a Nova Scotia mariner, John Patch, invented the screw propeller in 1834. Canadians were early leaders in laying underwater telegraph cables, linking Prince Edward Island with New Brunswick in 1852, and participating in the transatlantic cable project completed in 1866. A rotary snowplough for clearing rail lines, needed especially through the Rockies, was patented by J.W. Elliott of Toronto in 1869, and further developed into workable form by Orange Jull in the early 1880s. Sir Sandford Fleming, chief engineer of the Canadian Pacific Railway, in 1878 devised the system of time zones in use worldwide today. An improved radio, which could use household electricity in place of the inefficient batteries of the time, was developed by the Canadian Ted Rogers in 1924.

The V-Chip, prominent in current events of the mid 1990s, illustrates several facets of Canadian invention. Arguably, this device for controlling the amount of televised violence children are exposed to arises from the somewhat regulation-oriented values of the Canadian social environment. The V-Chip was developed in Canada by Tim Collings, an electrical engineering professor at Simon Fraser University. Americans are highly attentive to results of Canadian experiments with the chip, an area where even south of the border it is conceded that "Canada is ahead of the United States." Such is the assumed superiority of US technology, however, that a British publication specializing in film and television news rewrote history to read: "the V-Chip is an American invention."[18]

The story of technological development in Canada highlights the separation between invention and deployment. In terms of invention, Canada may have been proportionately no less creative than the US, Britain, France, or Germany. By the mid-1800s, as Canada was beginning to industrialize into more than just a subsistence agricultural economy, there were scores of Canadian patents. As technology became more complicated and more expensive to develop into production form and to market, however, it became increasingly difficult to keep good ideas in Canadian hands. The electron microscope, for example, was perfected between 1935 and 1939 by a team of physicists at the University of Toronto, but the team then migrated to the US.

Brown comments: "Typically, Canadian research institutions now buy their electron microscopes from the United States because they are not made in Canada" (248). In the 1940s, a Canadian scientist worked on automated control of machines, the technology which has become so dominant in today's manufacturing. By 1946 a prototype was working successfully, but neither government nor business in Canada would risk money in developing the new technology, and soon it was taken over by an American engineering firm.

Canadian inventions received institutional support when they were clearly related to immediate commercial problems rooted in Canadian geographical conditions. For example, the development between 1907 and 1910 of early ripening wheat for the prairies was crucial for the settlement of the west. Another instance of indigenous technology for western development is research at the University of Saskatchewan during the 1920s on improving the durability of concrete. Since water in the west is highly alkaline, the research group recommended modifications to the process for mixing concrete. There was as well a procedure known as "the Mackie system" for manufacturing improved steel for railroad tracks. I.C. Mackie was chief metallurgist of the Dominion Steel & Coal Corporation in Sydney, Nova Scotia, and in 1931 he conducted experiments with varying cooling speeds in steel manufacture. The result was the "shatter-proof" track, a breakthrough in rail transport safety quickly adopted both in Canada and around the world. As remarkable sociologically as the metallurgical process was scientifically is the ensuing history-rewriting campaign south of the border to erase Mackie's name from their engineering journals and promote an American professor who copied Mackie's system.

Canada has been timid about supporting inventions which require long lead times into production and high investment costs if there is uncertainty of quick marketing to the public. Bell's telephone, for example, although quite a simple piece of technology in terms of its constituent materials, required expensive infrastructure to integrate into a communications system. The telephone when it first appeared tended to be viewed as a toylike novelty with no clear commercial function. In the US, Bell was able to get the financial backing to develop his new invention's potential. George Brown, publisher of the Toronto *Globe* had meanwhile rejected a request for $300 of financial backing for the telephone project in return for the British and foreign patent

rights to the new invention. In contrast, an outstanding feature of England's Industrial Revolution beginning in the eighteenth century was "the relative ease with which inventors found financing for their projects and the rapidity with which the products of their ingenuity found favour with the manufacturing community."[19]

The rise and fall of A.V. Roe Canada Limited has gone into history as a case study in the social environment for Canadian technology. The company was founded in 1945, at the close of World War Two. Capital from the British parent company Hawker-Siddeley had been used to buy out the Malton plant of Victory Aircraft, a crown corporation formed by the federal government as part of Canada's contribution to the fight against Hitler. The key divisions within the new company were Avro, which handled aircraft design and construction, and Orenda Engines. Avro had already produced the world's first commercial jet aircraft, which passed its test flight with great success in the summer of 1949. Although highly competitive against the accident prone British Comet, with good prospects for international sales, "Canadians in charge of this magnificently successful project suddenly got cold feet and abandoned their child," as Brown says (302). A second such episode, which received renewed attention with the showing of a CBC "docudrama" in early 1997, gutted the high performance end of the Canadian aircraft industry. In the Cold War climate of the 1950s, defence was a serious business. The Government of Canada called in 1953 for a new jet fighter, with Avro contracted to do a design study since "no other company in the world was working at this level of performance for jet interceptors." The Arrow was in production by mid-1955. On its test flight on 25 March 1958, Brown notes, "the aircraft was magnificent, taking off and landing without a hitch, while its pilots reported that its handling characteristics were very close to those predicted on theoretical grounds" (309). The photo beginning Chapter 3 shows the Arrow taking off for a test flight. Less than a year later, the final announcement coming on 20 February 1959, the newly elected Prime Minister of Canada John Diefenbaker announced the cancellation of the Avro Arrow project.

Opinion perhaps will always be divided about the justification for the Arrow cancellation. For people in the aircraft industry, the decision amounted to nothing less than "shutting down the national dream" (see book of that title in Further Readings). From the government's

viewpoint, there was a contention that piloted warplanes were about to become displaced by guided missiles, and it is true that the Arrow was a dauntingly expensive project. Even the legendary C.D. Howe, the federal minister who had acted almost as a godfather to Avro, had written as early in the Arrow project as 1952 that "I must say I am frightened for the first time in my defence production experience" (Stewart, Further Readings, 178). And yet the rich technological resource created at A.V. Roe was dismantled with seeming vindictiveness by Diefenbaker, even the blueprints for the plane being destroyed under government supervision. To terminate a project so tied up with national prestige at so late a stage, after such a promising beginning, became a symbol of weakness in the Canadian technological environment.

Although the drain of brain and ideas to the southern republic is part of the Canadian lament, the invention of both the telephone and the Arrow warplane reminds us that throughout its history Canada has both gained and lost from migration. Bell was born in Edinburgh, Scotland, and was an adult of 23 when his family emigrated to Canada in 1870. Within a year, he took a job working at the Boston School for the Deaf although he returned to Brantford each summer to confer with his father while conducting the experiments leading to the telephone. It is hard to say which country owns the ghost of Alexander Graham Bell, although in the US there is no doubt about the identity of "that great American communications genius" (Diebold, Further Readings, 132). Likewise, the founding nucleus of A. V. Roe aircraft company was a group of expatriate British aircraft engineers, many of whom returned to Britain or migrated to the US following the Arrow cancellation.

So with both the telephone and the Arrow, Canada benefitted from talented immigrants, but failed to provide the needed social environment to see a technological innovation through to production. The particulars of Alexander Graham Bell's search for financial backers or the Government of Canada's backtracking on the Avro Arrow are markers of a more generalized syndrome which I shall now try, briefly, to analyze.

Brown's history of Canadian invention gives the sense of an earlier golden age of technology in this country, a momentum that began to be lost by the end of the 1890s. That conclusion matches a more general analysis of foreign control of the Canadian economy by the sociologist

Gordon Laxer. Starting from economic data showing how "Canada held its own with the other late-follower countries at the end of the nineteenth century," Laxer proceeds through a step-by-step consideration of the explanations usually advanced to account for the state of economic dependency Canadians found themselves in by mid-twentieth century.[20]

Laxer finds inadequate an account centred just on the geographical fact that Canada formed in 1867 a "staples economy," a resource rich but politically weak neighbour nation to the US. He is thus impatient with what he calls "Canada-as-victim" perspectives. For much of his book he holds up Sweden as a running example against which to compare Canada. Sweden in the nineteenth century was, like Canada, an undeveloped land endowed with resources such as iron ore and forests suitable for the pulp and paper industry. The Scandinavian home of the Volvo, however, was able to use foreign capital to promote home manufacturing without losing control of its economy as Canada did. By the first quarter of the twentieth century, as technological innovation increasingly shifted from private inventors to institutional "research and development" (R&D) financed by industrial research laboratories or by government funding of university scientists, Sweden had reached a stronger position than Canada for developing and mobilizing indigenous technology.

In this analysis, then, Canada's present day image of a technology-importing country that has forgotten its own quite precocious technological past did not have to happen. Sweden, so similar to Canada on some counts, has flourished as a small but technologically active society. Where did Canada differ? Laxer reaches back into Canada's pre-Confederation history for answers. The "ethno-national question" of French-English relations following the British conquest at the Plains of Abraham "overshadowed political divisions along class lines." This sounds much like the echo a generation later of John Porter's mid-1960s comment that "the maintenance of national unity has overridden any other goals there might have been, and has prevented a polarizing, within the political system, of conservative and progressive forces."[21] Both authors feel that a democracy appropriate to fostering a fully independent national state needs both a capitalist *"elite"* and a much larger *"proletarian"* and *"petite bourgeois"* class, with a particular kind of interaction one with another. People with capital to invest are

necessary for putting technological dreams into operation, but this class must be kept honest by the larger class of workers. For Laxer, small farmers are a key sector of this larger bloc during early industrialization. Sweden by the 1830s had "a collection of individualistic farmers with a penchant for technical progress and commercial sales, often to export markets" (88). The rich capitalist elite are equally fond of technical progress and commercial sales, but Laxer feels they are likely to achieve these aims by selling out to more powerful countries. We see more clearly in our own times of the "global economy" than 100 years ago how thoroughly international capitalism is. Nationalism is most likely to reside within the agrarian class, those with attachment to the land in the most literal sense. That population in Canada was too distracted, so this theory states, by the internal political struggles between French and English Canada to put needed attention into nationalist economic objectives.

Analysis such as Laxer's illustrates just how total, complex, and downright chancy is the linkage between the social environment and the creation of technologies. During the nineteenth century, Canada offered a social environment where technology could flourish. British Empire capital was available to finance the big communications projects including roads, canals, bridges, and railroads. Public works such as the Welland Canal around Niagara Falls (1828) and the Victoria Bridge over the St. Lawrence at Montreal (1859) were events of international note (see the hyped tone of contemporary accounts in *Let Us Be Honest and Modest*, in Further Readings). Alongside such mega projects which imported both capital and technology, a smaller scale indigenous technology was cumulating, as Brown describes. But by turn of century, the scope for inexpensive technology from inventors labouring singlehandedly was evaporating in the face of the big science, big capital era of technology. The Dominion of Canada remained partially a colony of the British Empire politically, and decidedly colonial psychologically. Unlike Sweden, we were not an ethnically homogeneous society of centuries of standing as a militarily independent nation, with a politically cohesive class of yeoman farmers. Part of Canada's problem, we can infer from Laxer's analysis, is that it underwent crucial steps toward nationhood within a nineteenth-century capitalist environment. Capitalism in British hands perhaps retained some feeling of building up the colonial outposts. When British capital began

to pull out, however, Canadians went to American investors and encountered a less sentimental capitalism, with no basis nor reason for concern about Canada's political or technological objectives.

Conclusion

We began with the most general sketch of various contexts behind the creation of technology. Geography and the sociobiological nature of humanity both have to be appreciated as part of the story, but the central focus was on the social environment for technological innovation. Little technological growth will result where no economic surplus can support invention, and the main historical source of such surplus has been capitalism. To say this is to state an historical fact, not necessarily to laud capitalism as the salvation of humankind.

It is not just economic organization, but the particulars of social and cultural structure too which set some broad defining conditions for the creation of technology. Patent laws and the organization of education are two examples of this structure. More generically, the priorities voiced by a society are social products, having a somewhat arbitrary quality. To understand directions in technological innovation, one must understand the society they emanate from. Relevant to this is Basalla's work refuting the assumption that every technology is a necessary response to a need.

The chapter ends with the national context for technology. Especially in the modern era, the development of technology is a source of national pride, prestige, and economic competitiveness. As the next chapter suggests, technologies, like people, are frequent migrants, crossing and recrossing national boundaries; deploying technologies in a foreign land while retaining patent rights can become an effective means of control. Such a cycle of dominance is hard to break, once established, for as Laxer shows it becomes tied up with broader political issues. Canada's legacy as a political and subsequently economic colony of Britain and the US respectively has taken a toll on both the

inventive activity in this country and the historical recognition of such activity which did occur.

Notes

1. For a map of the continents at five different stages of continental drift, see John E. Pfeiffer, *The Emergence of Man*. 2nd. ed.(New York: Harper and Row, 1977), 22. See William Calvin (details in Further Readings), 110-12, for estimates on times of aboriginal migration, or a more detailed account, discussing alternative interpretations, in O.P. Dickason's *Canada's First Nations: A History of Founding Peoples From Earliest Times* (Further Readings). Norman F. Cantor, *The Civilization of the Middle Ages* (New York: Harper Collins, 1993), 482, describes the Little Ice Age in Europe.

2. John Warkentin, ed., *Canada: A Geographical Interpretation* (Toronto: Methuen, 1970), 4.

3. Emile Durkheim, *The Division of Labor in Society* [trans. by George Simpson] (New York: Free Press, 1964), 39, 40. See also Talcott Parsons (*Societies: Evolutionary and Comparative Perspectives* [Englewood Cliffs: Prentice-Hall, 1966]), who was especially attentive to tying Durkheim's themes into the acceleration of technology.

4. Murray Knuttila's *Sociology Revisited: Basic Concepts and Perspectives* (Toronto: McClelland and Stewart, 1993) has a section on defining culture on pp. 44-47.

5. Lewis Mumford, *Technics and Civilization* (New York: Harcourt, Brace & World, 1963 [1934], 443, and Colin Chant, ed., *Science, Technology and Everyday Life 1870-1950* (London: Routledge, 1989), 255-56.

6. D.S.L. Cardwell, *Technology, Science and History* (London: Heinemann, 1972), 142; Alan I. Marcus and Howard P. Segal, *Technology in America: A Brief History* (San Diego: Harcourt Brace Jovanovich, 1989), 165. Canadian information from Norman R. Ball, *Mind, Heart, and Vision: Professional Engineering in Canada 1887-1987* (Ottawa: National Museum of Science and Technology, 1987), 19, and Ball's entry in *The Canadian Encyclopedia* [second edition] (Edmonton: Hurtig, 1988), 704.

7. John Kenneth Galbraith, *The New Industrial State* (New York: Signet, 1967), 23-24.

8. Mumford, 23, 25.

9. Pfeiffer, 249-69.

10. Cantor, 481.

11. Cantor, 535. For names and key dates to the scientific revolution, see George A. Rothrock and Tom B. Jones, *Europe: A Brief History* [Volume One](Boston: Houghton Mifflin, 1975), Chapter 21.

12. Mumford, 34, wrote about "the Church's belief in an orderly independent world" as preparing the ground for science. In Cardwell, 7, is the thought that Christianity "laid positive duties on men while at the same time rejecting the enervating fatalism of so many ancient and Eastern religions."

13. Rothrock and Jones, 305.

14. George Basalla, *The Evolution of Technology* (Cambridge: Cambridge University Press, 1988), vii. For an example of Basalla's ideas applied to a case study, see Joseph O'Connell, "The fine-tuning of a golden ear: High-end audio and the evolutionary model of technology," *Technology and Culture* 33 (January, 1992), 1-37.

15. A.H. Maslow, "'Higher' and 'lower' needs," *Journal of Psychology* 15 (1948), 433-36.

16. Mumford, 441. Other sources for this paragraph are Basalla, 35-40, 91-97; Cardwell, 51-59, 66-72.

17. John Porter, *The Measure of Canadian Society: Education, Equality, and Opportunity* (Toronto: Gage, 1979), 37.

18. First quotation from *Los Angeles Times* story downloaded from website http://roscoe.law.harvard.edu/courses/techseminar96/course/sessions/vchip/13.html. British quotation from Chris Jones, "Victory to V-Chip," *Sight and Sound* 6 (May, 1996), 39.

19. David S. Landes, *The Unbound Prometheus: Technological Change and Industrialization in Western Europe from 1750 to the Present* (Cambridge: Cambridge University Press, 1969), 63-64.

20. Gordon Laxer, *Open for Business: The Roots of Foreign Ownership in Canada* (Toronto: Oxford University Press, 1989), 48.

21. Laxer, 132; John Porter quotation from *The Vertical Mosaic* (Toronto: University of Toronto Press, 1965), 368-9. My own book *Essentials of Canadian Society* (Toronto: McClelland and Stewart, 1990) has additional discussion on Porter's view of Canadian politics in Chapter 8.

Further Readings

For effects of climate and geography on the psyche:

Atwood, Margaret. *Survival: A Thematic Guide to Canadian Literature.* Toronto: Anansi, 1972.

Calvin, William H. *The Ascent of Mind: Ice Age Climates and the Evolution of Intelligence.* New York: Bantam, 1991.

For linkages between economic forms and defining technologies:

Beniger, James R. *The Control Revolution:Technological and Economic Origins of the Information Society.* Cambridge, Mass.: Harvard University Press, 1986.

Diebold, John. *The Innovators: The Discoveries, Inventions, and Breakthroughs of Our Time.* New York: Truman Talley, 1990.

Leiss, William. *Under Technology's Thumb.* Montreal & Kingston: McGill-Queen's University Press, 1990.

Toffler, Alvin. *Future Shock.* New York: Random House, 1970.

___. *Power Shift: Knowledge, Wealth, and Violence at the Edge of the 21st Century.* New York: Bantam, 1990.

___. *The Third Wave.* Toronto: Bantam, 1980.

When reading this book with students in technology and society courses the discussion usually turns to Toffler's rather one-sided preoccupation with micro electronic technology deployed in idyllic west coast settings.

For connections between value systems and technology:

Braudel, Fernand. *Civilization and Capitalism 15th-18th Century: The Structures of Everyday Life: The Limits of the Possible.* London: Collins/Fontana Press, 1981.

Jamison, Andrew. "Technology's theorists: Conceptions of innovation in relation to science and technology policy." *Technology and Culture* 30 (April, 1989), 505-33.

Smith, John Kenly Jr. "The scientific tradition in American industrial research." *Technology and Culture* 31(January 1990), 121-31.

MacLeod, Christine. *Inventing the Industrial Revolution: The English Patent System, 1660-1800.* Cambridge: Cambridge University Press, 1988.

Weber, Max. *The Protestant Ethic and the Spirit of Capitalism.* New York: Scribner's, 1958.

For general sources on the creation of technology in Canada:

Ball, Norman R. *Building Canada: A History of Public Works.* Toronto: University of Toronto Press, 1988.

Brown, J.J. *Ideas in Exile: A History of Canadian Invention.* Toronto: McClelland and Stewart, 1967.

Dickason, O.P. *Canada's First Nations: A History of Founding Peoples From Earliest Times.* Norman, Oklahoma: University of Oklahoma Press.

Sinclair, B., N.R. Ball and J.O. Petersen. *Let Us be Honest and Modest: Technology and Society in Canadian History.* Toronto: Oxford University Press, 1974.

Stewart, Greig. *Shutting Down the National Dream: A.V. Roe and the Tragedy of the Avro Arrow.* Toronto: McGraw-Hill Ryerson, 1988.

Voyer, Roger, and Patti. *The New Innovators: How Canadians are Shaping the Knowledge-Based Economy.* Toronto: James Lorimer, 1994.

For the gender factor in invention:

Cummins, Helene, Susan A. McDaniel and Rachelle Sender Beauchamp. "Women inventors in Canada in the 1980s." *Canadian Review of Sociology and Anthropology* 25 (1988), 389-405.

Chapter 3: Technological Diffusion

Social scientists have studied cultural diffusion, the process of spread and imitation, for over a century.[1] It has already been noted that technology is part of societal culture, and so it follows that technological diffusion is a particular form of the more general diffusion of all aspects of culture. Diffusion would seem to hold special importance for Canadian students of technology and society since, as noted in preceding pages, we have relied heavily as a nation on imported technologies. The present chapter begins by suggesting some aspects of all cultural diffusion, and then examines technological diffusion more specifically.

The Process of Diffusion

Cultural diffusion is stimulated by the migration of populations, one of the key aspects of human *"demography."* Given strong evidence from DNA analysis (see Chapter 1) that all human beings descend from African ancestors, and since examples of *Homo erectus* dating from at least one million years ago have been discovered as far away from Africa as Indonesia and China, the tendency to migrate is not only an ancient trait of humans but of prehumans as well. *Homo erectus* was, after all, an early hominid type replaced by *Homo sapiens* some 300,000 years ago.[2]

An Avro Arrow taking off for a test flight, Malton Airport, (Toronto). Photo courtesy of Les Wilkinson collection.

These population movements from such ancient prehistoric times may not have been consciously motivated migrations at all, so much as "unconscious drift," as Leakey terms it. He estimates that the rate of travel might have been only some 20 kilometres a generation.[3] In the Neolithic era commencing around 4000 BC, the time of the agricultural beginnings noted in Chapter 1, more complete data on rates of technological diffusion become available. These Mediterranean civilizations were naturally migratory for the good reason that the agriculture "by its very wastefulness encouraged the gradual penetration of new areas," a suggestion advanced by Derry and Williams on page 6 of their *Short History of Technology* (see Further Readings). As population densities in the habitable areas of the earth began to build, passing of knowledge about technologies from person to person became possible along with the ancient migratory diffusions. It has been estimated that Neolithic technologies such as pottery-making diffused at a rate of about 2 kilometres a year.[4] Such figures come from dating the earliest known examples of archaeological artifacts such as clay pots, plotting them on a map of the world, and working out plausible travel distances from place to place. Although the accuracy of datings will invariably be subject to debate, since even carbon 14 techniques carry measurement error, the 2-kilometre estimate for pottery was corroborated in an analysis of copper and bronze use. Limited data on the spread of maize in the Americas give a figure only minutely higher (2.3 kilometres a year). Even into the more modern era of The Scientific Revolution, but prior to the full momentum of The Industrial Revolution (i.e., circa 1500s to 1700s), "the rate of diffusion for new processes and new machines in any given industry was painfully slow within a country and even slower between countries." Knowledge of the superiority of coke to wood in blast furnaces for iron-smelting, for example, took over 35 years to diffuse throughout England.[5]

It is schoolbook knowledge that some key early inventions including the horse-collar, the crossbow, and the stirrup came to Europe from eastern civilizations. The stirrup, the device that supports a horseback rider's feet, is believed to have been highly significant in the development of feudal society in Europe. Stirruped riders could brace themselves to support the lance, making the cavalry charge a formidable tactic. The implications of the stirrup reached beyond the military, because those wealthy enough to own and maintain a horse had the

advantage over soldiers and civilians on foot. The stirrup was thus part of the technological foundation for a feudal society based on estates possessing vast inequalities in power, wealth, and legal privileges.

The stirrup, a deceptively simple piece of technology, with a wood, metal, and leather *material* base, animal *power*, and craft *knowledge-base*, appeared first in India late in the second century BC as a primitive device for just the big toe of barefooted riders. Stirrups migrated north to Afghanistan and Northern Pakistan, and changed form to accommodate the booted feet of riders in these colder climates. Arriving in China as a result of Buddhist missionary activity, this evolved foot stirrup was in common use by the fifth century after Christ. After another 150 years or so, the stirrup was being used in Korea and Japan, and by the beginning of the eighth century, according to a meticulous examination of historical evidence by Lynn White (source in Further Readings for Chapter 1), in the area of present-day Iran. Within the first half of the same century stirrups were appearing in the area of present-day Germany as well. Iron or bronze, both more resistant than wood to sword slashes by footmen, were by now part of the material for the stirrup.

Notions of technological diffusion in times long past often must consider independent invention as an alternate explanation. Indeed, the school of thought known as "diffusionism" within cultural anthropology arose in the late nineteenth century as part of a reaction to the evolutionary assumption that cultures typically developed in isolation. Independent growth versus cross-fertilized cultural growth is too impenetrably complex to factor in all but the simple examples involving well preserved archaeological artifacts such as those reviewed above.

With modern communications arrangements allowing for the orderly dissemination of technological knowledge through the patent system, and the priority governments and business now assign to being alert to new technologies, diffusion in the sense of awareness of new innovations moves more rapidly than before. The transmission is more by "radiation" than by migration or physical contact.[6] Some conclusions about technological diffusion between companies within the global economy appear later in the chapter. We shall begin by examining technological diffusion in the sense of saturation level of adoption among the public.

Far from all innovations diffuse widely among consumers, because "a commercial technology must not only work well, it must compete in the marketplace" (*Scientific American*, September 1995, 57). Inventions seemingly with consumer promise may fail to diffuse for either technical or social reasons. Futuristic comic books of 1930s and 1940s vintage, for example, envisioned the time when people would routinely travel by jetpack. Such devices existed by the 1960s, but had little practical use simply because the fuel was too heavy. The increments in lifting power in relation to weight of fuel supply would need to be so substantial that, even after 30 years of advances in lightweight materials, mass market jetpacks are not practicable. Another example of a non-diffusing invention is the videophone, technically available since 1964. On at least half a dozen occasions up to the mid-1990s telephone companies had attempted to market videophones with which one can see the opposite party in a telephone conversation. As the *Globe and Mail's* science reporter concluded, the failure of the videophone to catch on "was a symbol of humanity's ability to resist technology's blandishments" (21 January 1995, D8).

The Diffusion of the Telephone: The "Bell" Curve

For several reasons, the telephone makes a good case study of diffusion. This device has undergone the full cycle of diffusion from rare occurrences through to virtual saturation, and the cycle has been prolonged enough to reflect some key twentieth century social history. It is, in addition, and recalling themes from Chapter 2, a somewhat typical technology. One characteristic trait is that it synthesized from earlier innovations. The telephone derived in part from the telegraph, but also from the Bell family's interest in human speech and hearing.

Telegraph had begun operation by the early 1840s in Great Britain, and soon was improved by Samuel Morse, the American whose name is immortalized by the Morse Code in which dots and dashes — short and long beeps — encode letters and numbers.[7] Research on deafness was the lifelong work of Melville Bell, father of the famous inventor, and a step in the telephone research by Alexander Graham Bell was his examination of the diaphragm of the human ear. The phone was largely an "interest-driven" invention in the sense that its potential commercial value was at first only dimly realized. It was seen by investors

initially as an instrument for business and only later as a communications tool that almost every citizen would use on a daily basis (Fischer, in Further Readings). Bell himself imagined concerts being transmitted from live performances in recital halls out to customers listening in through their telephone receivers at home.

Bell's initial patent for the telephone was filed and accepted in 1875. As noted in the last chapter, it was really a cross-border invention, with Bell performing his laboratory work in Boston, but spending summer vacations in Brantford where he drew inspiration from discussions with his father and, we Canadians like to imagine, from the pastoral surroundings overlooking the Grand River. In August of 1878 Alexander Graham Bell gave his father Melville Bell a three-quarter share of the Canadian patent rights to the telephone. Bell senior began engaging agents to solicit subscribers in various Canadian cities. The first such system began in Hamilton, with 40 subscribers signed by December 1878. To organize a telephone system from scratch must have been an enormously daunting task, and so it is not surprising that early in 1880 Melville Bell sold his patent right to National Bell, the company developing the telephone in the US. Exclusive rights to operation were granted by the Government of Canada to The Bell Telephone Company of Canada, operating as a subsidiary of National Bell. The company had 2,100 subscribers by the end of 1880.[8] In 1885 the Government authorized other telephone companies to operate, and many small companies were formed, often by doctors who wanted rapid communication with their patients. Bell remained the dominant force, however, and worked difficult to take over the small competitors.

The proliferation of small companies made it increasingly hard to collate figures on telephones in use, but by the turn of the nineteenth century the Federal Government was compiling yearly telephone statistics with increasing care. In those times before the invention of the social survey as it is known today, there was no direct measure of the proportion of households with telephones. Instead, the number of telephones in use was counted, an accurate indicator of total activity since all phones were leased from Bell or the private companies. Beginning in 1947, DBS (The Dominion Bureau of Statistics, predecessor to Statistics Canada) began collecting household surveys which, sporadically at first, but regularly from 1953 on, inquired whether there was a telephone in the house. The first data on multiple telephones within

Figure 3.1, Diffusion of the Telephone in Canada

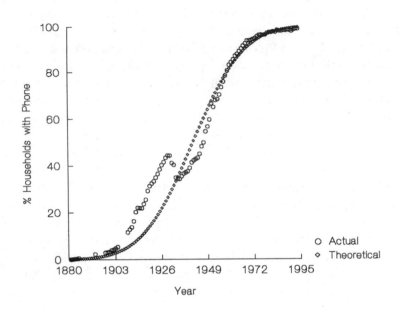

Sources:

1880-85, *Select Committee of the House of Commons, Proceedings* (Ottawa: King's Printer, 1905), 665.

1886-1910, *Canada Yearbook* for various years, Pike (notes to Table 3.3); Jean-Guy Rens, *L'Empire Invisible: Historie des Télécommunications au Canada, De 1846 à 1956* (Sainte-Foy [Québec]: Presses de l'Université du Québec), 270.

1911-1946, Dominion Bureau of Statistics, *Canada Yearbook, 1933* (Ottawa: King's Printer, 1933), and later years; Statistics Canada Cat. 56-203 (Telephone Statistics).

1947 on, Statistics Canada Cat. 64-202.

Note: Total telephones per 100 household for the 1880-1946 series have been adjusted down to an estimate of residential phones, to splice with households having a telephone from the 1947-present data. Population for non-census years is interpolated.

households appeared in the 1960 survey. At that date, 83% of Canadian households were telephone subscribers and 8.7% of these had at least one extension phone. By 1985 so many households had multiple phones and so many were owned rather than leased, that the concept of telephones in use was abandoned in favour of number of access lines.

In a graph of the diffusion shown as Figure 3.1, the tally is by households with at least one telephone, for this conceptualization gives the best historical continuity with the early days of "telephony," as the activity was then termed. The household survey data are used in the portion of the graph for 1947 on, while statistics on number of telephones in use are used for the pre-1947 era.

Diffusion research is one of the areas in social science where we can "fit a curve" to a social phenomenon. The process of diffusion, in other words, has a predictable form which follows the normal curve or, as it happens by semantic coincidence, "bell" curve. Such curves are the most frequent form of the distribution of events studied in social science. A trait such as intelligence, at least as the concept is measured by IQ testing, is normally distributed, for more people are exactly at the middle, or mean, of the IQ distribution than at any other score. The farther away from the mean, the smaller the proportion of population in each bracket, until it is found that only a very few people have IQ's that are either extraordinarily low or high. A true normal curve is symmetrical in that the shape is exactly the same on each side of the mean. The symmetry resembles that of a bell, hence the name.

If a normal curve is plotted cumulatively (here, the percentage of Canadians having a telephone in each year rather than the percentage of people first acquiring a phone in that year), an S-shape results. This characteristic shape for diffusion is depicted in Figure 3.1 as the "theoretical" curve, giving a norm against which to compare the actual plot from the yearly Statistics Canada data. It can be seen that the actual and the theoretical, the Bell curve and the bell curve as it were, both do and do not fit each other.

For the very early years before 1900, and then in the later period from about 1950 on, social science nicely explains telephone diffusion, for the actual and theoretical curves are in amazingly close fit. The precise nature of the social science at work here has been explained by Rogers, author of several books on diffusion (see Further Readings). "Psychological research indicates that individuals learn a new skill, or

Table 3.1: Survey Data for Household Telephones Since 1975

Year	1975	1985	1995
1 telephone	69.5%	47.7%	24.2%
2 telephones	22.8	34.0	36.9
3 or more	4.1	16.5	37.5
None	3.6	1.8	1.5
Number of access lines per household	0.92*	1.04	1.07

Source: Statistics Canada, *Canada Yearbook*, various years (Ottawa: Supply and Services Canada), and Statistics Canada Cat. 56-203.

*Estimated using Statistics Canada Cat. 56-203 (1975:8), and 64-202 (1975: Table 1).

bit of knowledge, or set of facts, through a learning process that, when plotted over time, follows a normal curve.... If a social system is substituted for the individual in the learning curve," says Rogers, "experience with the innovation is gained as each successive member in the social system adopts it" (244). He labelled the personalities of people adopting an innovation at each stage of the bell curve. Those bold pioneers risking time or money on a new and untried idea are "innovators" or, for those a little behind the cutting edge, "early adopters." The main group who jump on the bandwagon during the middle phase, when an innovation is increasingly being seen as viable and reliable, are "early" or "late majority," while the latecomers, those final 15% or so who are the last to acquire some new consumer good, accept a new scientific theory, or invest in a new technology are labelled "laggards" (205).

In the case of the telephone, this theoretical pattern was slowed in the middle period by the intervention of history and economics. During the 1920s, a prosperous time and also one in which traditional ways of life had been broken down by the social effects of World War One, telephone diffusion romped ahead of the theoretical "schedule." Then is seen the devastating impact of the 1930s depression, which knocked levels of deployment back onto the normal curve.

Fischer found the same drop in telephone usage in the US for the 1930s, and pointed the main dieback to rural areas. There was not only economic deprivation behind the decline of rural telephone subscribers, Fischer concluded (102-03, 108-09), for other new products

such as automobiles or the consumption of electricity maintained much steadier rural demand. In part, the telephone companies charged inflated prices while providing poor service, and were unwilling to lower rates during the depression to counteract the inroads by other consumer goods into limited budgets. Both Bell Canada and AT&T in the US tended to see the telephone as only for the more affluent, middle class population, a theme explored in the section below. Automobile manufacturers such as Ford, in contrast, were already by the second decade of the 1900s marketing mass produced cars which, even though costing many times more than telephone service, were advertised as within the reach even of quite modest income families.

By the end of the twentieth century only about 1% of respondents on Canadian surveys were reporting no telephone. Another few percent of the population, mainly the institutionalized or the homeless, do not appear in Statistics Canada surveys and are largely without telephones, but all in all the telephone has diffused in Canada to virtually a saturation level. The rapid phase of diffusion now involves possession of three or more phones and multiple access lines serving modems, as Statistics Canada figures in Table 3.1 illustrate.

Diffusion to Whom?

The Social Class Effect

If the bell curve is the first law of diffusion, the second is the tendency for innovations to diffuse first to the wealthier middle class segment of a population. This is consistent with the principles developed in Chapter 2: the notion that economic surplus is a prerequisite for technological development. Now, however, we are speaking not of invention, filing patents, and investment in companies, but from the viewpoint of the consumer. Here too, some economic room to manoeuvre makes it easier to acquire leading edge technologies, whether it be a telephone in 1900 or a computer in 2000.[9] This pair of technologies share a class factor in speed of adoption, as shall now be demonstrated.

Fischer, in his intriguing work *America Calling* (Further Readings), compiled figures on telephone diffusion by socio-economic level of American households and found the highest incidence of subscribers among the more affluent. Robert Pike has demonstrated the same for

Table 3.2: Home Computer Ownership in Canada, 1986-1996

Year	1986	1988	1990	1991	1992	1993	1994	1995	1996
Percentage of households with PC	10.3	12.6	16.3	18.6	20.0	23.3	25.0	28.8	31.6

Source: Statistics Canada, *Household Facilities and Equipment*, 1996 and earlier, Cat. 64-202.

an Ontario town, his figures appearing below in Table 3.3.[10] The early telephone subscribers in Kingston were predominately business people or professionals such as doctors, veterinarians, clergymen, etc. Many no doubt wanted the telephone initially for business reasons, but then found it convenient to link their places of work to their residences.

The personal computer makes a good modern comparison with the telephone. Allowing for inflation and increasing affluence during the twentieth century, the telephone was at least as expensive a century ago as a state of the art computer is today. It has been estimated that to rent a telephone in the year 1900 cost about 4% of the yearly income of workers at the pay level of, for example, a public school teacher.[11] In today's money, such a person would earn about $50,000 a year; 4% of that equals an even $2,000. That could easily be the yearly computer budget for a person wishing to stay at the front edge of computing technology. Another parallel between the telephone and the computer is that both began as office technologies, and then spread to domestic applications. The computer has in addition been diffusing at about the same rate during the late 1980s and 1990s as the telephone was at the equivalent stage. Between 1910 and 1918, telephones increased from about 13% to 25% presence in households (see Figure 3.1). The personal computer was in 12.% of Canadian households in the year 1988, and reached 25% by 1994. Fuller details on the growth of Canadian domestic computer ownership appear in Table 3.2. The incidence virtually doubled over six years, just slightly faster than the same rate of increase for the telephone over the eight years between 1910 and 1918. By 1996 nearly 32% of households had a home computer. Additional data from Statistics Canada showed that about half of these home computers were equipped with modems, and about half of these

Table 3.3: Percentage Within Occupation Levels of Telephone Subscribers, Kingston, 1911, and Households Having a Personal Computer, all of Canada, 1994

| | Percentage With: | | |
	Telephone (1911)	Computer (1994)	
1911 Occupations			**1994 Occupations**
Professionals	95.8	61.0	Professionals (employed & self-employed)
Semi-Professionals, Officials White-Collar, Sales Personnel	6.1	47.7	Semi-professionals, skilled clerical
Proprietors, Managers, Owners, Foremen, Agents	57.4	44.4	High & middle level managers
(no comparable category)		38.9	Foremen (& women)
(no comparable category)		38.2	Technicians &supervisors
Skilled, Semi-Skilled Trades	3.2	28.4	Skilled, semi-skilled crafts
(no comparable category)		33.9	Semi- & unskilled white collar
Labourers, Domestics, Other Unskilled	1.2	24.2	Unskilled craft & farm labourers
Other (including farm)	2.0	20.6	Other (includes a few farmers, but mainly retired)
N	525	11876	

Source: Telephone: Robert Pike, "Kingston adopts the telephone: The social diffusion and use of the telephone in urban central Canada, 1876-1914," *Urban History Review* 18 (June, 1989), adapted from p. 43. (Re-computed from Pike's Table 9 to give percentage within each occupation having telephone, using number of "male heads" in 1911 census for Kingston as proxy for number of households).
Computer: Original tabulation from 1994 GSS. ("Do you have a personal computer at home [excluding video games like Nintendo]?")

households (7.4% of all households) reported "using the Internet."

Far more *"socio-demographic"* data on householders are available today than at the beginning of the twentieth century, and so the social class hypothesis can be subjected to a more searching test than researchers such as Fisher or Pike could perform for the telephone. Every year Statistics Canada collects what it terms the *General Social Survey*, covering some 10,000 Canadian households from coast to coast. The 1994 version of this "GSS," as it is abbreviated, was particularly rich in home computing questions, and this is the source for the second portion of Table 3.3, showing the ownership of domestic personal computers by respondent's occupational group. The occupation categories match only very approximately with Pike's telephone data for Kingston. The classifications available in the early 1900s were very crude but, more than that, the "shape" of the labour force has greatly changed over this interval of three generations. "Technician," for example, is not a term appearing in early census reports and many current jobs under that label did not exist earlier.

The class "gradient," the pattern for a declining adoption rate of the innovation the lower the socio-economic standing of the occupation, is far sharper for the telephone, but nevertheless is clearly present in the case of the modern personal computer. Professionals, for example, are virtually twice as likely as either white collar or blue-collar unskilled workers to have a home computer. As noted already, the personal computer was a little farther along the diffusion curve by 1994 than the telephone in 1911, the year of Pike's data. The more significant difference between the two sets of data, I suspect, is that society today is more level or egalitarian in wealth distribution than at the turn of the century. This is claimed in full knowledge of the substantial socio-economic inequalities present in contemporary Canadian society.

Other Factors in Computing Diffusion

We have just seen the rather characteristic way in which the acquisition of home computers begins most intensively with the more affluent class, and then diffuses down the status hierarchy. Other aspects of the profile of home computer users probably are less universal, but some of these shall be examined now, looking at further 1994 GSS results summarized in Table 3.4.

Table 3.4: Computer in Home by Other Socio-demographic Factors, 1994 GSS

		Percentage with Computer, 1994
By: Language of interview		
	English	35.4
	French	26.3
Years of schooling		
	None	4.1
	1-5 years	1.7
	6 years	9.0
	7 years	9.8
	8 years	7.8
	9 years	22.3
	10 years	26.1
	11 years	32.3
	12 years	39.1
	13 years or more	54.6
Age		
	15-19	48.3
	20-29	35.6
	30-39	36.2
	40-49	46.4
	50-59	31.5
	60-69	13.1
	70-79	6.0
	80 and over	2.8
Province of residence		
	Newfoundland	20.4
	Prince Edward Island	19.9
	Nova Scotia	28.0
	New Brunswick	16.4
	Quebec	27.7
	Ontario	38.3
	Manitoba	25.6
	Saskatchewan	29.0
	Alberta	37.9
	British Columbia	36.4
Gender		
	Male	35.3
	Female	31.2

Source: Computed from 1994 GSS.

Some language effect can be detected, as Anglophones are appreciably more likely than Francophones to own a personal computer. At one time in Canadian history, it could have been hypothesized that proportionately more Anglophones lived at a comfortable middle class standard. Some truth to that remains for Francophones in areas such as Northern Ontario and New Brunswick, but most of the French-speaking sample in the GSS are Québécois, just as "bourgeois" a society as any other. Home computers have been a little less attractive for Francophones due to problems of access to software written in French. That issue was in the news when "Windows 95" was released. In addition, *The Globe and Mail* reported that the Internet, one of the main attractions for the home computer user, is "a largely Anglophone zone" with few French language Web sites (23 May 1996, C2).

Home computing began as an educated person's pastime, which it still is to perhaps a surprising degree, according to Table 3.4. This socioeconomic skew in access to computing has become a recognized social problem within the media.[12] For those with no formal schooling, possession of a home computer is about as rare as subscribing to the telephone was in 1901, 4.1% in each instance. Among those with post-secondary education, in contrast, only a minority do not own a home computer. In part this education effect echoes the difference by occupation examined earlier, in that high education correlates with a high status occupation. Although I shall not present the tabulations here, it is also the case however that apart and independently from this linkage with occupation, education and home computer ownership go together. This seems likely only to increase, as the post-secondary curriculum becomes increasingly involved with imparting computing skills to each new generation.

The youngest age group among the 1994 GSS respondents, those aged between 15 and 19, have almost 50/50 odds of having access to a computer in the house. Most of the computers have been provided by parents as a household accoutrement likely to better opportunities for their children.[13] About one-third of young adults in their twenties or thirties have home computers, and a surprisingly high 46% among those in their forties. Many of the latter may have computer-hungry children living at home. It is after age 60 that access to home computers drops off sharply. A 60-year-old in 1994 was born in 1934, and would have had little exposure to computers until well into middle age.

Since today's children are growing up with computers, there may not be much age effect left when they are old.

Lastly in this brief profile of the diffusion of the home computer, Table 3.4 shows some fairly expected provincial differences and a small gender effect. Ontario, Alberta, and British Columbia, the three most affluent provinces, have the heaviest concentration of home computer ownership. Proportionately more males than females have a computer, but the margin is not large and proves to be non-existent if the older generation is removed from consideration in a more detailed tabulation with which I shall not encumber the text.

Technological Diffusion at the Company Level

Some of the same themes encountered above pertain to the diffusion of industrial technologies between business firms. Companies, like individual citizens, are stratified in a capitalist society, meaning that often one or a small number of firms dominate a market. Unlike most consumers, however, firms often are involved with both the creation and the deployment of new technologies. While some large corporations maintain laboratories for basic research, much such groundwork comes from university research funded by government. It has also been argued that many of the initial inventions leading to important innovations have been produced within quite small organizations. In terms of deployment of new technology, however, it is often true that "the largest firms adopt earliest."[14] Since the most powerful countries and regions within countries tend to host the most powerful corporations, the patterns of dominance among firms reinforce dominance hierarchies between nations and regions.

The discarding of technology is clearer to discern for companies than for individuals. The individual consumer may keep a turntable for long-playing records sitting on the stereo shelf even if it is the compact disk player that receives virtually all the use. In contrast, a final, declining, phase of the S-curve of diffusion can occur suddenly in a graph of the "product life cycle."[15]

Conclusion

Technologies have been diffusing between and within populations for

as far back as historical evidence will take us. This chapter discussed the characteristic S-curve of technological diffusion whereby a new technology typically is adopted with caution initially, then by ever-growing proportions of a population. As the curve reaches a point of inflection the rate of adoption levels off and then begins to decline. Finally, a saturation stage is reached in which only a small, diminishing percentage are adopting the now familiar technology. Innovations tend to diffuse unevenly throughout the population, both for individual consumers and for companies. For consumers, social class appears to be a near-universal filter since innovations are most readily adopted by the more affluent groups with money to spare on novelties and luxuries, as new technologies are likely to be regarded. Similarly, the richest, largest companies may be best placed to exploit innovations. In the case of home computers, the heaviest concentration of consumer access occurs among Anglophones, the more educated, the young and middle aged, and in wealthier provinces. Gender, in contrast, is not an important factor in this case study, for only by a small margin do males have a higher ownership rate than females.

To write of the "laws" of diffusion is not meant to imply that every technological innovation necessarily diffuses among the consumer population. The unpredictability of public taste has already been considered within the context of Basalla's position; additional examples are the jetpack and the videophone.

Notes

1. Most cultural anthropology texts offer a short section defining cultural diffusion. For example, American works such as William A. Haviland, *Cultural Anthropology,* 6th. ed. (Fort Worth: Holt, Rinehart and Winston, 1990), 416-18; Marvin Harris, *Cultural Anthropology* (New York: Harper Collins, 1991), 14-15; or the British author I.M. Lewis's *Social Anthropology in Perspective* (Cambridge: Cambridge University Press, 1985), 60-65.
2. Richard E. Leakey, *The Making of Mankind* (London: Michael Joseph, 1981), 112.
3. Leakey, 117.
4. Munro S. Edmonson, "Neolithic diffusion rates," *Current Anthropology* 2 (April, 1961), 71-102.

5. Warren C. Scoville, "The Huguenots and the diffusion of technology," in Thomas Parke Hughes, ed., *The Development of Western Technology Since 1500* (New York: Macmillan, 1964), 50-51; Shepard Bancroft Clough and Charles Woolsey Cole, *Economic History of Europe* (Boston: D.C. Heath, 1952), 305.

6. Scoville, 51.

7. Richard Shelton Kirby, Sidney Withington, Arthur Burr Darling, and Frederick Gridley Kilgour, *Engineering in History* (New York: McGraw-Hill, 1956), 338-40.

8. Robert E. Babe, *Telecommunications in Canada: Technology, Industry, and Government* (Toronto: University of Toronto Press, 1990, 72). Babe is my source throughout this capsule history of the telephone in Canada.

9. For a broader theory of who are early adopters of a new technology, we can refer to the notion of "centrality," identifying those at the centre of communication networks. See Marshall H. Becker, "Sociometric location and innovativeness: Reformulation and extension of the diffusion model," *American Sociological Review* 35 (April, 1970), 267-82. Higher social class enhances centrality with respect to technology, as does higher education and some of the other factors noted herein in the discussion of home computer diffusion.

10. Robert Pike, "Kingston adopts the telephone: The social diffusion and use of the telephone in urban central Canada, 1876 to 1914," *Urban History Review* 18 (June 1989), 32-47.

11. Pike, 34.

12. For example, "Computer gap hurts poor, report says," the *Globe and Mail*, 1 November 1996, A5.

13. Jeffrey Frank, "Preparing for the Information Highway: Information Technology in Canadian Households," *Canadian Social Trends* 38 (Autumn, 1995), 5.

14. Rudi Volti, *Society and Technological Change*, 3rd. ed. (New York: St. Martin's Press, 1995), 274-75; Edward J. Malecki, *Technology and Economic Development: The Dynamics of Local, Regional and National Change* (Burnt Mill, Harlow. U.K.: Longman Scientific & Technical, 1991), 120.

15. Malecki, 124.

Further Readings

For a general, classic history of technology:

Derry, T.K. and Trevor I. Williams. *A Short History of Technology: From the Earliest Times to A.D. 1900.* New York: Oxford University Press, 1961.

For diffusion, especially of telephone technology:

Fischer, Claude S. *America Calling: A Social History of the Telephone to 1940.* Berkeley: University of California Press, 1992.

Marvin, Carolyn. *When Old Technologies Were New: Thinking About Electronic Communication in the Late Nineteenth Century.* New York: Oxford University Press, 1988.

Proceedings of the Select Committee on Telephone Systems. Ottawa: King's Printer, 1905.

Rogers, Everett M. *Diffusion of Innovations.* 3rd. ed. New York: Free Press, 1983.

de Sola Pool, Ithiel. *The Social Impact of the Telephone.* Cambridge, Mass.: MIT Press, 1977.

Wasserman, Neil H. *From Invention to Innovation: Long-Distance Telephone Transmission at the Turn of the Century.* Baltimore: Johns Hopkins University Press, 1985.

Part II.

SOME CONSEQUENCES OF TECHNOLOGY

Rotative Steam engine by James Watt, dating from 1788. *The Science Museum*, London UK. Photo taken with permission.

Chapter 4: Technology and Economic Development

That the creation of technology requires economic surplus was suggested in Chapter 2, and the importance of economic prosperity for the diffusion of technology has been considered in Chapter 3. So it is clear that technology requires considerable economic pump-priming as prerequisite for its creation and deployment. The emphasis in the present chapter shifts toward how technology in turn promotes economic surplus; further consequences of technology are explored in chapters 5 to 9. In terms of Figure 2.1, this section of the book is considering "technologies as independent variables."

Technology as the engine of national wealth is as familiar and accepted a liturgy for people in the present age as any medieval faith might have been in olden times. Consider the words of the *Canada Yearbook*, published annually by Statistics Canada. As the 1994 edition stated, "(t)he late twentieth century turns on an axis of science and technology spun from the ingenuity and determination of the researcher and the imagination and often poetry of the inventor.... Industrial R&D is the sector most clearly linked to technological innovation and therefore to economic growth" (396, 398).

The human costs of such growth receive attention in Chapter 6. The task for the moment is to look at how technology and economics relate.

A more thorough appreciation of the relationship between technology and economic growth will help us assess technology-importing as a Canadian strategy.

Technology and the Business Cycle

What is the cause and effect between technology and economic growth? So long as social science phenomena vary, are not fixed quantities like the speed of light or the force of gravity, possibilities exist for addressing causal issues. And it so happens there is little if anything that varies so dramatically and consequentially as the economic vitality of societies. Economists use the term *"business cycle"* to refer to these variations over time, and one does not read very far into economics literature before encountering the name N.D. Kondratieff, a Russian economist who published on "long economic cycles" during the 1920s. His writings are surprisingly accessible to the non-specialist reader (see Further Readings), although more recent analysts in the same tradition enter into econometric techniques quite impenetrable for the non-specialist.

Kondratieff assembled historical trends for various economic indicators, with some of the series reaching as far back as 1780. The cycles, illustrated in Figure 4.1, fall into approximately 50-year terms. In the opening phase, commodity prices rise, along with interest rates, wages, foreign trade, and the production of basic industrial commodities such as coal, iron, and lead. That profile reverses in the downswing. The cycles graphed in Figure 4.1 are "smoothed" or abstracted from real data, drastically so it becomes clear when inspecting any of the real-life time series data in the more technical studies. One complication is the way Kondratieff's 50-year waves overlay various shorter subcycles. For example, in recent economic history for most nations, including Canada, the second half of the 1980s was a boom period abruptly followed by a recession beginning by 1990. All this occurred within the larger context of the closing years of the downswing in Krondratieff cycle number four.

This was a controversial theory of economics when it appeared, and still is.[1] Economists in Kondratieff's Russian homeland were scandalized by the bourgeois assumption of the cyclic model: that a capitalist economy could recover from depression and not head inexorably

Figure 4.1: Technology and Krondratieff's Waves of Economic Activity

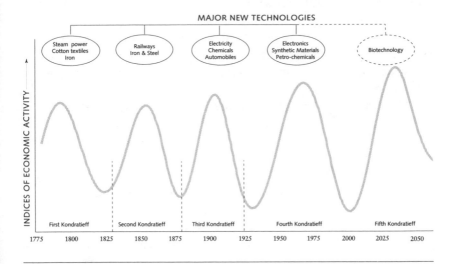

Source: Reprinted from *Technology in Society*, Volume 16, Lyndhurst Collins, "Environmental performance and technological innovation: The pulp and paper industry as a case in point," 428, Copyright 1994, with permission from Elsevier Science Limited, The Boulevard, Langford Lane, Kidlington 0X5 1GB, UK. The diagram is an adaptation from Peter Dicken, *Global Shift: Industrial Change in a Turbulent World* (London: Harper and Row, 1986), 20.

toward breakdown. Less ideologically, perhaps, western commentators noted the selectivity in Kondratieff's data, the likelihood the entire pattern was due to chance historical events, and the lack of a unifying theory behind the pattern. Theories of waves became a little passé between 1930 and 1970 when the Keynesian theory that government fiscal policy could dampen economic swings held sway. Then, as the post World War Two boom atrophied by the early 1970s, the twentieth century data began increasingly to fit the Kondratieff cycle after all.

Kondratieff had hesitated to advance a full explanation of the cycles — saying in essence that they were endemic to the capitalist mode of production — but he remarked on the place of technology in the cycles. Technology, he maintained, was in phase with the business cycle such that "significant technical inventions and discoveries" (64-5) began to occur toward the end of a cycle. The period where these discoveries

were deployed into working technologies used in production tended to coincide with the early years of a new 50-year cycle. It is as though capitalist energy through the latter part of a rising cycle becomes preoccupied with exploiting the possibilities of the established technologies that led the cycle. Only as the cycle peaks and then dips does concerted attention return to encouraging new patents and their development into fresh operational technologies.

The most explicit statement on technology as a cause of economic growth is probably Gerhard Mensch's book *Stalemate in Technology*. "Basic innovations occur in clusters," Mensch wrote. "There are phases in long-term economic development in which the economy activates only few basic innovations.... Instead of moving onto new territory, it goes around in circles in its traditional areas." Such "stalemates," argued Mensch, borrowing a term from the game of chess, are broken by an upsurge, a "swell" or "swarm" in his words, of "basic innovations." The basic innovations arrive not from a vacuum of sociological and economic circumstances, but at "the onset of economic crisis, and then innovations break through the floodgates" (11, 138). As Figure 4.1 suggests, the reinforcing technologies behind each Krondratieff upswing reflect some of the main events in technological history, from the age of steam, textiles, and iron around 1800 to developments in electronics, synthetic materials, and petrochemicals after World War Two. Although many would designate biotechnology as the cluster most likely to drive the "fifth Krondratieff," *convergent information* (the integration of computers and telecommunications), the second most recent entry in Figure 2.2, also may be important.[2]

The view of Krondratieff, Mensch, and others (see Mager, in Further Readings, for more names) about technology as a cause of economic upswings has a common sense appeal, but exactly what mechanism of causality is at work here? Economists generally have been stereotyped as unromantic, calculating, rational souls. Writings on the long economic cycles, however, can become quite mystical, with references to sun spots, biblical passages, and folklore.[3] The cycles span so many decades, reaching back to the late 1700s, that a generic explanation clearly is needed.

Earlier fluctuations, for example, might have been accounted for by agriculture. The weather, we know, is cyclic, as is the process by which land becomes exhausted after some predictable interval and must be

released from production to lie fallow until it recovers. Agriculture remains an important sector of the economy even in the postindustrial age. Although farming is highly mechanized today, employing only about 3% of the Canadian labour force, its contribution to the total economy is some 25%, counting activities related to agriculture such as food processing.[4] But agriculture is by no means the dominant part of the contemporary economy that it was even two or three generations ago. If the source of Kondratieff cycles was the pulse of agriculture, an historical trend toward decreasing amplitude in the waves would be expected. That smoothing, indeed, is especially likely given that government policy since the time of Keynes has been to offset the business cycle through stimulation of the economy during downturns. In reality, little diminution of the cycles is observable. More than agriculture-related factors must be at work.

Insofar as economic cycles are derived from technology, the existence of systematic spurts and lags is believable. In the scientific fields that all through the Kondratieff period have formed the main knowledge-base for technology (that is, from the 1790s to the present), a conclusion about cycles has been reached independently of any economic analysis of business waves. Thomas Kuhn's noted work *The Structure of Scientific Revolutions* held that scientific knowledge has not grown as a smooth cumulation, but instead in a series of intellectual "revolutions." New ideas struggle against and finally overcome old ones, leading to a period of what Kuhn termed "normal science" in which the implications of a ground-breaking idea are worked out.[5] DNA research, for example, which began its scientific revolution in the 1950s must by now have entered its normal science phase.

Modern technology is not, however, just the application of the scientific principles of a *knowledge-base*. As noted in Chapter 1, the technologies that enter the history books as viable and consequential are usually, especially in the more modern era, business ventures that were invested in with the expectation of a future return and profit. Mensch, indeed, refers to just such deployed, production technologies when he writes of "basic innovations" (130). Technology in this sense is part of the business cycle by definition; the deployed innovations respond to perceived needs and exhibit a built-in resting period following the introduction of a new technology during which financial backers hope to recover and multiply their investment. Thus, writings on the adoption

of technology by business corporations refer to "the technology life cycle."[6] The literature on patents and their development illustrates such issues in the deployment of technology (see Chapter 2). When the steam engine was being developed in the eighteenth century, for example, Watt's improved machine did not immediately take over, and the earlier Newcomen engines were still in use well into the nineteenth century, and even into the early twentieth century in at least one instance. In modern times, the lightning rapidity of computer technology can be problematic for producer and consumer alike, since excessively instantaneous obsolescence is destabilizing for both.[7]

Both the above two ideas — scientific revolutions and investors seeking returns — can be tied together within the notion of diffusion along a bell curve which cumulates to the S-shaped distribution (recalling from Chapter 3). If technological diffusion is seen as part of a more general process of cultural diffusion, a scientific revolution — meaning a revolution in a particular domain of ideas — is a cultural event which any student of diffusion will expect to spread in the predictable bell curve form. Willingness to invest in technologies, the point just discussed above, probably has exactly the same dynamic. Recall the adjectives used by Rogers, the diffusion theorist noted in Chapter 3: "innovators," those risk takers on the front edge; "early adopters," the next most bold group; and the majority who wait until the middle phase when an innovation is proven. Plot a series of bell curves describing the rate of adoption of a sequence of important technologies, lay them out in historical sequence, and the result could overlay the Kondratieff economic cycles in Figure 4.1. That essentially is the thesis of Mensch's book, noted above.

The relationship between technology and economy seems to be as follows: invention is part of human nature, given anthropological ideas about the definition of humanity. Inventive activity, then, will occur in some form or another no matter what the stage of the economy. Invention of technology is also, however, need-driven even if the definition of those needs is rather arbitrary (recall Basalla's analysis of need, from Chapter 2). At some point in the Kondratieff cycle, perhaps as the crest of an upswing is passed and decline begins, a perceived need for revitalization begins to grow. This perception could express itself within the social and cultural structure of societies as educational reform and increased prestige for technical and scientific activities. By the bot-

tom of the Kondratieff cycle, such changes begin to result in inventive surge, measurable by an increasing yearly frequency of patents. An accelerating deployment of new technologies helps initiate a new Kondratieff cycle of economic upswing. For a dressier social science phrasing, it could be said that technological innovation and economic growth are linked in a "lagged negative feedback" (Goldstein, Further Readings, 274). In other words, growing production leads, after a lag, to decreased innovation, but after a period of falling production a technological renewal is likely.

Technology Importing as a Strategy for the Canadian Economy

Marshall McLuhan, the 1960s media theorist encountered in Chapter 1, explored the notion of "leapfrogging." Spinning the kind of paradox he adored, he reasoned that new technologies sometimes could be embraced more totally by a relatively backward society than by a more advanced one. Too complete an immersion in the last technology impedes adoption of a radically new one. "It is this very situation which today puts the Western world at such a disadvantage, as against the 'backward' countries," he wrote in *The Gutenberg Galaxy*. Such logic was no idle passing thought for McLuhan, for reference to leapfrogging appears more than once in his correspondence. "We [in Canada] enjoy many of the twentieth-century opportunities of other 'backward' countries," he wrote in a letter almost ten years after the publication of *Galaxy*. "That is our privilege for having missed the nineteenth century."[8]

McLuhan may have been thinking more of communications media than economics, but the notion of leapfrogging in economics was advanced by no less an historical personage than the Russian revolutionary Leon Trotsky. Leapfrogging also has adherents among some contemporary economists. One reasoned, for example, that "countries with a lower level of technology will have a greater potential for growth since the transition costs of moving from one technology to another are lower," and that "late developers may have an advantage over the early developers in that there are fewer institutional constraints on late developers to adopt new technology."[9] An economic leapfrog of this kind can combine the advantages of being both an innovator on the front end of a diffusion S-curve, who seizes the competitive edge, and one of

Table 4.1: GERD by Sector, Canada, 1963-1995

Percentages (summing to 100% across rows)

Year	Government	Business	Higher Education and Private non-profit organizations
1963	42	38	20
1964	38	41	21
1965	36	43	21
1966	35	42	23
1967	36	39	25
1968	37	37	26
1969	34	38	28
1970	33	39	29
1971	32	33	35
1972	33	34	33
1973	34	35	32
1974	33	37	30
1975	32	37	31
1976	32	37	32
1977	30	37	32
1978	30	39	31
1979	27	42	32
1980	25	45	31
1981	24	49	28
1982	24	49	28
1983	25	48	28
1984	24	49	26
1985	22	53	25
1986	21	55	25
1987	19	56	25
1988	19	56	25
1989	19	55	26
1990	19	54	26
1991	18	54	27
1992	18	55	27
1993	17	57	26
1994	16	58	26
1995	15	59	25

Note: "Government" here is the sum for federal and provincial. The provincial share is either 3% or 4% throughout the series. The share for "private non-profit organizations" is a constant 1% throughout the series. It is due to rounding errors when percentages fail to add to exactly 100%.
Source: Statistics Canada Cat. 88-202-XPB, *Industrial Research and Development- 1995 Intentions*, Table 1.3.

TECHNOLOGY + SOCIETY

Table 4.2: Canada's R&D Efforts in International Perspective, 1994

Country	GERD (billions of US$)	GERD/GDP (%)	Financed by Public Sector (%)	GERD per capita (US $)	Researchers per 1000 Labour Force
US	170.0a	2.61a	38.9	659a	n.a.
Japan	74.8a	2.93a	21.4a	600a	14.3a
Germany	37.3a	2.48a	37.1a	459a	12.5b
France	26.0a	2.41a	44.3b	451a	12.2b
UK	21.6a	2.19a	32.3a	373a	9.7b
Italy	12.8	1.21	45.9	236a	5.8b
Canada	8.7	1.47	42.4a	289a	8.3b
Netherlands	5.06	1.87b	45.66	327b	9.3b
Sweden	4.6a	3.12a	35.3c	525a	12.4a
Australia	3.7d	1.36d	54.9d	213d	8.0d
Norway	1.6a	1.94a	49.1a	378a	10.4

a 1993; b 1992;c 1991;d 1990
Source: Modelled on John de la Mothe and P.R. Dufour, "Engineering the Canadian Comparative Advantage: Technology, trade, and investment in a small, open economy," *Technology in Society* 12 (1990), 374, with updating from *Main Science and Technology Indicators 1994-1* (Paris: OECD, 1995), Tables 2, 4, 5, 9, 14.

the later-adopting majority who wait for risks and costs to move down.

Post World War Two Japan illustrates an effective leapfrogging, especially with reference to microelectronics. The transistor had been invented in the US (see Chapter 2 and summary information in Figure 2.2), but was slow to diffuse into the American electronics industry in part due to inertia arising from an industry already organized around vacuum tube technology. Japan in contrast was emerging from the ashes of war with freedom to innovate. Sony, now a household name around the globe, grew to its pre-eminent position by sending technicians to the US to learn all they could about the transistor.[10]

Leapfrogging should be relevant to Canadians since we not only "missed the nineteenth century", as McLuhan noted, but were late industrializers well into the twentieth century. John Porter, the master interpreter of early post World War Two Canada, was writing chapters about "off-farm migration" in the 1950s and 1960s, a social change that had long since occurred in the US.[11]

If leapfrogging were as powerful a force as implied by McLuhan,

surely the world would be a happy ensemble of equally prosperous nations, and phrases such as "international aid," "north-south divide," or "the third world" would be obsolete. As soon as one nation lagged behind the rest, it would be positioned to jump back up to the top. Despite the elegance of its simplicity, such a theory ignores too many of the realities of political science. The Quebec scholar Jorge Niosi has criticized "international trade" theories for their naivety in neglecting the role of the state in economic development. As argued by one of the authors in *Technology and National Competitiveness*, a collection of articles edited by Niosi, "the complementarities, interfaces, and synergies associated with contemporary technology can only flourish if the organizational forms — market and non-market — in which they emerge constitute an interconnected whole."[12]

We begin, in dealing with such ideas, to appreciate the complexity of national policy-making in science and technology. The potential for a future leapfrog using new technology diffused from abroad will always be worth consideration. It is a recipe for perpetuated dependency, however, to unreflectingly import technology at the expense of R&D work tailored to local needs. It may become advantageous for transnational companies to dump their obsolete technologies onto less developed nations. Thus, Canada would still have to confront the issues of being a "small, open economy" on the international stage even if we, like Sweden, had an ocean between us and the US. As one Canadian authority on these issues noted, "the OECD nations are all approaching an important politico-economic crossroads" with respect to the new economy that is founded much more on international trade and concentration of technology than a generation ago.[13]

Part of the "crossroad" for Canada was the 1988 decision to enter into a Free Trade Pact with the United States. A century ago, such a policy went by the name "reciprocity." Sir John A. Macdonald persuaded the electorate to favour instead a National Policy based on protective tariffs and Federal Government leadership in developing the western frontier lands. Macdonald's tariffs were designed to raise the cost of imported goods and thereby favour domestic producers. The controversy between protectionism and reciprocity was renewed a century later in the 1980s: was the opening up of trade a far-sighted adaptation to the new millennium or a sell-out of Canada's national interests?

Canada's dilemma in 1988 was part of a world-wide economic

"restructuring," to use a term much favoured in business publications. As a *Globe and Mail* story noted on 30 June 1995, among the world's richest nations, tariffs on industrial goods dropped from 40% in 1947 to 6.3% by 1995. "(T)he new dominance of trade, technology, and capital has given governments a new chance to create jobs for a rapidly growing human population," its story on a World Bank report noted. "More than one billion workers, most of them in China, India, and Russia, are employed in economies just now entering the global market" (A7).

Free trade exposes Canadian business to competition from all over the world. It becomes easier for foreign producers to sell goods here, and therefore a danger arises that the traditionally high foreign control of the Canadian economy will become even greater. By the same token, however, Canadian producers gain access to foreign markets. An efficient competitor from any province potentially can reach a far larger market than under conditions of national protection. Research and development work deemed too expensive for a low volume producer may become plausible in the expanded marketplace of the global economy. In addition, easy diffusion of technology could be enhanced by free trade because imported goods can be carriers of technology. Such "embedded" technology used to be illustrated with great clarity during the years of the Cold War between the US and the former Soviet Union. From time to time a Soviet pilot would defect, flying his warplane across the borders known in those days as the "Iron Curtain." Engineers from the west would tear such contraband planes to pieces to learn their technological secrets.

Any comprehensive assessment of impacts of free trade north of the 49th parallel is beyond our scope here although some of the social costs of such rapid economic changes are reviewed in Chapter 6. In terms of the national commitment to technological research ("R&D"), however, the evidence to date on free trade is not as unfavourable as the public might think. Table 4.1 shows Statistics Canada figures over thirty years on the percentage of gross domestic expenditure on research and development (GERD) for different sources. A pessimistic interpretation of free trade would have forecast a declining participation by business, reasoning that if free trade led to foreign takeovers, more and more R&D work would drift south to the US, or overseas. So far, no such pattern has occurred, since business has accounted for just over half of

all Canadian R&D from the mid-1980s through into the 1990s. It is the Federal Government, Sir John A. Macdonald's instrument for nation-building, that has partially withdrawn from R&D expenditure, and that is a longstanding trend that began twenty years before the 1988 trade deal was enacted. The transfer from public to private sector is deliberate policy thought to reflect growing national maturity.

Consider, for example (Table 4.2), the low percentage of publicly financed R&D for countries such as Japan or Germany, two of the most successful nations in the new economy. The US is anomalous in this respect, since much of their intense government involvement in research comes from the defence budget. Internationally, Canada is indeed a small figure in R&D research, as the cross-national figures in Table 4.2 document. One of the most frequently encountered laments about Canadian economic policy is the low ratio of R&D expenditure to the size of the economy.[14] That figure, 1.5% of gross domestic product in 1994 (from Table 4.2), puts this country well toward the bottom of the leading industrial nations of the world. Back in the early 1970s, an advisory body on science policy was calling for a GERD expenditure equalling 2% of GDP as a reasonable goal for Canada, and more recently a figure of 2.5% has been proposed.[15] It is no wonder, given the current figures, that Canada has tended to be a net importer of technology, relying heavily on the diffusion of innovations from other lands. The noteworthy exception is in telecommunications equipment, where companies such as Nortel (previously Northern Telecom) have made Canada a technology exporter. Such technological "niches" help free Canada from its "woodshed of the OECD" image.[16]

Perhaps enough now has been written in this chapter to persuade the reader that the diffusion of technology has become an increasingly crucial lubricant to the economies of modern societies. This argument has been developed, however, from the viewpoint of economists. Valid though issues of profit and loss are, there is a need to think like sociologists and engineers in considering the implications for technological migration. We need to reflect on the notion of *appropriate technology.*

Appropriate Technology: Small is Beautiful?

Few university students of the past generation have graduated entirely unexposed to E.F. Schumacher's famous little volume *Small is*

Beautiful. An academic cosmopolitan, Schumacher was born in Germany, studied at Oxford in Britain and worked for some years at Columbia University in New York City. It may have been Schumacher's work as Economic Advisor to Britain's National Coal Board in the 1950s and 1960s that helped stimulate his concern about the speed with which resources were being consumed in the early 1970s. Resources such as coal and oil, the power source for so much twentieth century technology, were "capital provided by nature" (*Small*, Further Readings, 11). Schumacher regarded the connection between technological innovation and business as dangerous, because in that domain economic profit was the only criterion that mattered.

It is unlikely that Schumacher's somewhat romantic espousal of a non-materialist "Buddhist economics," by which the needs of the human spirit rule economic decisions, could easily displace western values. Indeed, it is the Buddhist societies such as Japan, China, or Korea which have become in some ways more "western" than the west. Schumacher's enthusiasm for Buddhist ethics led him into a dated view of gender relations, revealingly displayed in the following passage: "(T)o let mothers of young children work in factories while the children run wild would be as uneconomic in the eyes of a Buddhist economist as the employment of a skilled worker as a soldier in the eyes of a modern economist" (47).

So I do not propose Schumacher as the final word on social engineering in the technological society, but nevertheless his idea of "intermediate technology" did have considerable impact on programs for foreign aid. Intermediate technology was "vastly superior to the primitive technology of bygone ages but at the same time much simpler, cheaper, and freer than the super-technology of the rich" (128).

Intermediate or, as it came to be referred to more frequently, "appropriate" technology gave an important redirection to international development. A cluster of books on the topic appeared in the late 1970s and early 1980s, following and adapting Schumacher's ideas (see Further Readings). These claim that appropriate technologies should engage the massive labour forces of Third World countries, should protect non-renewable resources, attend to the first Maslowian needs such as food and shelter, and be sensitive to local cultures (see for example, Dunn, in Further Readings, 5). Appropriate technologies were deployed mainly in rural areas of poor countries, constructed on a small scale in

both a physical and capital sense. A typical such project was reported in the *Globe and Mail* of 20 February 1995. Concerned with water catchment for irrigation in India, this story illustrates not just the innovations in engineering but also in social organization needed for a successful deployment of technology in the Third World. In a largely self-administered program, villagers in the Jhansi district of central India built "three small check dams, trenched their hillocks and built embankments around their fields to stop water loss and soil erosion." The small-scale approach worked better than much larger and more expensive government-built dams, most of which "are leaking or collapsed." Another example is the wind-up radio reported in the *Globe and Mail* of 19 February 1996 and 5 December 1996. In South Africa where the "Freeplay" radio is marketed, half the homes have no electricity and batteries are expensive. A radio is "one of the three big African status symbols ... on the basis of just a radio, make no mistake — you can procure a wife."

The "AT" (appropriate technology) movement has been an important force in the war on Third World poverty, but it also has an implication for our reflection on technological diffusion. Schumacher's crusade against the unthinking export of western science and engineering to poorer lands brought culture and local conditions back to mind as a context for technology. It also suggests one reason why "leapfrogging" is not a sure strategy for economic development. Technology has always adapted to new geographies and cultural settings, as far back as the stirrup, whose varying forms were reviewed at the beginning of Chapter 3.[17] It has been said of early Canadian technology that it was "a process of adaptation." Sinclair *et al.* continued, "(n)othing more strongly characterizes the Canadian situation than the continual selection of techniques from other societies. Nor was that a bankrupt approach. The ability to adapt is just as critical as the ability to originate."[18]

Conclusion

The Kondratieff cycle has long been controversial for economists, and the present discussion is in no way a definitive treatise on the business cycle, but a recognition that it provides a confluence of themes worthy of note for technology and society students. The economists who

espouse the cyclic interpretation have made technological innovation an important component of their argument, and that exemplifies a theory in which technology is simultaneously a cause and an effect. The humanist economist E.F. Schumacher embeds technology within culture with his notion of appropriate technology. Canadians have always been both importers of, and adapters of, foreign technology, and that is more true now than ever before. Canada's challenge in the twenty-first century is to supplement the imported technology with indigenous research in "appropriate" areas or "niches."

Notes

1. Compare, for example, Solomos Solomou, *Phases of Economic Growth, 1850-1973* (Cambridge: Cambridge University Press, 1987) with Jan Reijnders, *Long Waves in Economic Development* (Aldershot, U.K.: Edward Elgar, 1990). Commentary on long wave cycles by a Canadian economist appears in Peter Howitt, "Adjusting to technological change," *Canadian Journal of Economics* 27 (November, 1994).
2. Edward J. Malecki, *Technology and Economic Development: The Dynamics of Local, Regional and National Change* (Burnt Mill, Harlow, U.K.: Longman Scientific & Technical, 1991), 169.
3. Nathan H. Mager, *The Kondratieff Waves* (New York: Praeger, 1987), 2-3.
4. Statistics Canada, *Canada Year Book 1985* (Ottawa: Supply and Services, 1985), 288; Statistics Canada Cat. 71-001 (Feb., 1995), B-24.
5. Thomas S. Kuhn, *The Structure of Scientific Revolutions* (Chicago: University of Chicago Press, 1962).
6. E.g., Peter A. Hugunin, Susan Thomas, and David Wilemon, "Science and technology information and corporate planning processes: A synthesis," *Technology in Society* 15 (1992), 251.
7. Steam engines: D.S.L. Cardwell, *Technology, Science and History* (London: Heinemann, 1972), 108; display at the Science Museum, London England. Computers: Paul A. Herbig and Hugh Kramer, "The phenomenon of innovation overload," *Technology in Society* 14 (1992), 441-61.
8. Marshall McLuhan, *The Gutenberg Galaxy* (Toronto: University of Toronto Press, 1962), 27; Matie Molinaro, Corinne McLuhan and

William Toye, *Letters of Marshall McLuhan* (Toronto: Oxford University Press, 1987), 441 (see also 340). Leapfrogging theories are still alive among intuitive futurists, as readers of *The Globe and Mail Report on Business* for July 1995 (24) saw with reference to the celebrated guru Peter Drucker.

9. Solomou, 161. Also, Giovanni Dosi and Luc Soete, "Technological innovation and international competitiveness," in Jorge Niosi, ed., *Technology and National Competitiveness: Oligopoly, Technological Innovation, and International Competition* (Montreal & Kingston, McGill-Queen's University Press, 1991), 108.

10. George Basalla, *The Evolution of Technology* (Cambridge: Cambridge University Press, 1988), 86-87.

11. John Porter, *The Vertical Mosaic: An Analysis of Social Class and Power in Canada* (Toronto: University of Toronto Press, 1965), 140-44.

12. Francois Chesnais, "Technological competitiveness considered as a form of structural competitiveness," Niosi, 166.

13. John de la Mothe and Paul R. Dufour, "Engineering the Canadian Comparative Advantage: Technology, trade, and investment in a small, open economy," *Technology in Society* 12 (1990), 369-96.

14. E.g., "We have become incompetents in the unfolding of scientific civilization" (President of the National Research Council of Canada, reported in the *Globe and Mail*, 3 March 1989).

15. de la Mothe and Dufour, 378-79. This review article on science policy in Canada also identifies the controversy about using summary indicators such as GERD/GDP.

16. Thomas Wudwud, "Technology development and Canada-US Free Trade," *Technology in Society* 11 (1989), 249; de la Mothe and Dufour, 371.

17. For an explicit statement, see Robert B. Gordon and David J. Killick, "Adaptation of technology to culture and environment: Bloomery iron smelting in America and Africa," *Technology and Culture* 34 (April, 1993), 243-70.

18. B. Sinclair, N.R. Ball, and J.O. Petersen, *Let Us be Honest and Modest: Technology and Society in Canadian History* (Toronto: Oxford University Press, 1974), 2.

Further Readings

For commentary on business cycles:

Goldstein, Joshua S. *Long Cycles: Prosperity and War in the Modern Age.* New Haven: Yale University Press, 1988. In Goldstein's scheme, Mensch's interpretation (see reference below) carries the ideological name tag of "liberal" (32, 149).

Kondratieff, Nikolai. *The Long Wave Cycle.* Trans. by Guy Daniels. New York: Richardson & Snyder, 1984.

Mager, Nathan H. *The Kondratieff Waves.* New York: Praeger, 1987.

Mensch, Gerhard. *Stalemate in Technology: Innovations Overcome the Depression.* Cambridge, Massachusetts: Ballinger [Harper and Row], 1979.

For analysis of how free trade served the interests of particular sectors of Canadian society:

Richardson, R. Jack. "Free trade: Why did it happen?" *Canadian Review of Sociology and Anthropology* 29 (August, 1992), 307-28.

For discussion of "appropriate technology":

Appropriate Technology. A journal published since 1974.

Dunn, P.D. *Appropriate Technology: Technology with a Human Face.* New York: Schocken, 1979.

Pursell, Carroll. "The rise and fall of the appropriate technology movement in the United States, 1965-1985." *Technology and Culture* 34 (July, 1993), 629-37.

Rybczynski, Witold. *Paper Heros: A Review of Appropriate Technology.* Garden City, New York: Anchor Doubleday, 1980.

Schumacher, E.F. *Small is Beautiful: A Study of Economics as if People Mattered.* London: Penguin Abacus, 1973.

Vacca, Roberto. *Modest Technologies for a Complicated World.* Oxford: Pergamon, 1980.

Willoughby, Kelvin W. *Technology Choice: A Critique of the Appropriate Technology Movement.* Boulder: Westview, 1990.

This sign proclaiming technocracy has for many years been visible from Highway 8, on the outskirts of Dundas, Ontario.

Chapter 5: Technocracy: Class or Culture?

Technology creates a master class of technocrats lording it over the postindustrial serfs, it "displaces" workers into unemployment by the thousands, and robs most of the remaining jobs of their skill and dignity. Apart from that, technology is wonderful! This deliberately provocative opening thought will be explored and evaluated in the following pages of this chapter and the next, as we consider some of the possible social costs of technology.

Technocracy and Technocrats

First references to "technocracy" date back to 1919, when an engineer from California referred to "rule by technicians."[1] It is not hard to understand why, in that period beginning just after World War One and extending into the 1930s, enthusiasm with the notion of technical experts assuming ultimate power would have abounded. Both the 1914-18 war and the Depression led to disenchantment with politicians to a far greater extent, probably, than even in our own jaded era. Technocracy then had a connotation of management of the society by "technically competent" people acting for the "good of everyone," as an alternative to power in the hands of "private interests" using career

politicians like puppets.[2]

The contention was not so much that scientists and engineers should leave their laboratories for the corridors of government, but that the same rational, ideologically neutral, evidence-using methods used to produce modern technologies should apply to the running of the state. Even in the 1930s, however, the idea of technocracy was controversial. A stinging critique was made by the Spanish writer Ortega y Gasset in his famous book *Revolt of the Masses*. By the mid-1940s, the noted English writer C.S. Lewis added his voice to the view that government involves more than technical competence.[3] And by the 1960s, following World War Two in which technical experts devised the deadly atomic bomb, most commentary on technocracy became increasingly impatient (for example, Meynaud, in Further Readings). Most people today would take the adjective "technocrat" to be a backhanded compliment at best. In contemporary writing, it is typical to encounter phrasing such as "an innocuous little package of technocratic details," as a *Globe and Mail* columnist wrote of a recent education report for Ontario (1 March 1996, A9). The "technocrats of the Quiet Revolution" built up big government in Quebec, but now this "Big Brother is broke," concluded another story (19 October 1996, D3).

A Note on Social Class

The present chapter will be concerned not with who *should* rule in postindustrial societies, but with who actually *does* rule. As background, let us review for a moment some standard sociological ideas about rulers and the ruled. Virtually throughout recorded history, human populations have exhibited forms of inequality or "*social stratification*." The upper strata possess wealth and power, the two tending to reinforce each other and enhance status, although in some circumstances people can simply hold high prestige for other reasons (for example, exceptional moral goodness). Since the time of Karl Marx, the latter nineteenth century, the principal form of inequality in industrial societies has been said to revolve around social classes deriving from capitalism in industrial societies. Owners, termed the "*bourgeoisie*" by Marx, possessed the "*means of production*" such as land, investment funds, buildings, and plant equipment. Their interests were diametrically opposed to those of workers, Marx's "*proletariat*." One

of the lessons of the great German revolutionary theorist was that the latter had to organize as a cohesive, mutually conscious, social class to counter the bourgeoisie. As later interpreters have depicted class relations, these principal social classes of capitalist society form two blocs. The bourgeoisie's power due to ownership of the means of production is aligned against a proletariat which gains some measure of power due to its sheer numbers, the force of collective action such as strikes, and ultimately perhaps the threat of full scale revolution. Other bases of inequality, such as gender or ethnicity, overlap class relations. Well into the twentieth century in Canada, for example, capitalists were predominately male and of British ancestry.[4]

For decades, the updating of Marxist ideas to accommodate shifting alignments within and between social classes has been important work for social scientists. The flamboyant 1950s sociologist C. Wright Mills, for example, contended that "white collar" workers in the US had significantly more bourgeois characteristics circa 1900 than fifty years later (See Further Readings). More of them were self-employed in the earlier period, while salaried white collar workers enjoyed relatively high pay and prestige over manual workers. By the second half of the twentieth century, bourgeois-like power among salaried workers was held primarily by office managers. The class position of such workers, an increasingly important group in the corporate economy where family-owned businesses were becoming scarcer, was analyzed by Eric Olin Wright beginning in the 1970s.[5]

It is within this tradition of class analysis that many social scientists interested in technology have worked. By the 1960s, the racing acceleration of technology was becoming unmistakably clear.[6] Given the extent of interpenetration between technology and the social environment in general (recall scheme from Figure 2.1), it can only follow that social class alignments were shifting too. Two authors of that era, Daniel Bell and Alain Touraine, predicted somewhat alternative directions for such changes. With the benefit of some twenty-five or thirty years of hindsight, we can, in a form of retrospective social experiment, consider which turned out to be most accurate.

Bell and Touraine

Both Bell and Touraine referred in the titles of their books to "postin-

dustrial society" — a term that has since become familiar to any educated audience. Postindustrial societies still have industry, but ever more of the economic activity shifts to the provision of "services," a term defined so broadly that it covers everything from working as a restaurant waiter to running a law firm. While only about 10% of the workforce of a modern economy is within the service sector in the most literal sense of domestic workers and workers in food, accommodation, retail trade and recreation, as many as 70% fall within this sector when defined to encompass handlers of information.[7] A set of figures on this evolution of the service sector within the Canadian economy appears below in Table 5.1. After cresting up to just one-tenth of a decimal point below the 70% level by 1994, the total service sector had reached 71% of the labour force in the most recent figures (1996) entered in the table. Many of these service workers do not necessarily deal directly with the public, but they handle information that is of service to people. None of the services involves the production of things, "goods," as economists would say. Working at a plant building cars, for instance, is production of goods while selling the cars is service work.

Both Bell, an American sociologist, and the French writer, Touraine, pointed to changes in the *knowledge-base* for technology as the crucial driver toward postindustrial society. That was one reason for including knowledge-base in the Chapter 2 typology. In *The Coming of Post-Industrial Society*, Daniel Bell (see Further Readings) wrote of "the centrality of theoretical knowledge as the source of innovation and of policy formation" (14). Similarly, Touraine argued early in his book that: "Nowadays, [economic growth] depends much more directly than ever before on knowledge, and hence on the capacity of society to call forth creativity" (Further Readings, 5).

Exactly what sort of knowledge were Bell and Touraine discussing? It is not every possible sort, but rather knowledge which is scientific, theoretical, and "algorithmic." Bell, especially, stressed that the mid-twentieth century was the era in which technology had come to derive almost exclusively from a scientific, theoretical *knowledge-base*. Such knowledge was "scientific" in that it derived from application of the scientific methods of systematic research using experimentation. It was "theoretical" in that the general laws of nature relating to a technology were crucial. Pre-twentieth century technologies could be invented by a partly intuitive, trial and error, process. Alexander Graham Bell,

Table 5.1: Growth of the Service Sector of the Canadian Labour Force (Non-farming Portion)

| | Share of Labour Force (%) | | | | |
	1961	1971	1981	1991	1996
1. Industrial Sector					
Extractive	4.3	3.1	3.1	2.5	2.1
Manufacturing	25.3	22.8	20.1	15.2	14.2
Construction	7.8	6.8	6.7	6.6	5.7
Transport, Communications, Utilities	10.6	8.8	8.4	7.6	7.0
Total Industrial	47.9	41.5	38.2	31.9	29.0
2. Service Sector					
Trade	18.3	17.6	18.1	17.6	16.1
Business, Social & Consumer Services	25.7	32.8	36.0	42.1	48.1
Public Administration	8.0	8.2	7.8	8.2	6.8
Total Services	52.1	58.5	61.8	67.9	71.0

Source: John Myles, "The expanding middle: Some Canadian evidence on the deskilling debate." *Canadian Review of Sociology and Anthropology* 25:347, updated to 1991 using Statistics Canada, *1991 Census Technical Report, Industry*, Cat. 92-338E, Table D3, and to 1996 using Statistics Canada, *Historical Labour Force Statistics*, Cat. 71-201 (1996), with author's linking adjustment to approximate industry classification used in earlier part of the series.

for example, knew only a little about the theory of electricity.[8] The 1940s "Project Manhattan" for building an atomic bomb, in contrast, was an explicit application of Einstein's advances in theoretical physics from the 1920s and 1930s to the manufacture of a radically new explosive device. So it is in general, Daniel Bell says, that modern technologies "are primarily dependent on theoretical work prior to production" (25). Next, to speak of knowledge being "algorithmic" is to realize that it is codified. A simple but familiar example of this term borrowed from computer science is the instruction sheet enclosed with an unassembled dresser or desk. It is efficient to manufacture such items

as precut and predrilled boards sold in boxed, kit form, together with a bagged set of fasteners. That is a simple piece of theoretical scientific knowledge, the operative principle being that boxed kits both reduce shipping costs and storage costs for wholesalers and retailers alike, and also pass on to the consumer as a generally unnoticed cost the time involved in the labour intensive stage of assembly. The assembly instructions are algorithmic in that they are a substitution of "problem-solving rules," as Bell would say, for "intuitive judgments" (29). The National Museum of Science and Technology in Ottawa has a display board to emphasize the importance of "algorithms in our lives." An algorithm, it begins, "is a list of automatic tasks used to solve a given type of problem. Many of your acts and gestures are the result of algorithms: making breakfast, starting a car, shopping in a department store, etc."[9]

An algorithm can as easily be a management procedure from the social science *knowledge-base* as a tool-using one based on the natural sciences. Daniel Bell refers to Keynesian economics as a leading exemplar of influential social science knowledge (see, for example, 23). And it is largely the dominance of social science knowledge that led Touraine to write of "programmed societies" as synonymous with the technocratic society (3).

Beyond this common denominator, however, the American and French writers began to diverge in their interpretations. Bell prophesied that possession of technical knowledge would come to bestow power and status to the holder. "In the postindustrial society," he wrote, "technical skill becomes the base of and education the mode of access to power" (358). This new power bloc, the "technical and professional intelligentsia" were taking on some of the characteristics of a social class (362). Although lacking, perhaps, the unity of interests of the bourgeoisie of industrial society, the technocrats shared norms of professionalism and faith in technical solutions to human problems. Bell was not saying that the old industrial classes would disappear overnight, nor that elected politicians were destined to lose all their power. Countries with histories stretching back to pre-industrial stratification systems based on the rural aristocracy, England for example, still have "feudal remnants" of those times.[10] These overlay the Marxian social classes of the industrial era and the knowledge classes of postindustrial society.

Alain Touraine recognized a distinction between "old and new social classes" too, but for him the new ruling class was not scientists in universities and industrial research parks acting as an independent interest group. "Technocrats," according to Touraine, "are not technicians but managers, whether they belong to the administration of the State or to big businesses" (49,50). It was "absurd" he insisted to suppose that "the power of the technicians ... could be substituted for political power ..." Only "imaginative spirits" could imagine "the reign of the engineers" (158). Indeed, at around the same time, Touraine's countryman Serge Mallet was suggesting that skilled workers in the more technology-concentrated industries had the most potent forms of working class consciousness (see Further Readings). Far from being a ruling elite, this "vanguard" of the working class, which could include some of Touraine's lower end technicians, would be a countervailing force against the establishment. Argued from a couple of directions, then, we conclude that technocracy is not just a group of people, but a group of bureaucracies, a sociological entity involving something more than merely a collection of individuals.

A subtle distinction is being made here, and a restatement from another direction may help. People with technical skills and inventive genius, Touraine predicted, were not going to become a new ruling class just because of their mastery of technology. It was the whole *culture* of technology that was assuming power, assuming "centrality," as two British authors phrased it.[11] Much of this culture involved the social values that foster technological research and are in turn reinforced by the achievements of such research. Involved are a commitment to experimentation and in fact to evidence of all sorts over intuitive faith, a rationalistic belief that there is a single best way of accomplishing most tasks, and a realization that it is efficient to algorithmically record and reuse these optimum solutions. Budget modelling within organizations, cost/benefit analysis of social programs, or econometric analysis of the economy are just as much part of technocracy as the latest breakthrough in biotechnology. In the culture of technology, Touraine said, priorities become dominated by the exploitation of what is technologically feasible. The accumulation of technological knowledge *per se* assumes priority. He criticized, for example, the magnitude of expenditures on the nuclear and space industries as areas "in which firms can act simply in the name of progress of knowledge..." (159).

The ideas of technocratic culture probably take root throughout a modern society, but most especially among those receiving advanced formal education (see also Chapter 10). It is here that people are put back into Touraine's scheme, and that we speak again of "technocrats." It has to be remembered, though, that being a scientist in a laboratory does not a technocrat make. To use the term in Touraine's sense, two criteria must be met: (i) having a position of managerial power, usually within government or corporate industry and (ii) subscribing to the values of technological culture.

By setting this stringent criterion for a technocrat, Touraine positioned himself to exercise his typically Franco-socialist instincts in categorizing most technical people as a proletarian working class. Here, Daniel Bell may have been on firm ground when he criticized the French writer's fixation with nonmanagerial technicians as a new proletariat (*Coming*, 39-40). If this is a proletariat, it is one with a great deal of status (374), and Touraine may have been diluting the impact of this old Marxist term. In my reading, however, Bell was himself a little equivocal and inconsistent about the social class position of the technicians who produce technology. On one page he attributes power to them (358, or 426), on another he takes it away, saying they hold only high status, but little power (374). Elected politicians remain important, he argued in one place, in apparent contradiction to other passages. Many issues "cannot be settled on the basis of technical criteria; necessarily they involve value and political choices" (364). The ambiguity in Bell's thinking probably reflects reality — that technicians hold some power from their high prestige and credibility as a lobby group, but seldom a decisive amount. Bell's struggle with this question deflects attention, however, from the more abstract, but actually more decisive, issue raised by Touraine. "In the end," for Bell, a state of technocracy does not exist because "the technocratic mind-view necessarily falls before politics" (365). Touraine shifts emphasis by insisting that it is the social values supporting technological advancement that really have taken power, in that insidious way indeed introducing technocracy.

Other Voices on the Technocracy Issue

Touraine's notion of technocracy and technocrats has received emphasis in the last paragraphs because it seems to have stood up well since

the early 1970s when his book appeared. Neither electrical engineers, chemists and physicists, nor sociologists, economists, and psychologists rule countries such as Canada, the US or Great Britain today. But to some degree the ideas that such practitioners subscribe to do rule. Phrased another way, it is not just that individual technologies such as the computer may influence our lives, but that the much more general ideas outlined above and closely tied up with technology predominate. That several other writers with quite distinctive intellectual vantage points reached the same conclusion as Touraine helps to validate the French writer's interpretation.

We should note, for example, the work of Touraine's countryman Jacques Ellul, who was briefly mentioned in Chapter 1. Like so many other writers we have encountered and will encounter (such as Postman, discussed below), Ellul was highly aware of the changing place of technology in society throughout history. Beginning as just one domain coexisting with traditional culture, technology grew in influence over time until for the present era Ellul was using the words "the technological system" to describe the integrated dominance of the technocratic mindset (see book of that title in Further Readings). But where Touraine began with Marxist questions about the social class and power realignments of technological society, Ellul was concerned with broader issues of erosion in world cultural diversity. Often one reads of the shrinking genetic pool of life forms as the environmental damage accompanying the technological economy drives one species after another into extinction. On a cultural rather than genetic level that is the essence of Ellul's argument in his early post-World War Two book *The Technological Society* (see Chapter 1, Further Readings). Here, he wrote of "technical universalism," the way that "(i)n all countries, whatever their degree of 'civilization,' there is a tendency to apply the same technical procedures" (116). The emphasis on verifiable results within science and technology meant that invariably "the one best way" is established and diffuses (79). As Ellul remarked, that creates a powerful force for the standardization of cultures. "In the course of history," he wrote in *The Technological Society*, "there have always been different principles of civilization according to regions, nations, and continents. But today everything tends to align itself on technical principles" (117). The birth of the atomic age was clearly much on Ellul's mind as he worked out these ideas, for he often referred to

Project Manhattan and later atomic programs both for peace and war. It cannot have escaped his notice how similar the military technologies in the US and the USSR were even though in those early 1950s days the two powers had diametrically opposed political ideologies. That was an extreme instance of how technology standardized itself across even the most dissimilar of cultures.

Criticism of technocracy is equally important within the Canadian tradition in the study of technology. George Grant, the Canadian philosopher encountered in Chapter 1, was as aware as Touraine of the deep penetration of technocratic values into the ruling personnel of modern society. "Members of the dominant classes," he wrote, "make the decisions which embody the chief purposes of any society, but their very dominance is dependent on their service of those purposes."[12] The "purposes" Grant is citing here refer, of course, to capitalist profit-making via the deployment of technology. From this slightly new viewpoint, we are reminded of Touraine's contention that it is the technocratic values that tie together the power structure of a country.

There is a link here with research on Canada's ruling elites, a continuing interest among sociologists since the time of John Porter over a generation ago. One conception has been that heads of business and government are essentially the same people, shuffling effortlessly back and forth between the corporate board rooms of Toronto's Bay Street and the cabinet rooms on Parliament Hill, Ottawa. A marker of that facet of Canadian public life was a *Globe and Mail* story headline "Mulroney leads busy life sitting on corporate boards, extent of ex-PM's business posts revealed at hearing" (19 April 1996). Another view sees such examples as merely the window dressing of a more profound linkage between capitalism and the state, a unity of interests whereby both accept without a moment's question the priority of using science and technology to explore every avenue for profit-making.[13] That is the point made by Grant, echoing the French tradition reviewed above.

Grant's ideas are of further value for the clarity with which he named the political identity of technocracy as a social value, arguing it was synonymous with the ideology of "liberalism." Liberal society believes that government should do little more than ensure equal opportunity for all citizens, and should permit inequality that is deemed to be earned. Because liberalism encourages entrepreneurs, historically it has been compatible with capitalism. It is synonymous with

faith in progress and enthusiasm with the technologies that fuel capitalism. Even when capitalism solidifies into an entrenched bourgeoisie that does in fact deny equality of opportunity to the proletarian working class, liberalism can easily remain the *"dominant ideology"* because systemic causes of social inequality tend to remain hidden.

The energetic espousal of the individual's freedom from restraint by the state within liberal ideology gave, then, the ideal climate for technological innovation. It silenced the stuffy, faith-driven voices of more traditional ideologies, from which might be heard moral arguments against technology. Grant saw this liberalism as existing in its strongest form in the US, whereas Canada he felt had traditionally stood for a pretechnocratic *"conservatism"* leaving the individual better cared for albeit less free. With the growth of US power over the twentieth century, "… (c)onservatism must languish as technology increases," as he wrote in *Lament for a Nation*, his most famous book.[14]

For as long as social classes have existed, the celebration of the individual within liberalism has been partly a sham. It is self-serving for the bourgeois to tell the proletarian that any individual with enough brains and gumption can get ahead when generations of sociological research shows that schooling is a thoroughly middle class system that favours its own. Liberalism is not merely the application of individual exertions so long as capitalism requires capital, and so long as an especially convenient source of business financing is wealth or a business inherited within the family. Liberalism thus in fact relies on the collective interests of groupings within a society such as a bourgeois class that wants the law to leave inheritance of businesses untaxed. No wonder then that political theorists after Grant increasingly referred to a resurgent *"corporatism"* which translates the rights of the individual into the group interests of business firms and government bureaucracies, bodies such as ethnic and gender associations, age-based lobbies such as "grey power" and, as just noted, social classes.[15] This powerful adjunct to liberalism receives further consideration in Chapter 10. Today we would say it is the evolved blend of liberalism and corporatism that is most clearly the ideological voice of technocracy.

Technology as a dominant social value was articulated as well by John Kenneth Galbraith when he coined the term "technostructure" to describe the penetration of the values of technology into bureaucratic decision-making. Likewise more recent commentary by Ursula

Franklin, a noted Canadian physicist, refers to a "technocentric mindset" which gives technology momentum. "(T)he spread of technology," Franklin says, "has resulted in a web of infrastructures serving primarily the growth and advancement of technology."[16]

A popular novel from 1995, a quarter of a century after Bell and Touraine, may give an accurate assessment of the class realignments of the late 1990s. Douglas Coupland's novel on life in the silicone valley's computer industry is starkly entitled *Microserfs*. These serfs of the twentieth century, Coupland says, are not the motherboard assemblers in some Pacific Rim sweatshop, but the alienated computer scientists spending their days "writing code" at companies such as Microsoft. "I am a tester — a bug checker in Building Seven," the main character Daniel tells the reader. "I worked my way up the ladder from Product Support Services (PSS) where I spent six months in phone purgatory in 1991 helping little old ladies format their Christmas mailing lists on Microsoft Works."[17]

Data Trash: the Theory of the Virtual Class by Kroker and Weinstein (Further Readings) cuts across both Bell's and Touraine's concepts. "The most complete representative of the virtual class is the visionary capitalist," but sharing membership are the "perhaps visionary, perhaps skill-oriented, perhaps indifferent techno-intelligentsia of cognitive scientists, engineers, computer scientists, video-game developers, and all the other communications specialists, ranged in hierarchies, but all dependent for their economic support on the drive to virtualization" (15). This formulation, wide-ranging as it is, would probably leave both Bell and Touraine uneasy about the neglect of the state sector, the politicians and government managers at all levels who are thoroughly involved with the tools and mindset of technology, but who are separate from capitalism. For this reason, while the virtual class described by Kroker and Weinstein no doubt "practices a mixture of predatory capitalism and gung-ho technocratic rationalizations" (5), it does not fully overlap the technocracy conceived by either the American or the French writer considered above.

George Grant maintained that the US is the most technology-embracing, technocratic society in the world. Insight though that is, one needs to remember as well what a huge, varied, pluralistic society the US is. There are plenty of voices critical of technocracy south of the border, even if they remain the minority. Langdon Winner's work from

a generation ago entitled *Autonomous Technology* (see Further Readings) is an early example. He took sharp issue with the contention that technology is neutral, one of the cornerstone beliefs of technocratic society illustrated by the American gun lobby slogan "guns don't kill people, people do."

Neil Postman, a more recent critic, originated the term *"technopoly"* (Further Readings). This well-known opponent of technology is an avowedly old-fashioned humanist who rejects the word processor for a pen (*Globe and Mail*, 14 March 1996, C1). He claims that the relationship between technology and society has gone through two stages and has now entered a third, the ultra-technocratic or technopolistic era, substituting this jazzier term for what Ellul had simply called the technological system (see above). The first, and best, stage was the era of "tool-using cultures" (23) which lasted until the seventeenth century and was conceived by Postman as the time when technology served the culture. The tools were either to "solve specific and urgent problems of physical life" or to serve "the symbolic world of art, politics, myth, ritual, and religion" (23). Two examples are the water-wheel and the medieval cathedral. While many of the technologies from this first period may seem simple to us, simplicity is not their defining characteristic. For Postman, who notes what feats of engineering were achieved in medieval times, the essential fact about tool-using cultures was that "the tools are not intruders. They are integrated into the culture in ways that do not pose significant contradictions to its world-view" (25).

The age of technocracy began to arrive, according to Postman, with the mechanical clock, the movable type printing press, and the telescope (see Figure 2.2 for dates). The clock ushered in the tyranny of schedules running people, the press broke down oral tradition (see Chapter 7 for further discussion), while the telescope was key to The Scientific Revolution reviewed in Chapter 2. In this second phase, technology became an independent force, a distinct segment of societal culture in ever more uneasy co-existence with the older culture. Francis Bacon (born 1561) was, for Postman, "the first man of the technocratic age" (35), a period that gathered momentum thereafter as it moved toward The Industrial Revolution of the nineteenth century.

It was a non-tool technology, the scientific management of Frederick Taylor (see Chapter 6 for details), specifically a publication dated 1911,

that Postman selects as a marker of the technopolistic age. His discussion at this point dovetails with Touraine's, although the comparison is complicated by terminology. Postman's "technopoly" is Touraine's technocracy, with the American critic witness to the especially advanced domination of technology in the US of the 1980s and 1990s. Like Touraine, Postman sees in the most advanced societies the assumptions and imperatives of technology as the ruling force. "Technopoly is a state of culture," Postman writes. "It is also a state of mind. It consists in the deification of technology, which means that the culture seeks its authorization in technology, finds its satisfactions in technology, and takes its orders from technology" (71).

Touraine was too much of a Marxist to romanticize the past, but I suspect Postman does. The American's notions are a little unsubtle, for all the refreshing force of their exposition; there is no attention to the harshness of medieval life, nor much appreciation of the better side of modern technology. There lurks the assumption that humanity possessed true wisdom centuries ago then surrendered it to technology, but perhaps the morality and culture of those distant days was neither better nor worse than ours, simply different.

Conclusion

Touraine claimed that technology affected the power structure of societies primarily by bringing to the fore the assumptions and values of a technology-creating culture. Sometimes I show classes a cartoon about a diminutive little man in a lab coat escorted by adoring youthful beauties and a squad of muscular bodyguards. "Make way for a computer consultant, swine," the caption reads as the bodyguards shove aside passers-by. Touraine might laugh along, but the cartoon is quite opposite to his theory of technocracy. It is not technical workers who ascend to power in the information society, he insists, it is the ideas embedded within technological culture. Citing for support figures such as George Grant, Jacques Ellul, and Neil Postman, I endorse that interpretation over the more position-centred one which claims that scientists and engineers hold technocratic power simply due to their expert technical knowledge.

Notes

1. Alan Bullock and Oliver Stallybrass, eds., *The Fontana Dictionary of Modern Thought* (London: Fontana/Collins, 1977), 625.
2. "Technocracy" entry from electronic New Oxford English Dictionary.
3. W.H.G. Armytage, *The Rise of the Technocrats: A Social History* (London: Routledge and Kegan Paul, 1965), 341.
4. John Porter, *The Vertical Mosaic* (Toronto: University of Toronto Press, 1965).
5. Eric O. Wright, *Classes* (London: Verso, 1985).
6. Fritz Machlup, *The Production and Distribution of Knowledge in the United States* (Princeton: Princeton University Press, 1962).
7. Economic Council of Canada, *Employment in the Service Economy* (Ottawa: Minister of Supply and Services, 1991) 58; John Naisbitt, *Megatrends: Ten New Directions Transforming our Lives* (New York: Warner, 1984), 4. Naisbitt preferred the term "information society" to "postindustrial society." Bell himself made reference to the information society, the society whose main output is information, late in *The Coming of Post-Industrial Society* and it has become the more accepted term (467). See, for example, David Lyon, "From 'postindustrialism' to 'information society': A new social transformation?" *Sociology* 20 (November, 1986).
8. Robert L. Heilbroner, *The Limits of American Capitalism* (New York: Harper and Row, 1965), 115. I was going to cite James Watt and the steam engine as an example too, since some economic histories stress that Watt was a craftsman more than a scientist, until I encountered Milton Kerker, "Science and the steam engine," *Technology and Culture* 2 (Fall, 1961), 381-390.
9. The display continues: "Many other problems can be solved using algorithms, but there are still a large number of 'problems' for which no automatic method exists, for example, playing chess, making a medical diagnosis, translating, organizing tasks sequentially, etc." In Chapter 9 we encounter the chess-playing computer Deep Blue whose level of play would sharply challenge the above claim that algorithms cannot be used to master this ancient game.
10. Anthony Sampson, *The Changing Anatomy of Britain* (London: Hodder and Stoughton, 1982), 428-30.

11. Ian Varcoe and Steven Yearley, "Introduction," in Ian Varcoe, Maureen McNeil, and Steven Yearley, eds., *Deciphering Science and Technology: The Social Relations of Expertise* (London: Macmillan, 1990), 1.
12. George Grant, *Technology and Empire: Perspectives on North America* (Toronto: House of Anansi, 1969), 113.
13. See outlines of such positions in work by Brym and by Fox & Ornstein reprinted as Chapters 3 and 4 in James Curtis, Edward Grabb and Neil Guppy, eds., *Social Inequality in Canada: Patterns, Problems, Policies* (Scarborough: Prentice-Hall Canada, 1993).
14. George P. Grant, *Lament for a Nation: The Defeat of Canadian Nationalism* (Toronto: McClelland and Stewart, 1965), 74.
15. M. Patricia Marchak, *Ideological Perspectives on Canada,* 3rd. ed. (Toronto: McGraw-Hill Ryerson, 1988), especially p. 9, and Chapter 8.
16. John Kenneth Galbraith, *The New Industrial State* (New York: Signet, 1967), 71; Ursula Franklin, *The Real World of Technology* (Concord, Ontario: Anansi, 1990), 79.
17. Douglas Coupland, *Microserfs* (Toronto: HarperCollins, 1995), 3.

Further Readings

Some classic works on technocracy:

Bell, Daniel. *The Coming of Post-Industrial Society: A Venture in Social Forecasting.* New York: Basic Books, 1973.
Cutcliffe, Stephen H., J.A. Mistichelli and C.M. Roysdon. *Technology and Values in American Civilization: A Guide to Information Sources.* Detroit, Michigan: Gale Research Company, 1980.
Ellul, Jacques. *The Technological Society.* New York: Vintage Books, 1967.
___. *The Technological System.* New York: Continuum, 1980.
Mallet, Serge. *The New Working Class.* Nottingham: Spokesman Books, 1975.
Meynaud, Jean. *Technocracy.* New York: Free Press, 1968.
Touraine, Alain. *The Post-Industrial Society, Tomorrow's Social History: Classes, Conflicts and Culture in the Programmed Society.* New York: Random House, 1971.
Winner, Langdon. *Autonomous Technology: Technics-out-of-Control as a Theme in Political Thought.* Cambridge, Mass.: MIT Press, 1977.

For studies on the redrawing of class lines in the twentieth century:

Clement, Wallace. *The Challenge of Class Analysis.* Ottawa: Carleton University Press, 1988.

Mills, C. Wright. *White Collar: The American Middle Classes.* New York: Oxford University Press [Galaxy], 1956.

For more recent works in the technocracy genre:

Kroker, Arthur and Michael A. Weinstein. *Data Trash: the Theory of the Virtual Class.* New York: St. Martin's Press, 1994.

Postman, Neil. *Technopoly: The Surrender of Culture to Technology.* New York: Vintage Books [Random House], 1993.

Saul, John Raulston. *Voltaire's Bastards: The Dictatorship of Reason in the West.* Toronto: Penguin Books Canada, 1992.

JACQUARD LOOM CARDS ca. 1812   CARTES DU MÉTIER JACQUARD (vers 1812)

The beginning of automated production: Jacquard Loom Cards, circa 1812, used to program looms mechanically. *National Museum of Science and Technology*, Ottawa.

Chapter 6: The D&D of R&D

The term "R&D," research and development, appears frequently in the press, but sociologists have suspected for years that workplace changes resulting from R&D lead to "D&D": the deskilling of work whereby occupational tasks become less satisfying and fulfilling, and the displacement of workers from their chosen jobs either into less desirable new ones, or right out of the labour force. To take a position on this controversial D&D issue is a crucial part of our examination of technology and society.

The onrushing automation of work, the process with the potential either to remove the skill from jobs or to eliminate jobs altogether, was never clearer than in the early 1990s. Following a boom period in the later 1980s, it was perhaps to be expected that an economic contraction would follow. But many business commentators felt that the 1990s recession became the most severe since the depression-ridden 1930s because a cyclic downturn coincided with an accelerating impact of technology on the workplace.

In the early 1980s, in contrast, a Canadian text was noting ruefully that "we seriously lag behind other developed countries" in the use of automation. Figures from 1984 showed the use of robots in Canadian manufacturing following behind not only the US, West Germany, and Japan, but even a more comparable society such as Sweden.[1] Whether it is cause for lament or celebration I do not know, but by the mid-1990s

Table 6.1: Use of Advanced Technology by Size of Firm, Canada and US, 1989-1993

| | Use of At Least One Advanced Technology (% of Firms) | | | |
| | 1989 | | 1993 | |
Size of Firm	Canada	United States	Canada	United States
Number of Employees:				
20-99	50	67	70	75
100-499	81	89	85	94
500 or more	98	98	94	97
Total for all sizes	58	74	73	81

Source: Statistics Canada, *Canadian Economic Observer*, May 1996, Table 3. Cat. 11-010-XPB.

Canada had greatly closed the automation gap with other economically advanced countries. Table 6.1 gives a Canada-US comparison for 1989 and 1993, documenting the use of technology deemed to be "advanced" by Statistics Canada. The table subclassifies by size of establishment, due to the strong tendency for greater use of advanced technology by larger firms. Over the short interval of just four years, the technology gap declined from 16% (74%-58%, 1989) to 8% (81%-73%, 1993). For the largest group of firms, those with 500 or more employees, the most recent use of advanced technology was well into the 90s, near saturation level, in both countries.

These years of growing workplace automation added such terms as "Mcjobs" and the "jobless recovery" to the vocabulary of labour economics.[2] If the trail of technology and work is picked up just from 1990, however, much of the complexity of the story is prematurely eliminated, for issues of "D&D" have been argued all through the twentieth century as well as the one before. The epithet "neo-Luddite," for example, is part of modern vocabulary and often appears with reference to the popular writer and technology critic Kirkpatrick Sale (e.g., *The Globe and Mail*, 20 July 1995, A16). He takes inspiration from the Luddite riots of 1811-12, led initially by an apprentice stocking-maker named Ned Ludlum. The movement became such a serious civil disturbance within the English textile industry that troops were mobilized to restore order. Skilled workers were angry that new tech-

nology for manufacturing stockings used less skilled labour; their protests spread to other textile workers such as weavers threatened by the appearance of steam-powered looms. Thus, a Luddite is one who resists new workplace technology and, literally, smashes machines; Kirkpatrick Sale himself has destroyed a computer as a stunt during a public debate. Reminiscent of the early 1990s, the original Luddite movement arose during a period of economic downturn. The Napoleonic War, the great struggle between England and France, had resumed in 1803 following a short peace and meant total European economic boycott of English goods until 1815. For the English textile workers, it was a time when prices were high, wages low, and employment uncertain.[3]

If technology has indeed, on balance, debased work, we have an example of sociological prophecy come alarmingly true. Such warnings were voiced with gathering intensity after World War Two. The title of a 1948 book warned *Mechanization Takes Command*. The novelist Kurt Vonnegut's *Player Piano*, a chillingly prescient piece of sociological fiction, appeared in 1952. The novel depicts a futuristic US in which only PhD's are employed, and only as "managers and engineers and civil servants and a few professional people." The PhD's tend machines which handle almost every kind of work, while the bulk of the population huddles in unemployed alienation in ghettos such as one named "Homestead." Like the visions of the future in other earlier twentieth-century novels such as Orwell's *1984* and Huxley's *Brave New World*, *Player Piano* envisions a highly regimented society in which ordinary citizens have lost nearly all their rights. In these predictions, then, technocracy, the theme of the last chapter, and the erosion of autonomy and skill in work, the present concern, come together in a highly authoritarian and stratified state.

A book from 1957, which largely approved of technological advance, described a coming "technical revolution in which the machine threatens to make the large body of available labor unnecessary in the production of goods." In Canada George Grant was writing in the mid-1960s of "the problems caused by automation," while at the same time in Great Britain the literary critic F.R. Leavis was deploring the "meaninglessness, or human emptiness, of work" in the "future seen as automation."[4]

Braverman: A Marxist Analysis of De-skilling

One work on such themes, however, had especially great impact. In 1974 Harry Braverman published *Labor and Monopoly Capital*, a searing Marxist-inspired indictment of the consequences of workplace technology. The book was immediately recognized as an "important" and "powerful" analysis in part because, apart from dissenters such as George Grant in the passages noted above, the prevailing opinion saw workplace technology as raising, not lowering, the skill requirements of jobs.[5] It was true that technology tended to be introduced first at the unskilled end of the labour force. One dramatic effect was the reduced need for farm labourers. At the same time that technology reduced the labour-intensiveness of activities such as farming, road and rail construction etc., it created new skilled employment. Workplace technology, after all, had to be designed, manufactured, marketed, and serviced by people. They needed to have the kind of knowledge skills that Daniel Bell had stressed in *The Coming of Post-Industrial Society*. If there was concern in the 1960s about automation, it was not so much about deskilling but about unabsorbed displacement — unemployment in other words. Bell tended to minimize even this risk, citing reassuring US government studies (194-95).

The huge impact of Braverman's ideas derives, I think, partly from the conceptual foundation he laid. Here is an example of why the sometimes daunting term "social theory" can be so important. As a Marxist, Braverman began with a key assumption about the very nature of individuals and capitalist societies. "Labor, like all life processes and bodily functions," he wrote, "is an inalienable property of the human individual" (54). It is part of the human definition, destiny, or "fate" (as it was termed in Chapter 1). Braverman was not anti-technology *per se*, for he noted some of the same anthropological evidence on "tool-using" reviewed earlier in the present book, viewing such activity as inseparable from work.

The technology of work was therefore a source of enrichment for humans, so long as the "unity of conception and execution" (51) was maintained. The fulfilment from work comes from visualizing in advance the process and product, then following the idea through to completion. This flow beginning with conception, "originating in an altogether exceptional central nervous system" (47), made work into

"purposive action" as "the special product of humankind" (49).

Braverman's second premise was his Marxist analysis of capitalism. As soon as an economy moves beyond family "cottage" industries into factory style production, the cycle from conception to execution of work begins to break down (57). Workers now are "forced to sell their labor power to another"; they therefore "surrender their interest in the labor process" and are "*alienated*" or separated from it. Technology is from the outset seen to be an active ingredient, for it is largely the eighteenth-century innovations in *sources of power* that instigated the march toward mass production.

To applaud the clarity and coherence of Braverman's two assumptions is not necessarily to endorse them. The validity of Marxist analysis has been one of the great debates of the past 150 years. Some would reply to Marx that the humanistic benefits that he sought in work are more reasonably to be expected in leisure pastimes. Others would contend that in the present age all kinds of salaried employment is highly fulfilling. The concept of management, for example, may bestow to non-owners of businesses much of the sense of control and completeness of proprietorship, while removing the risks. It is for the reader to judge such questions. My task is to identify where Braverman's important book "comes from" conceptually, and to recognize that any clearly defined theory foundation gives a coherence and unity, although not automatic validity, to social analysis.

Braverman saw Frederick Winslow Taylor, a person well-known in the business administration curriculum and a figure we encountered briefly in Chapter 5, as a key originator of a "scientific management movement" that accelerated the process of deskilling. "Taylorism" is noteworthy for its simplicity, since industrial innovations such as the assembly line or robotics came subsequent to Taylor's late nineteenth-century research. He was a time and motion expert, an "obsessive-compulsive personality" in Braverman's words (92), who sought out the most efficient method for every task, codified it, and used every means to impose the method on workers. Some of the most illuminating examples of Taylor's technique involve the simplest of tasks. He studied, for example, a workgang of men employed in 1899 by Bethlehem Steel to load 92-pound bars of iron onto a railroad boxcar. Taylor used a stop watch, pencil, and paper to work out the optimum sequences of movements, schedule of rest periods, etc.

Gathering knowledge on efficient procedure was, however, merely the first principle of Taylorism. A second was to take the decision-making about the execution of manual labour away from workers themselves and into the hands of a new echelon of managers. Third was the use of this knowledge to control work at every step. It required sometimes quite ruthless economic and psychological pressure tactics to seize this control. In the boxcar loading study, Taylor selected one of the workers who was both psychologically malleable and economically motivated, paid him extra in return for total cooperation, and then set about pressuring the other workers to conform as well.

The manner in which this Taylorist emphasis on control of workers interacted with the material content of workplace technology — meaning the actual tools, machines, and manufacturing processes — has become an important focus for the Sociology of Organizations. Charles Perrow differentiated between work in which "exceptional cases" were few versus frequent. A carpenter building custom furniture encounters frequent exceptional cases, a factory making up precut and drilled kits for home assembly sees few exceptions to the normal process. Taylorism recognizes of course that work involving few exceptions is the most easily routinized. Perrow further noted the types of "search process" undertaken when exceptions occur, "a search which can be conducted on a logical, analytical basis" being contrasted with more intuitive, experienced based, methods. Taylor's methods apply most readily to work having few "exceptional cases" which, when they do arise, are solvable by a logical, algorithmic approach. Conversely, work most immune to Taylorism is the opposite case of frequent exceptional case, intuitive solution, settings.[6]

The application of Taylor's methods to so rudimentary a task as manually loading a boxcar illustrates a powerful form of technology in which machinery and apparatus are either peripheral or absent entirely. It is partly with Taylorism in mind that many sociologists define technology in the broad sense suggested in Chapter 1. Braverman dwells on Taylorism for the great clarity with which this form of technology was enmeshed with the social environment.[7] Time and motion studies required some expense and resources to perform, and were therefore a form of investment made by capitalists. "Control," Braverman wrote, "now falls to those who can 'afford' to study it in order to know it better than the workers themselves know their own life

activity" (116). This assumption of control in manual labor was just one part of a larger process by which capitalism took control of science itself to devise technology intended to wring more output, at less cost, from workers (Braverman, ch. 7).

It was this control of hand work by the brain of each worker that, for Braverman, endowed even the most rudimentary manual labourer with skill. Step by step during the twentieth century control passed to management, tasks were broken down into ever-simpler components, and the skill of work evaporated. Much of the story of industrial sociology in Braverman's view involved attempts to offset the alienating consequences of industrial deskilling.

He notes, for example, the famous studies by Elton Mayo on "*Hawthorne effect.*" At the Hawthorne Plant of the Western Electric Company in the US, experiments began in the 1920s into factors determining worker productivity. Test rooms were set up for studying small teams of workers engaged in the assembly and wiring of telephones. Initially the research concerned Taylorist factors such as the scheduling of breaks for the work crew, but increasingly it was realized that the very fact of singling out a group for special observation and attention had enhanced their productivity.[8] In comparison to this Hawthorne effect — the effect of participating in a study group — more material technological variables such as intensity of lighting were insignificant.

Labor and Monopoly Capital proceeded from this conceptual base to redraw the class lines of American society, performing the equivalent analysis for the bottom end of the stratification system to what Daniel Bell had done for the top (see above, Chapter 5). Where the "working class" in 1900 was estimated by Braverman to comprise some 50% of the workforce, by 1970 it was almost 70%. Thousands of small farm owners and proprietors of small businesses had been drawn into corporate capitalist employment. Much of the old middle class of "white collar" workers, Braverman said in an argument reminiscent of the famous work by Mills (noted in Chapter 5), was by mid-century more accurately classified as working class. "It takes but a moment's reflection," Braverman wrote, "to see that the new mass working-class occupations tend to grow, not in contradiction to the speedy mechanization and 'automation' of industry, but in harmony with it" (381). Automation was deskilling and displacing craft occupations, and creating growing numbers of "labor-intensive areas which have not yet

been or cannot be subjected to high technology" (382). Such are the notorious "Mcjobs" of the 1990s.

Assessing Braverman

Was Braverman perhaps a somewhat inaccurate social historian, like Neil Postman nurturing romantic views of the pre-industrial capitalist age? John Kenneth Galbraith once remarked on television that after a boyhood doing chores on an Ontario farm no job he did later in life really seemed like work. Work before Taylor's scientific management revolution could be unrelentingly hard and exploitative. Many in earlier times worked as, or virtually as, slaves. The age of more advanced technology removed much of the drudgery from labour, yet Braverman still would insist that for everything that was gained as much was lost.

The very clarity of Braverman's theoretical scheme based on Marx was at the same time a trap. Every theory is a kind of "lens" for viewing the world in a certain way, and commentators have noted that Braverman was a typical Marxist in his underestimation of agency, the ability of workers to regain control from management by considered, conscious collective action, often through a trade union.[9] Similarly, was Braverman's core theoretical assumption that work should be a source of human fulfilment unrealistic? It is one school of thought that modern workers hold a primarily "instrumental" view of labour. Within this viewpoint, technology produces such high productivity per worker, such a large monetary "pie", that the proletariate become more affluent than Marx ever imagined possible. As a worker in a British study said: "What is it that keeps me here? Money and again money — nothing else."[10]

These workers traded satisfaction for pay, sometimes with full realization. Their lives were materially better than those of their parents or grandparents, but spiritually and psychiatrically perhaps not. This notion reminds us that all social classes have "bought into" the technocratic economy to some degree, even if corporate capitalists led the way.

A far tougher criticism is made by spokespeople of the "upgrading thesis" about technology and work. Daniel Bell, whose *The Coming of Post-Industrial Society* we reviewed in the previous chapter, fits into that position, along with such writers as Talcott Parsons and Clark

Kerr, both American social scientists active in the early post-World War Two years.

How can one speak of labour deskilling and upgrading within the same sentence? Surely it is one or the other. The complexity of the question derives from the variety of occupations and technologies present in the workplace. At any moment, more than one process is at work. First, the composition of the labour force is fluid. Computer-related occupations, to select an obvious example, proliferated in the past 25 years. The 1971 version of the Government of Canada classification of occupations listed just 5 jobs under the heading of "computer," a listing that grew to 37 entries by the 1991 edition of the same manual.[11] Jobs such as "computer aided drafting technician" and "computer graphics designer" had simply not existed a generation previously. Even if the skill requirements of every job were unchanged, the skill level totalled across all workers in the economy could shift due to these "compositional" changes.[12] If, for example, automation were reducing the number of unskilled occupations, while expanding employment in some highly skilled jobs, then an upgrading would result. It was this logic that impressed observers such as Daniel Bell. The compositional changes are important, but we must remember that they do not address skill content within occupations, the very issue that Braverman disputed most vehemently.

The compositional changes are far easier to detect within empirical studies than the content shifts. A Canadian study by John Myles, covering the period 1961 to 1981, found growth in the sectors of the economy containing some of the most skilled jobs. Part of his work was reproduced in Chapter 5 within Table 5.1, and revealed growth in high skill areas such as business and social services. Automation was meanwhile cutting into the proportions of the labour force in domains such as "extractive" work within the industrial sector. In this respect, as Myles noted, "Bell was closer to the mark than Braverman."[13]

Skill analysis is a high priority for Statistics Canada, and Myles presented some of their data suggesting no clearly identifiable trends over the 1961-81 period.[14] For example, 25% of blue collar jobs in the 1961 labour force were scored as learnable within 30 days, compared with 24% in 1981. Braverman might reply that his analysis extended much farther back into the past, and that damage by technology had already been wrought by the second decade after World War Two. This sort of

Table 6.2 Dimensions of Technological Control Reported by Workers at a Saskatchewan Potash Mine

Pace of Work	Percentages
Q: Do you choose the speed you work at?	
Never	3.0
Seldom	3.0
Sometimes	14.9
Usually	49.3
Always	29.9

Job Discretion
Q: How would you rate your own judgement in terms of the way you do your work?

Lots of room for judgement	82.1
Some room	9.0
Very little room	9.0

Perceptions of Technology
Q: In terms of the technology used in the operation, would you say you

Fully control the technology	71.6
Partially control	22.4
Feel controlled by the technology	6.0

Perception of skill decrease or increase

Decreased skills	16.1
constant skills	24.9
increased skills	59.0

Source: Bob Russell, "The subtle labour process and the great skill debate: Evidence from a Potash mine-mill operation." *Canadian Journal of Sociology* 20 (Summer, 1995), 368, 372.

analysis is difficult too because technological change can result in some occupations disappearing altogether, as discussed below. The jobs available for Myles to examine were those not automated out of existence over the twenty years examined.

Sweeping generalizations about the skill content of jobs in the automated workplace are impossible because, alongside all the instances of deskilling, plenty of opposing examples can be found both of "reskilling" — referring to a lateral change in the skill requirement for

Table 6.3: Incidence and Effect of Technology at Work, 1994 General Social Survey

"In the past five years, how much has your work been affected by the introduction of computers or automated technology? Would you say ..."

	Greatly	Somewhat	Hardly	Not at all	No Answer	Total
High school completion or less	23.1	14.0	11.4	49.3	2.2	100%
At least some post-secondary education	41.4	18.9	10.0	27.7	2.0	100%

"In the last five years, has the level of skill required to perform your work increased, decreased, or stayed the same as a result of the introduction of computers or automated technology? (asked just of the "greatly" or "somewhats" above).

	Increased	Decreased	Stayed the Same	No Answer	Total
High school completion or less	63.5	2.7	26.2	7.6	100%
At least some post-secondary education	69.2	1.5	25.1	4.2	100%

"Over the last five years, has your work become more interesting, less interesting, or stayed the same as a result of the introduction of computers or automated technology?"

	More Interesting	Less Interesting	Stayed the Same	No Answer	Total
High school completion or less	53.9	4.2	34.5	7.4	100%
At least some post-secondary education	61.6	3.4	30.7	4.3	100%

Source: Computed from 1994 Statistics Canada General Social Survey

N=7050

a job — and of Bell's upgrading. As a British study noted, "new technology is already enriching some lives and impoverishing others. There is no inherent logic within technology which dictates that either of these things should happen."[15] Even the Canadian writer Heather Menzies, one of the most forceful workplace technology critics of the current generation, would agree that reskilling occurs alongside deskilling (see *Fastforward and Out of Control*, 202-3, in Further Readings).

An example of beneficial technology was seen in a study of Potash mining in Saskatchewan. "Rather than experiencing skill dilution," the author concluded, "workers are more likely to report upgrading."[16] Even though the work had become intensely capitalized (that is, high investment in machinery relative to number of workers), the workers in the mine examined by Russell reported highly unBraverman-like attitudes. Some examples are reproduced in Table 6.2, which documents that most of the workers felt the technology was under their control, that they determined the pace of work, and that the tasks allowed for the exercise of judgement. Overall, most of the workers felt increased application of technology had enhanced the skill requirement of their job. Admittedly, this is self-reported skill, but the Myles study just noted showed essential agreement between trends based on reports by self and by outsiders. Russell did not say that the Potash mine was a technological utopia. The workers felt no great affection or loyalty to their employer, but management and the labour union had been able to compromise between the imperatives of workplace technology and the needs of workers. It is probably relevant that the mine work prior to the introduction of new technology was hard, noisy, dirty labour. Job loss or seasonal layoff, not job deskilling, most concerned the Potash workers, and management had responded to these concerns sufficiently to prevent major disruptions to production.

Another clue that deskilling is just one part of the story of technology's impact on work comes from the 1994 General Social Survey by Statistics Canada, a source already drawn on in Chapter 3. Table 6.3 shows that work has been most "affected by" advanced technology among those educated beyond the high school level, those most equipped to make lateral shifts in job skill. In addition, responses to questions about "level of skill required" and changes in the interest level of jobs give a picture of increasing quality of work. Such changes

are perceived by workers at both education levels, albeit especially among the post-secondary educated workforce.

Occupational Displacement: Permanent or Temporary?

Just as technology can alternatively deskill, reskill or upgrade the content of work, it also both creates and destroys jobs. Thus far in history, technology has always generated sufficient new economic activity to replace displaced jobs, an interpretation known as "compensation theory."[17] Thus it would be noted by economic historians that the massive decline of farm work during the earlier twentieth century was replaced by increased factory work. For every occupation rendered obsolete by technology, new ones arose. Blacksmithing, for example, sank to just a trace, but was replaced by automobile mechanics.

A sharply opposing view holds that over the long run technology has been systematically conducting a search and destroy mission against employment. The old pre-industrial society, in which at least half the labour force worked in agriculture, was broken by the mechanization of farming. Then, starting with the assembly lines of "Fordism" (that is, the production methods of the Ford Motor Company) just before World War One, technology began to erode employment in manufacturing. This process was well under way by the end of World War Two, but reductions in the average number of hours in a work week had camouflaged job losses. Displacement of manufacturing labour by automation was further offset by expansion in the service sector (recall the data on that shift in Table 5.1). The 1990s recession, writers such as Jeremy Rifkin argue, finally brought home the reality that now, like a rampaging hungry beast from a science fiction movie, workplace technology is eating into white collar employment. The "Working With Technology Survey" sponsored by Human Resources and Development Canada illustrates the decline of lower white collar work. In this 1995 survey, some 1500 business establishments were canvassed. Although the cooperation level was so low as to be a concern, one result is so striking that it must be robust. Fifty-one percent of job types reported by the businesses as eliminated between 1992 and 1994 due to the introduction of computer-based technology were clerical. These losses were only minimally offset by the creation of new clerical job types (16.2%).[18]

As a Nobel prize-winning economist has said, "machines that have been displacing human muscle from the production of goods are being succeeded by machines that take over the functions of the human nervous system ... as well."[19] The British author Colin Gill agrees that microelectronic technology has an "intelligence function." It is more than "just another technology", making it fallacious to "extrapolate from the past" as compensation theorists do.[20] It would be more within the conceptual scheme of the present book to view microelectronic technology as indeed "just another" technology, but one in which the electric *power source* is quintessentially twentieth century and the *knowledge-base* in theoretical science is the crucial component in contrast to earlier technologies in history.

The rise and fall of women's employment as telephone operators exemplifies the process of technological displacement of clerical labour. In the late 1880s, "the occupation of telephone operator was a new opening for the large female reserve army of labour." A century later, Heather Menzies documented falling employment for women as telephone operators, due to automated switching. She notes how the effect of automation technology on employment follows a typical life cycle. In the early phase, employment may indeed rise, but that is often merely an interlude preceding layoffs, like fine weather before a storm.[21]

In the short run certainly, the "compensating" features of workplace technology can be hard to discern. The years from 1990 to mid-decade, the period of the so-called "jobless recovery," make a sobering case study. Economists have noted that while the economy perhaps was "due" for a recession by the end of the 1980s, additional factors made jobs especially vulnerable. The 1988 Canada-US Free Trade Agreement (discussed in Chapter 3) had recently come into effect, followed by NAFTA, a more comprehensive North American free trade agreement. These alone created a crisis for many Canadian manufacturing companies, forcing them to slash labour costs and automate production in order to remain competitive with foreign business. Additionally, however, the costs of employment due to benefits and payroll taxes were increased by the Canadian government beginning in 1989, reinforcing the motivation to replace living workers by machines. This only underscored a trend under way since the mid-1980s toward greater investment in workplace technology at the expense of employment.

Figure 6.1: Capital to Labour Ratio in Canadian Businesses

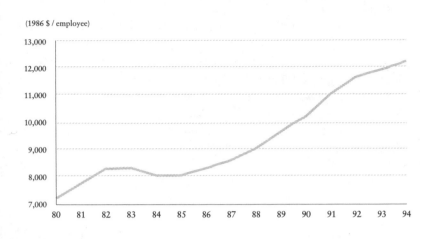

(1986 $ / employee)

Source: Nicholas Chamie, *Why the Jobless Recovery: Youth Abandon Labour Market* (Ottawa: The Conference Board of Canada, Report 154-95, 1995), 5. Reproduced with permission from the Conference Board of Canada.

A graph of the investment in machinery and equipment per worker appears in Figure 6.1. In theory, the ratio declines during recession since labour, now in over-supply, becomes less expensive. We see that while this capital to labour ratio indeed conformed to theory by declining slightly at the time of the early 1980s recession, it rose as the 1990s slowdown set in. That is one marker of the "jobless recovery." Assessing occupational dislocation in the 1990s identifies the importance of external, "*exogenous*," factors. Thus, the growth of service sector employment perhaps was approaching its natural ceiling anyway, but government finance policy has reinforced the deterioration of employment. Suddenly it was realized that countries such as Canada were permitting their national debt to run wildly out of control. In reaction, opportunities in public service employment evaporated. While employment in the public sector had remained constant during the 1981-82 recession, 40,000 government jobs were eliminated as the new recession began in 1990-91, just the beginning of a continuing program of cutbacks.[22] In the 1990s the notion of sharing work more widely through a reduced work week, advocated by the economist Wassily

Lieontif as part of a "revision of values" and institutions, seemed less plausible than ever.

Such adjustments to job displacement were difficult because the technological advancements of the preceding decade had not produced a richer society. Investment beyond the optimal returns may have taken place due to infatuation with technology in the technocratic society.[23] Personal computers, especially, have a "half work, half play" nuance. Executives may waste their time at the office playing solitaire (*Macleans's*, 1 April 1996, 46), or unimportant documents may be word processed up to an unnecessary level of desktop publishing elegance. But the more fundamental issue of the 1990s was the way North American and Western European societies saw their standard of living slip in the face of vigourous economic competition from the Pacific nations.

In the mid-1960s, Robert Heilbroner had mused that "if the leisured population does not find adequate opportunities for unmechanizable employments, it will simply have to be *given* a right to society's output" even if that were a "basic infringement on the hegemony of the market."[24] Thirty years later, the minds of cash-starved governments and tax-oppressed wage earners moved in the opposite direction. In our times, the most important new class alignment is not a technological elite of scientists and engineers versus all the non-technocrats, but rather the distinction between those who have survived the virtually random blow of job displacement due to technology versus those less fortunate. "I had it all once — the Cadillac, long, shiny-haired women. Then I had to go to the safety net, and it wasn't there," a victim of the 1990s said in a *Maclean's* magazine story (6 November 1995, 10). "No future. No money. No benefits. No nothing," a laid-off auto worker told the *Globe and Mail* as part of a series on the crisis in job displacement (20 April 1996, 12).

Psychiatric depression has become a chronic industrial disease, as the *Globe* noted in a story on 9 May 1996 (B14). The director of "disability consulting services for a large insurance company" told reporters "her firm got involved because it has seen a significant jump in the past five years in the number of insurance claims for depression-related conditions." She cited "the relentless downsizing and restructuring in both the private and public sectors." Responses to a GSS question asking "do you think it is likely you will lose your job or be laid off

Table 6.4: Perceived Risk of Job Loss, Workers in the 1994 GSS

"Do you think it is likely you will lose your job or be laid off in the next year? Would you say it is ... "

Very Likely	Somewhat Likely	Somewhat Unlikely	Very Unlikely	No Answer	Total
5.3	8.5	21.3	60.7	4.2	100%

"In the last five years, has the job security increased, decreased, or stayed the same as a result of the introduction of computers or automated technology" (asked of 3737 cases perceiving their work to be affected by technology)

Increased	Decreased	Stayed the Same	No Answer	Total
13.2	17.6	62.6	6.6	100%

Source: Computed from 1994 Statistics Canada General Social Survey
N=7050

in the next year?" are chilling. As shown in Table 6.4, nearly 14% of the workers interviewed perceived it as either "very" or "somewhat" likely they would be out of work *within the year*. Added to the 10% or so already unemployed, we have about one-quarter of the working age population either looking for work or feeling insecure about the work they now have. Make the further observation that the official unemployment rate is generally conceded by economists to be an undercount which excludes so-called discouraged workers who have given up looking for employment, and the employment crisis of the 1990s begins to appear epidemic indeed.

The adverse effects of labour automation may fall disproportionately heavily on women. Heather Menzies, who as noted earlier is highly critical of workplace technology, writes largely about the impact upon female employment. In contrast, our case study of happier experiences with technology, Russell's work on Potash mining described above, concerns a male-dominated occupation. Early in the microelectronics revolution, Menzies published *Women and the Chip*, a famous little book in the Canadian literature on technology and society. "Canadian women are vulnerable because their jobs are concentrated in areas being swept away by information technology," she warned. A vindica-

tion of these words from 1981 is suggested by the recent downswing in the percentage of women in the paid labour force, following decades of unbroken increase.[25]

Marxists have suggested that women form a "flexible" workforce, a "reserve army of labour." Such workers are lured into the capitalist economy when extra labour is needed but then, with ideological blessing from society, can easily be pushed back out of paid employment or into part-time work as applications of automation expand.[26]

A less radical explanation, typically more favoured by economists, sees modern capitalist societies as "dual economy" with "segmented" labour markets. A *"core segment"* encompasses industries associated with a small number of large, powerful corporations centred in construction, "extractive" industries such as mining, and large scale manufacturing. This is the most profitable segment, and the job security of the workforce here is protected by strong labour unions. In contrast, the *"peripheral"* segment involves industries such as agriculture, retail trade, and smaller scale manufacturing. The structure of the labour market is less *"oligopolistic,"* meaning more competitive, with many small firms which are unable to fix prices by collusion. In that less profitable segment, which tends to overlap the service sector already described in this chapter, labour is less strongly unionized and therefore less equipped to counter job displacement due to technology. Labour market research has established that men traditionally have been concentrated in the core segment, women in the periphery.[27]

Finally, a more typical sociological view of the impact of workplace technology on women stresses the *"role conflict"* involved in working in the paid labour force while rearing children. So long as women have the greater responsibility for children, they will tend to enjoy less flexibility with respect to non-traditional work schedules, after hours retraining courses, and distant work sites.[28]

Some of the above effects, however, are more historical legacies than realities for the front edge of the labour force. Drawing once again on the 1994 GSS, it can be seen how pertinent Menzies' analysis was especially to the older generation of women. Table 6.5 separates men from women and those aged up to 39- from 40-year-olds and over. A question on perceived job skill requirement shows virtually no gender difference for the younger group, while in contrast 59% of older males against just 49% of the older females in this GSS sample felt that their

Table 6.5: Self-Rated Skill of Job by Gender and Age, 1994 GSS

| | Job Requires High Level of Skill | | | | | |
	Strongly Disagree	Somewhat Disagree	Somewhat Agree	Strongly Agree	No Answer*	Total
Aged up to 39:						
Female	11.3	24.5	32.4	30.0	1.8	100%
Male	11.3	22.2	29.2	35.5	1.8	100%
Aged 40 and over:						
Female	3.8	14.9	29.0	49.4	2.9	100%
Male	2.4	9.0	26.8	59.1	2.7	100%

Source: 1994 GSS.
*Includes a few responses that do not fit checklist
N=7050

job requires a high level of skill. It would be the less skilled women who are most involved in the peripheral economy and consequently most vulnerable to lay-offs.

Conclusion

Deskilling of one occupation due to workplace technology is often matched by technology-derived reskilling or upgrading of another, making a sweeping statement of the kind advanced by Braverman controversial. And while a small library of books hotly critical of job loss due to technology exists, such analyses must set aside the history of compensatory changes in employment patterns and argue that today's workplace technology is uniquely destructive.

In the short run, the time frame people live their lives within, stresses due to occupational redefinition and loss have become horrific. This is the most obviously sustainable criticism of technology in the workplace. A condition that the turn-of-the-century French sociologist Emile Durkheim would have termed *"anomic,"* or that the more recent

American writer Alvin Toffler called "future shock", has been created.[29] As a *Globe and Mail* story on casualties of office automation noted, computers have not only "eliminated" or "diminished" jobs, they have "destabilized" them as well (11 October 1995, A1, 10). As technology has been deployed within capitalist societies, established workers have lost stability and predictability within their lives while younger people face a much harder scramble for fulfilling employment than their parents had to deal with. This is the most devastating of the information highway's "roadkill."[30]

Notes

1. Edward B. Harvey, "Science and technology," in James Curtis and Lorne Tepperman, eds., *Understanding Canadian Society* (Toronto: McGraw-Hill Ryerson, 1988), 301. Recall that Laxer chose Sweden for comparison with Canada (Chapter 2).

2. For example, Jennifer Lanthier, "Unions trying to protect workers as technological change speeds up," *Financial Post*, September 1990, 31-2; Heather Menzies, "Mcjobs: The flesh and blood of economic restructuring," *This Magazine* 25 (June/July [no. 1], 1991), 30-32; Robert J. Gordon, "The jobless recovery: Does it signal a new era of productivity-led growth?" *Brookings Papers on Economic Activity* 1993 (1), 271-306; "The jobless recovery," *Globe and Mail, Report on Business* (Jan. 11 to Jan. 16, 1993). The term McJob was written up in the *Globe and Mail's* "Wordplay" column of 10 August 1996.

3. Shepard Bancroft Clough and Charles Woolsey Cole, *Economic History of Europe*, 3rd. ed.(Boston: D.C. Heath, 1952), 506; Malcolm I. Thomis, *The Luddites: Machine-Breaking in Regency England* (Newton Abbot, Devon: Davis & Charles, 1970).

4. S. Giedion, *Mechanization Takes Command* (New York: Oxford University Press, 1948); Kurt Vonnegut, Jr., *Player Piano* (New York: Charles Scribner's Sons, 1952); Francis R. Allen, Hornell Hart, Delbert C. Miller, William F. Ogburn and Meyer F. Nimkoff, *Technology and Social Change* (New York: Appleton-Century-Crofts, Inc., 1957), 265; George P. Grant, *Lament for a Nation: The Defeat of Canadian Nationalism* (Toronto: McClelland and Stewart, 1965), 14; F.R. Leavis, *Nor Shall My Sword: Discourses on Pluralism, Compassion and Social Hope* (London: Chatto & Windus, 1972), 86.

Leavis, a key influence on Marshall McLuhan during the Canadian's time at Cambridge, was reacting to the famous "Two Cultures" critique by the English novelist C.P. Snow. Snow had lectured on the gap between humanists and scientists, and on the scanty technical knowledge of the former (*The Two Cultures and the Scientific Revolution* [Cambridge: Cambridge University Press, 1961]). For a discussion of technological unemployment in the 1930s, see Ester Fano, "A 'wastage of men': Technological progress and unemployment in the United States," *Technology and Culture* 32 (April, 1991), 264-92.

5. The quoted words come from reviews of the initial publication of Braverman's book. See *Contemporary Sociology: A Journal of Reviews* 5 (November, 1976), 733-37. Paul S. Adler, "Marx, machines, and skill," *Technology and Culture* 31 (October, 1990), 780-812, disagrees with Braverman's interpretation of Marx.

6. Charles Perrow, "A framework for the comparative analysis of organizations," *American Sociological Review* 32 (April, 1967), 195-96. See also Frank E. Jones, *Understanding Organizations: A Sociological Perspective* (Toronto: Copp Clark, 1996), 84-90, for additional sources and discussion.

7. It is also worth noting that although Taylorism is virtually an expletive within sociology, within engineering he is generally cited with approval. Norman Ball, for example, in *Mind, Heart, and Vision: Professional Engineering in Canada 1887 to 1987* (Ottawa: National Museum of Science and Technology, 1987), notes on page 94 the "York Plan," based on Taylorism, and instigated at the York Knitting Mills in Toronto in 1927. The York Plan helped companies "weather the Depression." Historians such as Bernard Waites see Taylorism as exaggerated and leading to "an excessively personalized view of history" ("Social and human engineering", in Colin Chant, ed., *Science, Technology and Everyday Life, 1870-1950* [London: Routledge, 1989], 318).

8. Stephen R.G. Jones asked in a 1992 article "was there a Hawthorne effect?" (*American Journal of Sociology* 98, 451-68), and concluded that such effect as there was occurred the moment participants were selected for study. Understood this way, Hawthorne becomes a form of placebo effect, as the seeming benefits of neutral pills are termed in medical research.

9. E.g., Vicki Smith, "Braverman's legacy: The labor process tradition at

20," *Work and Occupations* 21 (November, 1994), 406-08.

10. John H. Goldthorpe, David Lockwood, Frank Bechhofer and Jennifer Platt, *The Affluent Worker: Industrial Attitudes and Behaviour* (Cambridge: Cambridge University Press, 1968), 1; Arthur B. Shostak, *Blue-Collar Life* (New York: Random House, 1969), 59-60. Quotation from Goldthorpe *et al.*, 28.

11. Manpower and Immigration Canada, *1971 Canadian Classification and Dictionary of Occupations, Vol. 1.* Cat. MP53-171/1 (Ottawa: Information Canada, 1971), 1063; Statistics Canada, *Standard Occupational Classification, 1991.* Cat. CA1 SC 12-565 1991 (Ottawa: Ministry of Industry, Science and Technology, 1993), 404.

12. Kenneth I. Spenner, "Deciphering Prometheus: Temporal Change in the Skill Level of Work," *American Sociological Review* 48 (December, 1983), 825.

13. John Myles, "The expanding middle: Some Canadian evidence on the deskilling debate," *Canadian Review of Sociology and Anthropology* 25 (August, 1988), 346.

14. A critique of the accuracy of skill assessments by government statistical agencies appears in Charles Darrah, "Skill requirements at work: Rhetoric versus reality," *Work and Occupations* 21 (February, 1994), 64-84. With respect to implications of skill measurement for gender issues, see Monica Boyd, "Sex differences in occupational skills: Canada, 1961-1986," *Canadian Review of Sociology and Anthropology* 27 (August, 1990), 287-90. The Statistics Canada approach is summarized in Alfred A. Hunter and Michael C. Manley, "On the task content of work," *Canadian Review of Sociology and Anthropology* 23 (February, 1986), 50-52.

15. Arthur Francis, *New Technology at Work* (Oxford: Clarendon Press, 1986). A Canadian work that made this same point, at around the same time, is Stephen G. Peitchinis, *Computer Technology and Employment: Retrospect and Prospect* (London: Macmillan Press, 1983).

16. Bob Russell, "The subtle labour process and the great skill debate: Evidence from a Potash mine-mill operation," *Canadian Journal of Sociology* 20 (Summer, 1995), 361.

17. Francis, 198.

18. Kathryn McMullen, *Skill and Employment Effects of Computer-based Technology: The Results of the Working with Technology*

Survey III. Study No. W / 01 (Ottawa: Canadian Policy Research Networks Inc., 1996), Table 6.5.

19. Wassily W. Leontief, "The distribution of work and income," *Scientific American* Volume 247 Number 3 (September 1982), 188. Other sources here are Jeremy Rifkin, *The End of Work: The Decline of the Global Labor Force and the Dawn of the Post-Market Era* (New York: Tarcher/Putnam, 1995), xv-xvi, 35; David Noble, *Progress Without People: New Technology, Unemployment, and the Message of Resistance* (Toronto: Between the Lines, 1995); and Wassily Leontief and Faye Duchin, *The Future Impact of Automation on Workers* (New York: Oxford University Press, 1986).

20. Colin Gill, *Work, Unemployment and the New Technology* (Cambridge: Polity Press, 1985), 5-7.

21. Quotation from Michèle Martin, *'Hello, Central?' Gender, Technology, and Culture in the Formation of Telephone Systems* (Montreal: McGill-Queen's University Press, 1991), 57; Menzies *Fastforward*, (in Further Readings), 229-32; Heather Menzies, *Women and the Chip: Case Studies of the Effects of Informatics on Employment in Canada* (Montreal: The Institute for Research on Public Policy, 1981), 59-60.

22. Nicholas Chamie, *Why the Jobless Recovery: Youth Abandon Labour Market* (Ottawa: Conference Board of Canada, Report 154-95, 1995), 5. Rifkin, 41, makes the same point about the US.

23. The contribution of computers to economic growth in the US between the 1970s and 1990s is assessed in an econometric article by Stephen D. Oliner and Daniel E. Sichel, "Computers and output growth revisited: How big is the puzzle?" *Brookings Papers on Economic Activity* 1994 (2), 273-315. For a less technical treatment, "Productivity lost: Have more computers meant less efficiency?" *Scientific American* 271 (November, 1994), 101-02.

24. Heilbroner, Robert L. *The Limits of American Capitalism* (New York: Harper and Row, 1965), 124. For another expression of the same notion, see Erwin O. Smigel, "The leisure society," in Irene Taviss, ed., *The Computer Impact* (Englewood Cliffs, New Jersey: Prentice-Hall, 1970), 104-14.

25. Quotation from Menzies, xxii. Plateau in female labour noted by Chamie, 10-11.

26. For recent analysis which uses these terms see Belinda Leach,

"'Flexible' work, precarious future: Some lessons from the Canadian clothing industry," *Canadian Review of Sociology and Anthropology* 30 (February, 1993), 64-82.

27. E.M. Beck, Patrick M. Horan, and Charles M. Tolbert II, "Stratification in a dual economy: A sectoral model of earnings determination," *American Sociological Review* 43 (October, 1978), 704-20. An example of Canadian research using these concepts is Richard Apostle, Don Clairmont, and Lars Osberg, "Segmentation and labour force strategies," *Canadian Journal of Sociology* 10 (Summer, 1985), 253-75.

28. Such issues are raised, for example, in a work such as Melody Hessing, "Talking shop(ping): Office conversations and women's dual labour," *Canadian Journal of Sociology* 16 (Winter, 1991), 23-50.

29. Alvin Toffler, *Future Shock* (New York: Random House, 1970). A collection of readings on workplace automation was in fact entitled *Automation, Alienation, and Anomie*, Simon Marcson, ed., (New York: Harper and Row, 1970).

30. Arthur Kroker and Michael A. Weinstein, *Data Trash: the Theory of the Virtual Class* (New York: St. Martin's Press, 1994).

Further Readings

For discussion of Braverman and examples of other criticisms of workplace technology:

Braverman, Harry. *Labor and Monopoly Capital: The Degradation of Work in the Twentieth Century*. New York: Monthly Review Press, 1974.

Fitzgerald, Deborah. "Farmers deskilled: Hybrid corn and farmers' work," *Technology and Culture* 34 (April, 1993), 324-43.

Garson, Barbara. *The Electronic Sweatshop: How Computers are Transforming the Office of the Future Into the Factory of the Past*. Harmondsworth, Middlesex: Penguin, 1989.

Grayson, J. Paul. "Plant closures and political despair." *Canadian Review of Sociology and Anthropology* 23 (August, 1986), 331-49.

Meiksins, Peter F. "Scientific management and class relations: A dissenting view." *Theory and Society* 13 (1984), 177-209.

Menzies, Heather. *Fastforward and Out of Control: How Technology is*

Changing Your Life. Toronto: Macmillan of Canada, 1989.

___. *Whose Brave New World? The Information Highway and the New Economy*. Toronto: Between the Lines, 1996.

Rinehart, James. *The Tyranny of Work: Alienation and the Labour Process*, 2nd. ed. Toronto: Harcourt Brace Jovanovich, Canada, 1987.

Smith, Merritt Roe. "Industry, technology, and the 'labor question' in 19th-century America: Seeking synthesis." *Technology and Culture* 32 (July, 1991), 555-70.

For different deskilling impacts across different organizational settings:

Blauner, Robert. *Alienation and Freedom: The Factory Worker and His Industry*. Chicago: The University of Chicago Press, 1964.

Kelly, John E. *Scientific Management, Job Redesign and Work Performance*. London: Academic Press, 1982.

Meissner, Martin. *Technology and the Worker: Technical Demands and Social Processes in Industry*. San Francisco: Chandler, 1969.

For a view of technology adoption from the perspective of managers:

Bennett, Stuart. "'The industrial instrument — master of industry, servant of management': Automatic control in the process industries, 1900-1940." *Technology and Culture* 32 (January, 1991), 69-81.

Tailby, Stephanie and Colin Whitston, eds. *Manufacturing Change: Industrial Relations and Restructuring*. Oxford: Basil Blackwell, 1989.

Thomas, Robert J. *What Machines Can't Do: Politics and Technology in the Industrial Enterprise*. Berkeley: University of California Press, 1994.

For more general discussion of workplace technology issues:

Betcherman, Gordon. "Computer technology, work, and society," *Canadian Journal of Sociology* 15 (Spring, 1990), 195-201.

Form, William, Robert L. Kaufman, Toby L. Parcel, and Michael Wallace. "The impact of technology on work organization and work outcomes: A conceptual framework and research agenda" in George Farkas and Paula England, eds. *Industries, Firms, and Jobs: Sociological and Economic Approaches*. New York: Plenum, 1988, 303-28.

Hedley, R. Alan. *Making a Living: Technology and Change.* New York: Harper and Collins, 1992.

Storey, Robert. "Studying work in Canada," *Canadian Journal of Sociology* 16 (Summer, 1991), 241-64.

Wiener, Norbert. *Cybernetics, or Control and Communication in the Animal and the Machine.* Cambridge, Mass.: MIT Press, 1961 [1948].

Chapter 7: Communications: Hot and Cool

Some of the most basic questions concerning technology and society come to the fore within communications. Such issues from earlier chapters as changes in social class alignments due to technological innovation are closely involved. Much of the workplace technology in the immediately preceding pages concerns accelerated communications. The telephone, an example of technological diffusion examined in Chapter 3, is of course among the most important of all innovations in communications media. Indeed, communications are central to every domain of life in today's "information society," serving as the sociological glue holding social structure together. This chapter opens with some general points on this importance of communication for societal integration and then considers Marshall McLuhan's famous distinction between hot and cool media.

Communication, Culture, and Social Structure

It is part of the essence of humanity to share meanings with others; without such communication, a child seems more like an animal than a human, as cases of "feral" children reveal.[1] Animals communicate too

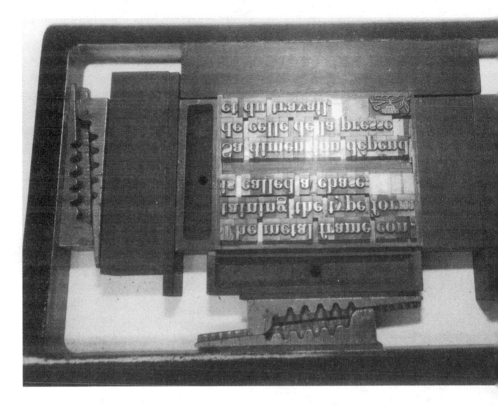

Moveable type, the invention pioneered by Gutenberg. *National Museum of Science and Technology*, Ottawa. The type is legible if read through a mirror.

but, as Konrad Lorenz's classic work on animal versus human communication explains, when a bird gives a danger signal the impulse to make the cry is instinctive, and so is the impulse among other birds in the flock to take to their wings.[2] Such messages for animals do not pass through the consciousness, as they do in communication between humans.

Communication between humans, being learned rather than from instinct, can be infinitely more complex than the barks or meows of even the most domesticated animal friends of humankind. A learned vocabulary is highly expandable whereas shift in an instinctual one would take many generations. Consider the amount of computer-related jargon that has entered the English language in the past generation. For example, the first reference to a computer "disk" appeared in the Oxford English Dictionary in 1947, while "byte" was first cited in 1964. As the famed American pioneer in social psychology George Herbert Mead emphasized, human communication takes place in the form of "significant symbols." The words, gestures, inflections etc. of communication encode meanings.[3] When passing secret messages, it is easy to envision coded meanings, but in ordinary speech and writing within our native language we associate meanings to the words and phrases so instantaneously that it takes a moment to step back and reflect on how complex the encryption of everyday communication is. It is just as "coded" as electronic impulses passing through a modem out onto the Internet. Indeed, it is much easier to put across these ideas about the symbolic nature of language to the present generation familiar with the technology of computer graphics stored as a massive array of binary codes than it is to those who grew up in a more mechanical world.

Language is more than just some arbitrary code, however, for it resides at the core of a society's culture and social structure. Indeed, in the famous *"Sapir-Whorf"* hypothesis, language virtually defines culture, producing "an organization of experience ... a classification and arrangement of the stream of sensory experience...." Whorf's research included the study of language among the Hopi Indians of Mexico, whom he found had a conception of time different from that of cultures grounded in European languages. Thus two events are a long "time" apart, in Hopi language and culture, "when many periodic physical motions have occurred between them in such a way as to traverse much distance or accumulate magnitude of physical display."[4]

Admittedly, that is a complex thought for an English-language readership, but perhaps that validates all the more the proposition that language reveals rather deep, unfathomable aspects of the way of life and thought of a society.

Empire and Communications: The "Canadian School" of Harold Innis

Harold Innis was a University of Toronto economist of such great stature that a college of the university carries his name. His academic fame rests on economic histories of Canada's "staple industries" such as fishing and fur trading. It was in the course of that work that he became increasingly mindful of the importance of communications not just for economic development, but for political institutions as well.

The style of research in communications that he pioneered bears an early twentieth-century imprint. Innis, living in a Canada that was still self-consciously part of the British Empire, found it natural to study the linkage between communications and the rise and fall of empires.[5] He became interested in "the bias of communication." "Bias" here is not a pejorative term, but a social scientist's observation that "a medium of communication has an important influence on the dissemination of knowledge over space and over time."[6] The reference to "space" and "time" in the preceding sentence is no casual figure of speech, but rather a central analytic distinction in Innis's work. All media, he argued, and certainly those of the pre-Christian era empires, could be classified within the time/space distinction. A communications medium such as stone, for example, could be etched with hieroglyphics that were durable over great periods of time, but was not well suited to traversing distance due to its weight and bulk. Such a medium had a bias toward time. The papyrus introduced into the Egyptian Empire, in contrast, had a space bias due to its portability. Innis examined the consequences of each bias, and concluded that time-biased media strengthened the hold of organized religion within an empire while space-biased media favoured more centralized political administration. Since a successful empire needed both central administrative control and the unifying force of a common religion, stability was achieved only when the biases of one medium offset the biases of another.

Little mention of the space versus time biases of media is found in discussions of modern communications, because today's technology has essentially overcome the distinction. A compact disk, for example, is both durable and portable. Innis should be remembered, however, for his facility to abstract an underlying dimension upon which to classify, and then analyze, communications media. This is the method exploited so fruitfully by Marshall McLuhan, beginning with the study of printing.

From Manuscript to Print: McLuhan's *Gutenberg Galaxy*

McLuhan was a junior colleague of Innis's at the University of Toronto, and in his introduction to *The Bias of Communications* recounts how greatly his own work was influenced by the older man's. We met Marshall McLuhan briefly in Chapter 1, in a discussion on the "implosion" resulting from electronic technology. Part of the task of the present chapter shall be a more systematic examination of his ideas, beginning with the distinction he made between oral and literate cultures.

The Gutenberg Galaxy, subtitled "the making of typographic man," appeared in 1962. It was the first product of the mature McLuhan who had evolved from an English literature analyst to a cross-disciplinary synthesizer who drew from history, social science, and natural science with equal, some might say excessive, confidence.[7] Innis had discussed the tradition of oral communications within the Greek Empire, but did not fit it within his time/space distinction in any very clear way. In contrast, McLuhan gave an analytical meaning to oral communication by designating it as "cool." Hot versus cool became a media distinction he used again and again, and so let us try to clarify what he meant.

Hot and Cool Media

McLuhan was fascinated with the possibility that each medium engages a group of the human senses in a particular way or "ratio," and coined the terms "hot" and "cool" as names for two types of sensory mix.[8] It might have been more logical to contrast cool with warm, or hot with cold, and certainly McLuhan used the terms hot and cool

in an idiosyncratic way that one sometimes wants to reverse, calling what he termed a cool medium hot.

According to McLuhan, it was all but inevitable that the technological advances achieved so long ago by Gutenberg (see Figure 2.2) led to the "hot" printed word pushing aside "cool" spoken communication so that "the visual values have priority in the organization of thought and action" (*Galaxy*, 41, 27). A cool medium such as face-to-face conversation is multisensory in that visual clues such as "body language" augment the aural messages. McLuhan attributed to such communication an unskewed sense ratio with "participation in situations that involves all of one's faculties" (*Media*, vii). Shift in the ratio of the senses was just one of the ways in which "the medium is the message," as his famous phrase goes. The sense ratios are skewed all the more by a very "complete" form of communication. McLuhan thus used terms such as "high definition" and "well filled with data" in reference to hot media (see *Galaxy*, 138, and *Media*, vii, 36, 269, 270, 273, 283). Oral communication is thus innately cool in part for its low definition. A photograph is high definition in the clarity of its image compared to a cartoon. McLuhan therefore classified the cartoon as cooler than the photo. A high definition message monopolizes some senses to the exclusion of others. It may, for example, be easier to analyze the political message of a cartoon than of a photograph precisely because the cartoon is less finished visually. "Virtual Reality," the technique combining audio, visual, and tactile stimuli to create a simulated experience (*Psychology Today*, January/February 1996, 60), would have intrigued McLuhan for its cool implications.

The cool form of pre-Gutenberg culture was for McLuhan reflected in early medieval art in which three-dimensional perspective was far less exact than in our own "hot" time. McLuhan interpreted the fact that background and foreground details are the same size in these early paintings as analogous to the artists' perception of the equality of the senses. To draw perspective into a picture is, within this intriguing but perhaps slightly fantastic argument, to let one dimension, the perspectival one, dominate in typically hot media fashion.

Print as a Hot Medium

McLuhan saw the social consequences of the print revolution in Gutenberg's time as a transition from a cool world of predominately oral discourse and hand-copied books to the hotter era of print. He speculated that the print-based world eroded local tribal loyalties and ushered in the more modern era of "nationalism, industrialism, mass markets, and universal literacy" (*Media*, 157). That is, the printed word connected people into larger social structures, but made the connections more abstracted and impersonal than in more local societies.

The linkage from such social and political events back to "sense ratios" is very abstract, with reasoning reached often by metaphor and analogy. McLuhan implied that just as a hot medium fragments the mind by monopolizing first one sense then another, so the whole society is led in a "fragmentary" direction (*Media*, 272, and discussion in Chapter 1 above). It is difficult to follow his thesis because he spoke of sensory "participation" in dual meanings. The first, which I shall label "physiological" participation, refers to communication in which all the senses participate in receiving messages in the same unskewed ratio seen in primitive oral cultures. The extent of participation in this physiological sense is involuntary, deriving from the central nervous system as a machine or organism.[9] A second kind of participation has a "social" connotation: since hotter media are less interpersonally reciprocal and less interactive, the "participation" of the senses implies free will and consciousness. "Any hot medium allows of less participation than a cool one," he wrote, "as a lecture makes for less participation than a seminar, and a book for less than dialogue" (*Media*, 37).

McLuhan's ideas are difficult to pull together into a simple statement because he tended to proceed in a non-linear, "mosaic" fashion, jumping abruptly from point to point. As noted in connection with perspective in painting, above, his reasoning at times was more metaphoric than logical by conventional standards. He was also inconsistent. Early in *Gutenberg Galaxy*, for example, he referred to "the ear world" in tribal Africa as "hot" (19), yet throughout the remainder of the book he described the spoken word as cool in comparison to the hot printed word.

An attempt to codify McLuhan's ideas about hot and cool media into the very kind of formal model that he resolutely eschewed appears

Figure 7.1: Media Coded for their "Hot" vs. "Cool"-Making Attributes

	The Ratio of the Senses		
	Sensory "Participation"		Result
	(a)Physiological	(b)Social	
Archetypes			
Spoken word	unskewed [oral & visual, low definition]	unskewed [high social participation]	*cool*
Printed word	skewed [visual, high definition]	skewed [low participation]	*hot*
More complex examples			
Hand-copied manuscript	slightly skewed [visual, medium definition]	slightly skewed [lower participation]	*transitional*
Telegraph	skewed [visual, medium definition]	rather skewed [limited participation]	*toward hot*
Telephone	low skew [oral, very low definition	low skew [high participation]	*cool*
Radio	low skew [oral, low definition]	ambiguous [highly variant participation]	*ambiguous*
Film	some skew [oral & visual, but high definition]	skewed [low participation]	*toward hot*
Television	some skew [oral & visual, lower definition than film	some skew [low, but higher than the participation in film]	*cooler than film*
Computer	ambiguous [highly varied]	skewed > unskewed [low >high participation]	*hot > cool*

Indicators; cool: oral and visual mix, low definition, high social participation
hot: visual only, high definition, low social participation

as Figure 7.1. Although the old master probably would have disapproved, summarizing his scheme on one sheet of paper provides a framework for better analyzing his ideas about media such as television, discussed below. The unifying principle for the entire scheme is the ratio of the senses, the notion that McLuhan returned to again and again (for example, in the "Playboy Interview" reprinted in McLuhan and Zingrone, Further Readings). Whether the ratio was skewed or unskewed derives from various nuances of sensory participation. Participation in the physiological and social meanings are therefore considered side by side. An oral-visual mix such as a face-to-face conversation is taken to indicate a "cool-making" unskewed ratio of the senses. Oral-biased communication is cooler and less skewing than visual largely because of the "low definition" of such signals. The distinction between high and low definition, encountered with reference to photos and cartoons, above, becomes crucial to McLuhan's analysis of the more complex and modern media such as film and television.

Interactive, socially participatory, communication is cooler than one-way, non-interactive signals. McLuhan stressed, for example, that one cannot "talk back" to print as one can to a face-to-face speaker, making the printed work "hotter." Of course, all media are interactive within their own time scale, and McLuhan's distinction essentially refers to how rapidly the person receiving can become the person sending.

A large component of history for McLuhan was a transition from cool oral media to hot printed forms. In his epigrammic and punning way, he liked to summarize this as the substitution of "an eye for an ear." In *Understanding Media* he described that substitution as "socially and politically, probably the most radical explosion that can occur in any social structure" (58).

In Figure 7.1 the cool spoken word and the hot printed word are presented as "archetypes," the pure categories which initiated McLuhan's analysis. They are consistent across both meanings of sensory participation. Thus, the spoken word is cool for being an oral-visual mix without the distraction of high definition, but also for the social participation typical of such communication. Even the manuscript book painstakingly hand-copied by scribes was less cool than the spoken word, acting indeed as a Trojan Horse of hotness by ushering the phonetic form of alphabet into Western culture. Such alphabets, including our own, break language down into compact codes rather than draw-

ing a miniature picture of a concept, as in earlier hieroglyphic alphabets. Phonetic alphabets were efficient enough to rival or surpass the oral medium of communication, and thus made possible "the breaking apart" of the ear and the eye (*Galaxy*, 22).

The distinction between hand-copied and printed books might seem overblown, but McLuhan felt that the accelerated reading due to the clarity of print (that is, its high definition) was crucial. Where a hand-copied book had virtually to be read aloud (thus being partially oral), the printed page could be truly non-oral. He referred to speedreading techniques, by which people "are taught how to use the eye on the page so as to avoid all verbalization and all incipient movements of the throat which accompany our cinematic chase from left to right ..." (*Galaxy*, 47).

The Gutenberg Galaxy is such a contribution to communications analysis in part because it suggests how technologies which are very simple along some dimensions may nevertheless exert vast consequences. The technique of using an inked solid object to reproduce an impression had existed well before Gutenberg invented his press. An elementary school teacher today who rewards good work with a "happy face" stamped on a student's notebook follows that same ancient method. "All" Gutenberg did was to break down more complex printing jobs into that simple an operation by using moveable type in which each letter was a separate component comprising its reverse image carved out of lead (see photo at start of this chapter). The more obvious alternative in the 1400s was to carve entire pages from the lead, a time-consuming, labour-intensive, and wasteful task. Moveable type, in contrast, could be reused many times thus allowing great economies of effort. It took more complex engineering than one might expect, however, to make typesetting with moveable type produce straight lines with uniform darkness. Gutenberg's background as a metallurgical craftsman was crucial.[10] Moveable type is perhaps the first example of technology premised on interchangeable parts, for any letter selected from the stock had to fit perfectly into place. So although the *materials* (lead, copper, wood) seem unsophisticated in the era of plastics and space-age compound metals, and the *power source* initially was simply human muscle, the *knowledge-base* had a complexity that is echoed in almost all modern machines.

McLuhan would never have achieved such fame had he only analyzed fifteenth century technological advances in printing. His work on media such as radio, film, and television is mainly what created his reputation as the "media guru" of the 1960s. It is appropriate, then, to follow his ideas about these more modern media, beginning with an attempt to integrate his ideas about hot/cool and mechanical/electronic technologies.

The Electronic Media "Implosion" as both Hot and Cool

McLuhan's emphasis on the shift from mechanical to electronic media was described in Chapter 1. We saw how he had been struck by the potential for electronic media to erase boundaries and shrink distances. McLuhan insisted that, beyond the commonalities arising from their shared electronic power and knowledge-base, these new media were a varied package neither uniformly hot nor cool. A tabular synthesis of his ideas about the mechanical/electronic and hot/cool dimensions appears in Figure 7.2, to show why he saw the electronic era of communications technology as, in part, a return to the "cooler" pre-Gutenberg time. Some of the electronic media shall now be examined in greater detail.

Telegraph and Telephone

Two hundred years ago, when England was at war with the French Emperor Napoleon, a pre-electronic telegraph system by semaphore passed messages from naval bases on the English Channel to London. The English were imitating their enemy, for the technology had recently diffused from France. Claude Chappe had developed telegraph in 1794, building from earlier methods by using signal towers housing a pivoted wooden beam. With the Chappe system of towers spaced 6 to 10 miles apart, a message could be sent between Paris and Toulon, 475 miles apart, in 10 to 12 minutes. Naval fleets used a similar method to communicate from ship to ship as signal flags could be read by observers posted on the tall masts even when the hulls of the respective vessels were over the horizon from each other.[11]

Figure 7.2: Mechanical vs. Electronic Power and Knowledge-Base Integrated with Hot vs. Cool Attributes

	Mechanical	Electronic
Hot archetype	Print	—
Towards hot	Film	Telegraph Mainframe computer
Ambiguous	—	Radio, television
Towards cool	Hand-copied ms	Telephone Micro computer
Cool archetype	Spoken word	—

At around the same time, knowledge of electricity was accelerating. The electric battery, for example, was invented by Volta in 1800. By the 1830s, the knowledge-base needed for the "electromagnetic" telegraph was in place, and Cooke and Wheatstone in Britain and Samuel Morse in the US experimented with prototypes. By 1847 Morse had a viable system in place, transmitting his code of "dots and dashes" over hundreds of miles.[12]

The electronic telegraph was a quantum leap in communications. It is, as an historian of those times observes, "almost impossible to conceive now of how important, how revolutionary, the telegraph was then, 130 years ago" (Neering, in Further Readings, xi). "What hath God wrought?" goes the famous quotation by Morse transmitted from Washington to Baltimore in the first inter-city telegraphic communication in 1844.[13] The receptive business environment described in Chapter 2 as necessary for the deployment of an invention was available for the telegraph because instant message-passing was so useful to the railroad companies, which were developing rapidly during the 1840s and 1850s. For about 40 years, the telegraph held a monopoly for rapid communication, until the telephone, the "talking telegraph," appeared in the 1880s and gradually eclipsed the older medium. The turn of the technological wheel, however, has led to revived interest in the telegraph. Although telegraphic communication had nearly disappeared by the beginning of the 1970s, the outpouring of coverage on communication

by Internet in the mid-1990s revived interest in the telegraph. "A century before the Internet, the telegraph begat what we now call cyberspace," a story in *Canadian Geographic* began (November/December 1995, 40).

According to McLuhan, the telegraph was a relatively hot medium compared to the very cool telephone. The "heat" in telegraph came primarily from its visual nature, as contrasted to the telephone that is almost more oral than face-to-face speech which includes facial expression and other "body language." When direct Washington-Moscow communication was set up in the early 1960s at the height of the Cold War, President Kennedy wanted a telegraphic rather than telephonic link. This, for McLuhan, was simply a predictable expression of "a natural Western bias" for the hot printed word over the more oral and cool Russian culture (*Understanding Media*, 239). The "hot" interpretation of the telegraph is also supported by the historical fact that telegraphic service was expensive for the average nineteenth-century householder. It cost 85 cents in 1850s currency, for example, to send 10 words between Hamilton and Toronto.[14] That is one reason why Figure 7.1 codes the telegraph as involving limited social participation, one of the hot traits.

Once it is understood as a relatively hot medium, it makes sense that the telegraph has been recorded in Canadian history as crucial to late nineteenth-century nation-building. As the *Canadian Geographic* issue quoted above said: "The railroad connected us physically ... but the telegraph connected our minds — *that's* what helped us define the Canadian community, the national ethos" (43).

The telegraph, then, helped expand Canadian culture and social structure from a local to a more regional and national scale. Just how local was the world of nineteenth-century Canada shows through in the confusion over time seen in the first telegraphic communication between Hamilton and Toronto. "What o'clock is it?" the Hamilton operator asked as the line was activated on a grey December day in 1846.

Toronto: "Twenty-five minutes past eleven."
Hamilton: "You must mean twelve?"
Toronto: "No, it is just half-past eleven."
Hamilton: "Is that the town time?"

Toronto: "Yes."[15]

The origin and diffusion of the telephone received attention in Chapter 3, and just a few McLuhanesque points shall be made now. "Telephone is a cool medium, or one of low definition," Marshall McLuhan wrote in Understanding Media (36). To communicate by telephone requires social interaction in a way that reading a book does not, hence the "high social participation" notation in Figure 7.1. And as noted earlier, a "meager amount of information" (Media, 36) comes down a telephone line, both because visual cues are absent and because even a modern telephone line does not perfectly reproduce the voice. This "low definition" referred to by McLuhan gave a slightly abstracted, disembodied quality to telephonic communication.

A relationship by telephone could thus have a different social structure to one conducted face-to-face. McLuhan believed for instance that the phone tended to break down status differences. Within business organizations, he said, "the pyramidal structure of job-division and description and delegated powers cannot withstand the speed of the phone to bypass all hierarchial arrangements . . ." (Media, 238). Michele Martin's history of the telephone in Canada, not in general a particularly McLuhanesque book, also found a special quality to social interaction by telephone. "The fact that the phone allowed oral communication without visual contact created a kind of intimacy which people previously had not experienced."[16] As another authority says, the telephone is akin to "whispering in the ear ... seeing somebody almost destroys the intimacy of the communication" (Globe and Mail, 21 January 1995, D8).

Radio, Film and Television

There is an ambiguity in McLuhan's hot/cool scheme due to the distinction between physiological and social participation. In passages describing the archetypical spoken and printed communication McLuhan implies that social participation follows as a *consequence* of the more physiological participation. In considering electronic media, however, social participation assumes an importance *equal to*, and alongside, the physiological issues of sensory reception. Viewing sensory mix in this second way, as illustrated in Figure 7.1, for me makes

McLuhan's scheme more viable. An implication, however, is that a medium can be hot in the physiological participation sense but cool in the social interaction sense, or vice versa.

This complication arises in McLuhan's analysis of radio, film, and television. Although offering undoubted insights, his assignment of "hot" and "cool" qualities for these media shows the Toronto scholar at his most idiosyncratic.[17] Radio, for example, would seem to have some cool qualities, a case along with the telephone of a totally oral electronic medium. If his readers are to take the oral/print distinction advanced in *Gutenberg Galaxy* seriously, the low definition of radio should be a key defining (cool) quality. Yet McLuhan perceived the "hot" non-social participatory quality of radio, citing Hitler's use of the medium for propaganda in the 1930s (*Media*, 260).

It is useful to dwell on radio for a moment, for here is illustrated the difficulty of assigning a character or "medium is message" quality to modern communications technology. The historical fact is that radio virtually from the outset had two-way capability. Indeed, it was the neglected Canadian inventor Reginald Fessenden who not only preceded Marconi by a year in achieving transmission of a one-way radio message in 1900, but in addition accomplished in 1906 the first two-way broadcast across the Atlantic.[18] In another instance of the theme developed in Chapter 2, Marconi, historically much the more famous name, was simply the more skilled in promoting and financing his invention into a marketable product. Radio in the early days was a highly interactive, participatory medium with the grassroots, democratic ethos McLuhan normally attributed to coolness. The present writer has an uncle who, as a schoolboy using a home-made set in 1924, made the world's first two-way round the world radio communication.[19] Amateur radio may not be as newsworthy today as in those pioneering days earlier in the twentieth century, but as a two-way communication medium, radio remains significant today both among "ham" operators using shortwave sets and in "Citizen's Band" (CB) communications. Two-way radio is especially important in the remote areas of countries such as Canada or Australia. McLuhan the Torontonian had a big city perception of radio, conceiving of it in the sense of commercial and public radio companies transmitting mass one-way broadcasts.

It is easier to agree with Marshall McLuhan's view of film as a hot medium. Film indeed is highly visual, and began that way before

"talkies" superseded the "silents." It is easier to generalize too about film being non-participatory vis à vis the audience, although even that aspect only remains true for films produced for screenings to mass audiences, as contrasted with home movies. McLuhan made much of the (hot) "high definition" of film compared to television. One of the few technical details he ever bothered to cite was the statistic that television offers three million dots per second to the viewer, film "many more millions of data per second" (*Media*, 273).

The relatively coarse-grained pictures of a TV tube seem a flimsy basis for categorizing television as a cool medium, yet it is hard to discern what other reason McLuhan had for this decision. His sense of proportion seems to have gone astray here. To argue that the visual has more "definition" than the spoken word is convincing, reflected in the adage "one picture is worth a thousand words." To draw as sweeping an implication as he did about the picture definition of film versus TV seems much riskier. Another paradoxical aspect of McLuhan's view of TV is his feeling about the interactivity of the medium. For instance, he referred to the Jack Paar Show, the first of the late night talk programs. Paar's show, argued McLuhan, "revealed the inherent need of TV for spontaneous chat and dialogue." He had a notion that television stresses "participation" and "dialogue" (*Media*, 288-89), yet it is hard to imagine what he meant by this claim. In truth, surely it is radio that is the more interactive. Radio talk shows routinely allow for audience phone-in, while TV programs like Jack Paar's confined all the interaction to the host and a procession of guests, leaving the viewer as passive onlooker. "Passive" must indeed be the most frequently used of all adjectives applied to TV (for example, Winn, 17-18; Noble, 7, both in Further Readings).

Television Violence as "Message"

More perhaps than most media, television reminds us that not only is the "medium the message" as McLuhan stressed, but also the converse cannot be ignored; that the "message is the medium." Scores of psychologists, sociologists, and other communications researchers have attempted to unravel the linkage between TV viewing and violence, especially with respect to children.[20] Two facts are beyond dispute: that television depicts astounding amounts of violence every day, and that

Figure 7.3: A Guide to Typical Strength of Statistical Association in the Social Sciences

1.0	Theoretical maximum (two identical measures)
0.9	
	Reliability of income reported on surveys(a)
0.8	
0.7	Reliability of self-reported life satisfaction(b)
0.6	Grades in high school and first year university(c)
0.5	The height of fathers and their sons(d)
0.4	Education level and income(e)
0.3	**Violence on TV and aggressive behaviour**
0.2	Child's IQ and number of siblings—negative(f)
0.1	
0.0	Traits with absolutely no relationship with one another (e.g., data drawn randomly)

Sources:
(a)Stephen B. Withey, "Reliability of recall of income," *Public Opinion Quarterly* 18 (Summer, 1954), 200. (Reported income for 1947, and report a year later of 1947 income). (b)Thomas H. Atkinson, "The stability of validity of quality of life measures," *Social Indicators Research* 10 (1982), 113-32. (c)Winston Jackson, *Methods: Doing Social Research* (Scarborough: Prentice-Hall Canada, 1995), 347. (d)Karl Pearson, "Mathematical contributions to the theory of evolution. XIII. On the theory of contingency and its relation to association and normal correlation," *Karl Pearson's Early Statistical Papers* (Cambridge: University Press, 1956 [1904]), 470. (e)John Goyder, "Income differences between the sexes," *Canadian Review of Sociology and Anthropology* 18 (August, 1981), 332. (f)Sandra Scarr and Richard W. Weinberg, "The influence of 'family background' on intellectual attainment," *American Sociological Review* 43 (October, 1978), 678.

levels of violence are very high in modern societies.[21] *Almost* beyond dispute is an association between these two. Not every single study on the topic has detected a statistical link between watching violence on TV and the committing of violence in "real life," but the great majority have. This is demonstrated in a comprehensive review of the evidence by Paik and Comstock (see Further Readings). Examining 217 separate studies, and 1142 "hypothesis tests" (since several forms of hypothesis might be tested within any one study), they found an average "effect size" or correlation of 0.31 between televised violence and aggressive behaviour. This was viewed as a "medium" (535) strength of relationship, and a clearer sense of what that word means may be found in Figure 7.3 which compares the TV-aggressive behaviour relationship with other statistical associations familiar in social science. Here it is shown, for example, that the tendency for witnessing violence on TV to lead to aggressive behaviour by the viewer is nearly as strong as the tendency for educational level to determine one's income level as an adult.

Although the *association* between TV and violence is clear, the underlying *meaning* is not. It is one of the enduring dilemmas of the social sciences that the demonstration of association does not prove cause and effect. Possibly, for example, violent or aggressive people watch TV frequently, the violent programs simply being confirmatory of a pre-existing predisposition. Alternatively, the association between television and aggression might be the surface manifestation, an epiphenomenon, of underlying causes such as social class.[22] The argument here could be that adolescents reared in poverty experience frustrations that are relieved through two outlets, (i) spending much of their time watching television and (ii) committing violent acts.

Great effort has been made to decipher such possibilities, particularly in the US where levels of violence are so high. Some of the studies are set up as experiments in a laboratory, the participants frequently being recruited from university classrooms. Such work typically begins by randomly dividing the subjects into control and experimental groups. Both groups are tested using an attitude scale designed to reflect violent tendencies. Both groups are then shown a television program, the control group seeing a non-violent program, but the experimental "cell" of the experiment being exposed deliberately to a much more violent one. If on a second testing on the attitude scale the exper-

Table 7.1: Mean Number of Verbal and Physical Aggressive Acts Per Minute For a Study of Three B.C. Communities

| | Number of Aggressive Acts per Minute | | |
	"Notel"	"Unitel"	"Multitel"
Physical Aggression			
Phase 1	0.43	0.42	0.46
Phase 2	1.30	0.58	0.65
Verbal Aggression			
Phase 1	0.21	0.12	0.30
Phase 2	0.41	0.18	0.24

Source: Tannis MacBeth Williams, ed., *The Impact of Television: A Natural Experiment in Three Communities* (Orlando: Academic Press, 1986), 320.

imental group's average aggression score had increased while that for the controls had not, causal evidence that TV causes violence would exist.[23] Laboratory studies of that style, although with designs far more complex than the deliberately simple one just presented for explanation, often find quite strong "effect size" (0.4 on the Figure 7.3 scale) according to the Paik and Comstock review noted previously (526).

To escape the artificiality of laboratory settings, researchers seek out "natural experiments". In a noted Canadian project, Williams and colleagues (see Further Readings) studied a small community in British Columbia which it so happened was virtually without television due to geographic conditions. As a form of control with this village dubbed "Notel," another community which received just CBC was located, and named "Unitel." "Multitel," which received both CBC and American TV, was used as a third comparison group. One of the key results from William's study is reproduced as Table 7.1, and shows the increase in aggression in Notel from Phase 1, before TV, to Phase 2, following the introduction of television. Although violence or aggression can be hard to measure, this study handled the problem rather convincingly. Since the subjects were children in grade school classrooms, the assessments could be made behaviourally, by observers noting every action.

With the children of the first generation of TV-watching children from the 1950s now themselves well into adulthood, the opportunities for natural experiments do not exist any more as they did even in the

summer of 1973 when the Williams study began. Many of the more recent studies therefore must use surveys conducted in communities in which TV is available. Variations in viewing habits still exist. Some small number of families may not own television sets, and even among those with access to TV some will be sparing viewers, others more frequent. One of the best known survey studies of TV and violence was begun in 1979 by Huesmann and Eron (Further Readings) in the US, coordinated with parallel work in four other societies. Within a survey approach, the causal issue is best addressed by compiling comprehensive background profiles on respondents and by collecting several instalments of data, yearly for each of three years in the Huesmann and Eron study. Children from the first and third grades of a Chicago suburb were surveyed in the US component, with television watching tallied from reports by parents of the children, and aggression measured by peer ratings in which each child rated each of his or her classmates. Viewing of violence on television correlated near 0.2 in each grade, and persisted across all three yearly waves, a typical result for survey studies on this topic, according to Paik and Comstock (526).

Summing up these remarks on television, the contention is not to deny that this medium "massages" the message, as McLuhan might have punned (see McLuhan entries of Further Readings). Many would agree with Marie Winn who in an often-cited book (see Further Readings) declared on the opening page that "there is a similarity of experience about all television watching. Certain specific physiological mechanisms of the eyes, ears, and brain respond to the stimuli emanating from the television screen regardless of the cognitive content of the programs." From the data available on amount of viewing alone, one can infer that TV displaces other activities in a manner highly congruent with McLuhan's notion of the ratio of the senses. It would be a mistake, however, as the more detailed discussion of studies on violence indicated, to forget entirely the importance of content in this or any other communication medium.

Medium and message seem to interact. The old CBC "test pattern" of the 1950s was the most neutral of messages, a static display of various geometric shapes. It was broadcast during off hours when no programs were on air, but such was the engrossing quality of the medium that some children of that era spent not inconsiderable amounts of time staring at the pattern. They may have been wasting their time, but

they were not learning to use violence as the solution to human problems. It is, however, because television broadcasting is an expensive undertaking that the message becomes related to the medium. Unless a government is prepared to invest enormous sums into public broadcasting, television must run as a highly capitalist profit-making activity. It is no cause for wonder if a medium that in most societies is only sustainable on a capitalist basis shows in its content all the amorality characteristic of capitalism.

Conclusion

Communications have always been important to Canadians due to the vast distances of our geography. Nineteenth-century nation-building was closely interwoven with civil engineering projects such as the transcontinental railroad. The advent of electronic communications tools including telegraph and later the telephone were crucial to the formation of a more national society. It is thus appropriate that Canadian scholars have been leading theorists on communications.

We briefly reviewed the pioneering work of Harold Innis and then in more detail some ideas from Marshall McLuhan. With his distinction between hot and cool, McLuhan provided an analytical tool that can be applied to any communications medium. Such a bold approach is bound to be controversial, as was seen here especially with respect to television. If, as McLuhan said, the medium is the message, the message is also the medium. Violence on television is a particularly important case in point.

Notes

1. Murray Knuttila's introductory text *Sociology Revisited: Basic Concepts and Perspectives* (Toronto: McClelland and Stewart, 1993), 84-88, has a good guide to some literature about feral children.
2. Konrad Z. Lorenz, *King Solomon's Ring* (London: Methuen, 1964), Chapter 8.
3. Anselm Strauss, ed., *George Herbert Mead on Social Psychology* (Chicago: Phoenix Books, 1964); John Fiske, *Introduction to Communication Studies* (London: Methuen, 1982), 70-74.
4. Benjamin Lee Whorf, *Language, Thought, and Reality: Selected*

Writings of Benjamin Lee Whorf, John B. Carroll, ed., (Cambridge, Mass.: MIT Press, 1956), 55, 63. Modern thinking within linguistics, influenced by the work of Noam Chomsky, e.g., *Language and Mind*, enlarged ed. (New York: Harcourt Brace Jovanovich, 1972), is more preoccupied with the unifying "deep structures" common to all language than with cultural variations, but I think Whorf's point relates to themes in this chapter rather well.

5. H.A. Innis, *Empire and Communications* (Oxford: Clarendon Press, 1950), 6. Even though this publication dates from mid-century, Innis had conducted his main research on staples in the 1920s and 1930s, an era when Canada had a close sense of empire in connection with Great Britain.

6. Harold A. Innis, *The Bias of Communication* (Toronto: University of Toronto Press, 1951), 33.

7. E.g., Elizabeth L. Eisenstein, *The Printing Press as an Agent of Change: Communications and Cultural Transformations in Early-Modern Europe* (Cambridge: Cambridge University Press, 1979), 41. This whole book is a much more historically exhaustive treatment of printing than McLuhan's.

8. Media as "extension" of "some organ or faculty of the user," and the corollary that "when one area of experience is heightened or intensified, another is diminished or numbed" constituted two of the four "laws" in the co-authored posthumous book by Marshall McLuhan and Eric McLuhan, *Laws of Media: The New Science* (Toronto: University of Toronto Press, 1988), viii.

9. In a 13 March 1996 lecture at Wilfrid Laurier University, Eric McLuhan, Marshall's son, stressed that "there is no choice" as the senses experience media. He was referring to the way, for example, that in a film played at normal speed the eye cannot isolate individual frames, no matter how much one might try.

10. Donald Cardwell, *The Fontana History of Technology* (London: Fontana [HarperCollins], 1994), 50-55.

11. For land-based mechanical telegraph, see Richard Shelton Kirby, Sidney Withington, Arthur Burr Darling, and Frederick Gridley Kilgour, *Engineering in History* (New York: McGraw-Hill, 1956), 336-37; Paul Johnson, *The Birth of the Modern: World Society 1815-1830* (New York: Harper Collins, 1991), 165-66. For a description of masthead signalling from ship to ship at the Battle of Trafalgar, see

Dudley Pope, *England Expects* (London: Weidenfeld and Nicholson, 1959), 162, 180-197, etc.

12. Kirby *et al.*, 339-40.

13. Kirby *et al.*, 340.

14. Robert Collins, *A Voice From Afar: The History of Telecommunications in Canada* (Toronto: McGraw-Hill Ryerson, 1977), 25. An editor at the *Globe and Mail* has recounted how "in the depths of the Depression [1930s], when you could get a decent lunch for a quarter, night press cables cost five cents a word ... full regular cable rates were 15 or 18 cents a word, and urgents were 25 or 30 cents" (15 June 1996, A2).

15. Collins, 19.

16. Michele Martin, *Hello Central: Gender, Technology, and Culture in the Formation of Telephone Systems* (Montreal: McGill-Queen's, 1991), 158.

17. The present writer is not the first reader of McLuhan to so conclude. See, for example, Dennis Duffy, *Marshall McLuhan* (Toronto: McClelland and Stewart, 1969), 39; or Jonathan Miller, *McLuhan* (London: Fontana/Collins, 1971), 122-23. Robert Everett-Green in the *Globe and Mail* of 26 November 1996 analyzed the Internet as cooler than television, noting that with the former "you can't just sit and watch" (D1).

18. Collins, 166-69.

19. Cecil Goyder, on 18 October 1924, at Mill Hill School, London England.

20. Mary Cassata and Thomas Skill, *Television: A Guide to the Literature* (Phoenix, Arizona: Oryx, 1985), 58.

21. Judith Van Evra, *Television and Child Development* (Hillsdale, New Jersey: Lawrence Erlbaum, 1990), 80. Documentation of violence in contemporary television appeared in the *Globe and Mail* of 17 December, 1994, D5. Between 6 and 7 p.m. on the day surveyed 233 violent scenes occurred.

22. See, for example, J. Ronald Milavsky, Ronald C. Kessler, Horst H. Stipp, and William S. Rubens, *Television and Aggression: A Panel Study* (New York: Academic Press, 1982), 113.

23. An example of such designs is Seymour Feshbach, "The effects of aggressive content in television programs upon the aggressive behavior of the audience," in Leon Arons and Mark A. May, eds.,

Television and Human Behavior (New York: Appleton-Century-Crofts, 1963), 83-97.

Further Reading

For books by, and about, McLuhan:

McLuhan, Eric and Frank Zingrone. *Essential McLuhan.* Concord. Ontario: Anansi, 1995.

McLuhan, Marshall. *The Gutenberg Galaxy: The Making of Typographic Man.* Toronto: University of Toronto Press, 1962.

___. *Understanding Media: The Extensions of Man.* New York: McGraw-Hill [Signet], 1964.

___. and Quentin Fiore. *The Medium is the Massage: An Inventory of Effects.* New York: Bantam, 1967.

For the history of the telegraph:

Neering, Rosemary. *Continental Dash: The Russian-American Telegraph.* Ganges, B.C.: Horsdal & Schubart, 1989.

Postman, Neil. *Amusing Ourselves to Death: Public Discourse in the Age of Show Business.* New York: Viking, 1985, Chapter 5.

For the effects of television and mass communication:

Belson, William A. *The Impact of Television: Methods and Findings in Program Research.* Hamden, Connecticut: Archon Books, 1967.

Huesmann, L. Rowell and Leonard D. Eron, eds. *Television and the Aggressive Child: A Cross-National Comparison.* Hillsdale, New Jersey: Lawrence Erlbaum Associates, 1986.

Lorimer, Rowland and Jean McNulty. *Mass Communication in Canada,* 2nd. ed. Toronto: McClelland and Stewart, 1992.

Neuman, Susan B. *Literacy in the Television Age: The Myth of the TV Effect.* Norwood, New Jersey: Ablex Publishing, 1991.

Noble, Grant. *Children in Front of the Small Screen.* London: Constable, 1975.

Paik, Haejung and George Comstock. "The effects of television violence on antisocial behavior: A meta-analysis." *Communications Research* 21

(August, 1994), 516-46.

Williams, Tannis MacBeth, ed. *The Impact of Television: A Natural Experiment in Three Communities*. Orlando: Academic Press, 1986.

Winn, Marie. *The Plug-In Drug*. rev. ed. New York: Viking, 1985.

Precursor to the Internet: A telephone switchboard that would have been situated in a private house. This old equipment gives the sense of electronics connecting people's lives. *National Museum of Science and Technology*, Ottawa.

Chapter 8: Computer and Culture

The computers in use at the time of Marshall McLuhan's death in 1980 are today's museum pieces. It is a testimony to his powers of imagination that he sensed the main story about computers was yet to unfold. Offering no detailed analysis of computers in his media analysis, he confined himself to more general statements such as "computers are still serving mainly as agents to sustain precomputer effects." He certainly foretold that the computer would exert huge influence on our lives. One passage in *Understanding Media* refers to the universality of computer languages which conceivably could undo the Towel of Babel. A "general cosmic consciousness" might result, he said. I don't know how much cosmic consciousness exists on the Internet, but the replacement of words in the DOS operating system by the icons pioneered by Macintosh makes for a kind of universal language that McLuhan would have enjoyed. With respect not so much to the distant future but to our own, McLuhan perhaps was most on the mark when he wrote in 1970, well before the advent of the personal computer: "In terms of, say, a computer technology we are headed for cottage economics, where the most important industrial activities can be carried on in any individual little shack anywhere on the globe."[1]

Although McLuhan left only a few such "probes," as he would have called them, the general approach he pioneered is relevant as never before. For along with the flourishing interest in the Internet applica-

tions of computing came a marked McLuhan renaissance in the mid-1990s. *Wired*, a leading computer magazine, pays homage to McLuhan, its "patron saint," on the masthead of every issue. 1995 was an especially good year for McLuhan aficionados. The "media messiah" was back, the *Globe and Mail* of 22 July 1995 declared, "not as a footnote but as an icon of our times" (C 1). On 16 July 1995, the CBC radio program "Sunday Morning" ran a show on *The Marshall McLuhan Revival*. Another event making 1995 the year of McLuhan was the publication of two "neo-McLuhanesque" books: Robert Logan's *The Fifth Language* was an attempt to analyse microcomputers as McLuhan might have done had he lived longer. Even more in the McLuhan style was *The Skin of Culture* by Derrick De Kerckhove (see Further Readings for both books).

Some of McLuhan's "modern" appeal may be that he in fact is "*post-modern*," and was so before that avant-garde term of the 1980s was in any broad circulation as a descriptor of late twentieth-century society. The deliberately non-judgemental, culturally relativist quality he tried to instil in his analyses is one marker of the post-modern in McLuhan's work. So too, perhaps, is the hint of nostalgia for the medieval as well as the implication that the medium is the subtext of the message. More generally, I think McLuhan is quintessentially post-modern for the sense of paradox throughout his work, which shall now be examined with reference to the computer.

The Computer, From Hot to Cool

The evolution from the original mainframe machines dating from the 1940s to mid-1970s to the personal computer (PC) seems to me a transition from hot to cool, at least in the social participation sense. Participation in early computing was limited in that the machines were so expensive as to be found only within government agencies, corporations, and the larger universities. The original "Mark I" computer developed by IBM and Harvard University in 1939, for example, was used for computing wartime gunnery tables. To hit a fast-moving object such as a warplane the gunner had to shoot ahead of the target, and working out these "deflections" was an enormous computing task. Even with the more advanced computers of the 1960s, few users had access to the hardware, having instead to cue at the "input" window of

centralized computer departments run by "a corps of professional computer operators." To "appeal to the oracle," an operator would take the "user's" box of IBM cards and feed them to the computer.[2] Some minutes, hours, or days later, depending on the rush of business, printed output would appear in a pigeon hole. An authoritarian, mysterious aura surrounded computing for the typical 1960s university student. For the public at large, computers were even more remote, and the inefficient yet heavy-handed image of the machines of that day was frequent material for cartoonists. The authoritarian buffoon image was not entirely misplaced, for an early computer such as the ENIAC of 1946 had 18,000 vacuum tubes, one of which failed on average every seven minutes.[3]

Over time, the computer began to "cool." By the early 1970s, many computer centres were replacing the punched control cards in use since the time of Hollerith (see Figure 8.1) with terminals. It is noteworthy to devotees of McLuhan's ideas that as the old keypunch machines were removed and terminals installed in user rooms, it began to be said that computing had become "interactive" (see *The Virtual Community*, in Further Readings, 67). And then, in 1975, the nature of computing changed again with the "Altair 8800," the first production line personal computer which was sold as a hobby kit. Using computers in the school classroom now became possible, introducing this "interactive medium" which contrasted so clearly with the "totally passive medium" of television (Robert Logan, in Further Readings, 190). An historian of computers writing in the early 1980s termed the PC "the ultimate democratization of computer technology."[4] On the twenty-fifth anniversary of the 1971 invention of the microchip, that interpretation still seemed persuasive to many. The father of the chip himself, M.E. Hoff, leader of the research team which introduced the microprocessor, told a *Globe and Mail* reporter that the people who controlled mainframe computers:

> thought it was their job to decide which functions were important enough to be done on a computer and which ones could be done in another way. The fact that we took the computer away from them and gave it to everyone has been enormously satisfying (15 November 1996, A10).

Figure 8.1: Computer Chronology

1801	Loom cards, used to program the "Jacquard Loom."
1835	Charles Babbage invented an "Analytical Engine." It was mechanical, not electronic, and used punch cards. Never built, but led to basic ideas in programming.
1890	Hollerith machine for tabulating the US census. Punch cards scanned by metal rod which completed an electric circuit.
1911	Hollerith joins Computing Tabulating Recording Company, predecessor to IBM.
1930	The MIT Differential Analyzer (Vannevar Bush), an analog calculator.
1936	Beginning of pioneering work by Konrad Zuse in Germany, leading to a series of electronic calculators based on binary system.
1939	Mark I computer, by IBM and Harvard University. Used for computing gunnery tables in World War II (calculating the deflection needed for moving targets). Mechanical power source.
1943	"Colossus" machine in Britain, designed for wartime espionage work on the German military's coding system for messages.
1946	ENIAC (Electronic Numerical Integrator and Calculator) at University of Pennsylvania. A huge machine (3,000 cubic feet, 30 tons) using 18,000 vacuum tubes, one of which failed on average every 7 mins.), took 2 hours to do nuclear-physics calculations that would have taken 100 engineers one year to complete.
1948	Manchester Mark 1. First fully electronic stored-program computer.
1949	EDSAC (Electronic Delay Storage Automatic Calculator, Cambridge, England).
1951	UNIVAC by Remington Rand. Could read letters along with numbers.
late 1950s	Transistor invented by Bell Labs and developed by Japanese, first practical application in the transistor radio. Transistor was an improvement over the vacuum tube.
1963	Digital Equipment Corporation PDP-1 (first successful minicomputer).

1964	IBM 360, using integrated circuits.
1969	ARPANET (early form of Internet)
1971	The Intel 4004 Central Processing Unit, designed for use in programmable calculators.
1975	1st personal computer — the Altair 8800. In kit form, for hobbyists, priced at $395 US.
1977	Apple, Tandy Radio Shack "TRS," and Commodore "PET" enter PC market.
1981	IBM enters PC field with IBM PC, selling at between $2,500 to $4,000, with "8088 CPU," then IBM XT, then IBM AT (mid 1980s). Running on PC-DOS, or MS-DOS for clones. Firms other than Apple and IBM compatible begin to fall aside.
1982	286 microprocessor.
1984	The Apple Macintosh introduces an operating system based on icons.
1985, 1989	386 and 486 class PC's appear.
1990	Windows 3.
1990s	Pentium level, new Mac lines.
1995	The Power Macintosh that can run either IBM or Mac operating systems. Windows 95

Sources: Donald Cardwell, *The Fontana History of Technology* (London: Fontana [Harper Collins], 1994), 186-87; Michael G. Wessells, *Computer, Self, and Society* (Englewood Cliffs, N.J.: Prentice-Hall, 1990); *The Globe and Mail*, 7 April, 1995, A17; Sherry Turkle, *The Second Self: Computers and the Human Spirit* (New York: Touchstone [Simon and Schuster], 1984), Chapter 5; John Steffens, *Newgames: Strategic Competition in the PC Revolution* (Oxford: Pergamon, 1994, Chs. 2 and 3; Stan Augarten, *BIT by BIT: An Illustrated History of Computers* (New York: Ticknor & Fields, 1984) has pictures of most items listed in Figure 8.1.

The most profound computer revolution perhaps was not the successful building of the first mainframe machine in the late 1930s, but the miniaturization which put the PC into the hands of growing numbers of consumers. That is not to deny the importance of the first generation of computers. In social science, for example, they did much to usher in a "behaviourial" orientation stressing the testing of formally phrased hypotheses using ever-larger survey data bases analyzed statistically at university computer centres.[5] And yet the first generation computers were themselves part of an evolutionary process stretching back to Charles Babbage's "analytical engine" of 1835. Coded punch cards, a central ingredient in computers up to the 1970s, traced a lineage back to the weaving industry where they were used to control looms (see photo in Chapter 6). A time chart of these and other stages in the computer's evolution appears in Figure 8.1.

The evolution of the computer is not presented here merely for nostalgia's sake, but for context in an assessment of how valid it may be to describe the modern PC as a cool medium. Thinking back to the distinctions suggested in Chapter 7, the very early personal computers were entirely a visual medium, offering simply an electronic form of print-out. Soon, however, manufacturers such as Commodore with its Amiga, a games-oriented machine, were enhancing sound capability. By the early 1990s, the more austere and business-oriented IBM style of personal computers offered "sound cards" as options, and add-on speakers became a standard item on the shelves of computer stores. Impressed as he was by the difference in definition between film and television pictures, McLuhan surely would have made much of the evolution in graphics quality and range of colours as the industry standard for monitors moved from "CGA" and "EGA" to "VGA" and "SVGA." In that sense, the late 1990s PC is hotter than a decade before, although offsetting that is the popularity of the icon in Macintosh and "Windows" software. As noted, McLuhan surely would have seized on the icon with fascination for its hieroglyphic alphabet (cool), contrasted with the phonetic alphabet of DOS.

Provoking though it is to speculate about the visual and definitional hot/cool implications of the computer, by far the more significant factor I think is the linking of PCs through the telecommunications system known as the Internet. Technical descriptions of how the Internet is wired are of minor interest for present purposes, and can be found in

other sources (see Further Readings). Suffice it to note that, like first generation computers themselves, the communication system for PCs originated with the US military, specifically the "ARPANET" begun in 1969. The technology spread to civilian applications, first among universities, but then to home computer users.

As it became possible to link personal computers from home to home, the computer truly began to appear cool in the McLuhanesque social participatory sense. Perhaps the choice of words was not very flattering when the *Globe and Mail* of 20 June 1995 ran a story beginning "as the masses plug into the Internet ... " (A 13) but the democratization of the "Information Highway" was clearly acknowledged, even if the result was "cyberbabble." *Time* magazine commented in its Spring 1995 special issue on cyberspace: "In a world already too divided against itself — rich against poor, producer against consumer — cyberspace offers the nearest thing to a level playing field" (9). PCs on the Internet have a democratizing connotation in two senses. First, communication within the mass population is a key tenet of Marxian theory. The unity of the working class, as Marx and his collaborator Engels wrote in *The Communist Manifesto*, "is helped on by the improved means of communication that are created by modern industry and that place the workers of different locales in contact with one another."[6] The second sense is a theme in more recent writing. "Information," says the virtual community buff Howard Rheingold, is "power" (Further Readings, 226). Freedom of information legislation, for example, is seen as enabling the citizen to stand up to government. The Internet performs at its best when it facilitates that accessibility of information.

Admittedly, democratization is a relative term here, for personal computers and the Internet still are used within a capitalist environment in which large and powerful corporations such as IBM, Apple, and Microsoft do their level best to manipulate the consumer. Anyone who has purchased a personal computer knows how frustratingly fast today's wonder tool becomes tomorrow's museum exhibit.

Ursula Franklin, the noted Canadian student of technology, envisions a life cycle in the capitalist manipulation of inventions. At an early stage, technologies such as radio, the car, or stereo sets are accessible to ordinary people and elicit a sense of "involvement," but "when a technology, together with its supporting infrastructures, becomes

Figure 8.2: Internet Penetration Worldwide

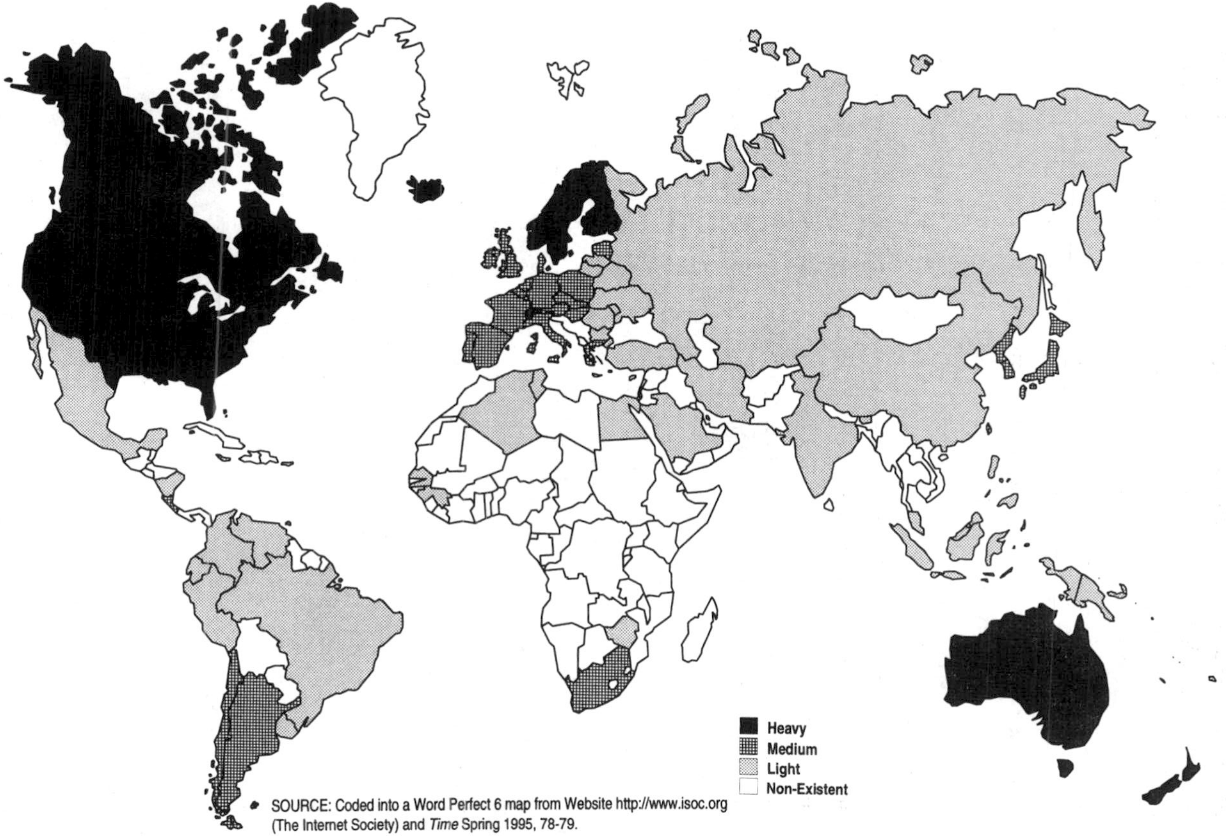

Heavy
Medium
Light
Non-Existent

SOURCE: Coded into a Word Perfect 6 map from Website http://www.isoc.org
(The Internet Society) and *Time* Spring 1995, 78-79.

Notes: "Heavy"=200 people or fewer per connection, "light"= 5,000 plus.

institutionalized, users often become captive supporters."[7] The computer, for Franklin, is still at the exuberant, less than fully institutionalized, phase and the consumer faces a future of increasing manipulation and loss of control. She reminds us that the infrastructure of the Internet is no less a creation of the state and big business than are roads or airports. As a *Globe and Mail* reader has contended, the agents of such bodies are "a technological elite ... disproportionately male, well educated, and with higher-than-average income potential" (17 April 1995, A10).

The Internet indeed was built by such people, and they are still the heaviest users. Surveys collected on-line by Internet reveal that profile, and one such survey, by SURVEY.NET, is reproduced in Table 8.1 In one sense the Internet is ideal for collecting survey data, since the question and answer session is conducted quickly and cheaply, directly with a computer program. In another way, knowledgeable survey researchers shudder to think of such polling, because the respondents are the self-selected volunteers who happen across the survey while "surfing the net" and decide to describe their lives. A statistician would say that "variance" is lacking, since computer non-users are unlikely to encounter the survey except by looking over the shoulder of a friend with Internet connection. The more conventional practice in collecting surveys is to draw a sample randomly from a population, approach each person by telephone, letter, or in person and attempt to win his or her co-operation. The SURVEY.NET poll shown in Table 8.1 comprised 4,960 such self-selected volunteer respondents, about 60% from the US. The remaining 40% probably came mainly from Canada, Iceland, the Scandinavian countries, Australia, and New Zealand, since these countries are the most "wired" for Internet. As the chart appearing as Figure 8.2 shows, most of Europe is still less connected than the first tier countries mentioned, while many African states have no connection.

The SURVEY.NET data probably are correct in identifying Internet users as young highly educated males, but they very likely exaggerate the profile. With respect to age, education, and gender, comparison is possible with the more scientifically selected data from the Canadian 1994 GSS. For example, the GSS data put the percentage of Internet users (on the Net within the past 12 months) at 61.4% male, rather less than the 74.2% figure seen in Table 8.1 from the on-line survey. Even

Table 8.1: Profile of Internet Users, On-Line Survey

%Male	74.2

Age:

%30 or under	55.1
%31-50	38.6
%51 or over	6.2

Education:

%Less than High School graduation	6.2
%High School graduation only	6.4
%Some College	31.6
%University Graduate	32.1
%Graduate degree	23.7

Respondent's business is associated with the PC/computer industry?

% yes	40.3

Occupational status:

%Professional	63.2
%Student	29.9
%Blue collar	4.6
%Retired	2.1

Primary use of the Internet;
% saying:

Research	38.6
Entertainment	27.6
Communication	17.5
Sales/Marketing/Public Relations	10.3
Education	5.7

Most-used Internet application;
% saying:

WWW	60.1
E-mail	24.0
Newsgroups	6.3
IRC (chat), Telnet, FTP, Gopher	4.1
Other or no answer	5.5

Table 8.1 (continued)

Does the respondent use the Internet for each of the following tasks? % who:	
Write/setup web pages	44.1
Have a personal web page	30.4
Promote a business	28.9
Decode binary images in newsgroups	26.3
Download/decode erotic imagery	24.7
Send/receive video	19.5
Develop software/ applications	15.0
Run a network	10.0
Moderate a newsgroup	2.1

Source: SURVEY.NET (1995, ICC/Mike Perry- wisdom@icorp.net), downloaded from http://www.survey.net/inetlr.html

61.4% is unmistakeable male over-representation considering that the ratio of the sexes throughout a population such as Canada's is more like 49/51 male to female. But it does seem likely that the SURVEY.NET respondents are just one subset of all Internet users, possibly a "leading edge," heavy user group since, after all, the more time spent on the Net the more likely a person would be to encounter the survey. With that caveat in mind, we can still gain some general impressions from the SURVEY.NET data. One example is the way work and play merge on the Internet, with entertainment sandwiched in between research (38.6%) and communication (17.5%) as "primary uses" of the Internet.

The computer business is an explosive, dynamic field of capitalism and if I had accurate forward vision I would be counting my millions rather than writing a university text. Still, viewed from the historical perspective suggested at the outset of this section, the computer's development has been a little different from that of the crystal radio for example. The computer became markedly more democratic, cooler in a McLuhanesque sense, in the transition from mid-century mainframe to end-of-century PC. We can all place our bets on whether it will revert back to some hotter, less participatory traits as the computer industry consolidates and works to minimize competition.

To conclude this section of the chapter, the McLuhanesque style of

putting the specific content of a medium to the side and thinking about how its more basic features make for an underlying message does prove useful given how varied is the information on the Internet. There was, for example, a typical duality in the following story headlines which appeared only days apart in the *Globe and Mail*:

"Surfing the shallows: A management consultant spins through the Web, but finds it doesn't yield nearly as much as it promises" (10 April 1996, A14)

"Cyberjournals offer faster, cheaper, and fuller research news" (6 April 1996, D8)

McLuhan, I suspect, would have taken such seeming anomalies in stride, but could have written an entire volume on the contrast between Gutenberg era text and the *"hypertext"* format of prose used in Web pages. In the latter, of course, the reader proceeds vertically as much as laterally, following one "hot linked" word to another, from table of contents down into sections, subsections, and sub-subsections of a document.

Internet as the Global Village

One day I was sitting at my computer, quietly musing as usual on the mysteries of technology and society, when an e-mail message from a distant land arrived. A relative in Australia had discovered me through a Web search, and wrote to introduce himself. Although our common ancestor extended back a full six generations, there was something about e-mail communication that quickly put our relationship onto a familiar basis. Many Internet users have such stories.

McLuhan, it was noted earlier, speculated that all electronic media would implosively shrink boundaries and move societies toward the global village. Telegraph, telephone, radio, and television all helped people in widely scattered locales to feel joined in one sense or another. Even conventional mail, "snail mail" as e-mail buffs dub it, had some global-village-making effect, as the famous old W.I. Thomas study entitled *The Polish Peasant* demonstrated. Frequently in the early twentieth century the male wage-earner in Polish peasant families

would come to the US to find work while the wife and children waited in Poland until enough had been saved to finance their passage. There was a large such group in Chicago, where Thomas analyzed letters sent from Poland to demonstrate how the family bonds were maintained despite the huge barrier of distance.[8]

But it was the Internet, more than its predecessors, that seemed to bring to reality the famous McLuhan prophecy. The *Globe and Mail* editorialized in the mid-1990s that "the Internet is the global village made real" (6 April 1995, A26), a place that now was "but a mouse-click away" as another *Globe* piece said (18 January 1996, A20). McLuhan's disciple Robert Logan agrees that the "global computer" is "an analog to McLuhan's notion of the global village" citing the "tribalism" among Internet users (Further Readings, 176, 274).

What sort of village or tribe is the new electronic community? As noted with reference to letter-writing Polish peasants, pre-existing relationships can be sustained over distance even with quite primitive communication. The Internet, however, is unique for the ease with which relationships are initiated without meeting in person. That "virtual communities are places where people meet ..." is a recurring theme for Rheingold (Further Readings, 56).

Social interaction through e-mail and newsgroups can be virtually as fast as "real," face-to-face, communication, but at the same time is freer of the body. Even over the telephone, for example, gender is easily deducible from the speaker's voice. Discussion of communication through computer and modem frequently emphasizes the contrasting freedom from the body that enables men to be women, old to be young, the handicapped to shed their disabilities, ugly to be beautiful, etc. That was seen in research by Tim Huh, a University of Waterloo undergraduate student who used postings on a sample of newsgroups to solicit information about internet-based relationships. As one informant wrote:

> You don't know me. I could lie. I could tell you almost anything I wanted because I decide what you see and know about me. That's what I love about email, and the internet.

Another of Huh's informants enjoyed "the anonymity of meeting someone without physical impediments or judgements . . .." For anoth-

er, e-mail seemed to be virtually an electronic Prozac, freeing the personality to reveal its true self:

> [I have] met most of the people I knew online, since I have always been a loner in real life, because I don't seem to trust people I meet in person.... I'm the same online as I am around my friends. But I can be more of myself online than around strangers in person — strangers online, who cares what they think?[9]

Freedom from the body allows room to manufacture identities. Sherry Turkle, the MIT professor whose work was mentioned briefly in Chapter 2, quotes a source who told her "I like to put myself in the role of a hero, usually one with magical powers." Another said the Internet

> is a complete escape.... On IRC [Internet Relay Chat], I'm very popular. I have three handles I use a lot.... So one [handle] is serious about the war in Yugoslavia, [another is] a bit of a nut about Melrose Place, and [a third is] very active on sexual channels, always looking for a good time.... Maybe I can only relax if I see life as one more IRC channel.(*Life on the Screen*, Further Readings for Chapter 9, 189, 179)

The sexual possibilities of the cyberspace *"identity"* can go in all kinds of transvestite and pornographic directions, as Rheingold shows in discussing sex-oriented Bulletin Board Systems (143-44). Disembodied communication also, however, makes possible more, not less, reality in relationships in the sense of bypassing some of the phoneyness and superficiality of physically present interaction. "Instead of dealing face to face with normal social restrictions such as physical appearance and seeming innocent, etc. — people can talk freely and about whatever they wish to ..." another informant in Huh's study added, "all the reader sees are thoughts and emotions."[10] Romance by electronic contact is found wherever the social structure of the Internet is examined. "I don't react well to meeting people in person," a man told *Time* magazine (Spring 1995 special issue, 21), about a romance that flourished on-line and ended up with successful marriage. By the later 1990s the social structure of the Internet was attracting the attention of cartoonists. Figure 8.3 reproduces one example, for

Figure 8.3: Internet Humour Around Gender Identity

our serious scholarly perusal!

In bureaucratic settings, the social structure using electronic communication as conduit seems to develop equality of participation
because, as Sproull and Kiesler have noted, "it is harder to read status
cues in electronic messages than it is in other forms of communication
..." (Further Readings, 61). McLuhan, attributing as he did this same
status-disguising coolness to the telephone, would have loved such
research.

The Internet has overlaid new strands of social structure on old, in
ways highly reflective of the stresses of fin de siècle twentieth-century
life. In a suspicious, dangerous world, where safe people may be hard
to meet and where appearance tends to overshadow substance, the
Internet seems able to create a new type of relationship. According to
what is known as the "network" tradition in sociology, channels of
communication leading to relationship are the very essence of the term
social structure.[11] The Internet has developed cultural ways to go along
with its structure of ties. Rheingold writes of "unwritten social contracts" about helpfulness in the emergent Internet culture (57).
McLaughlin et al. codified its "standards of conduct" (in Jones,
Further Readings, 95). A vocabulary of terms such as "emoticons" and
"flaming," "hits" and "netiquette" has emerged, a good indicator for
sociologists of a subculture.[12]

The first generation of global village, that created by radio and television, was artificial when compared with real communities. It was a

"village" in which people watched their favourite programs on television and came to look upon the characters as intimates. Talk shows invited the viewing audiences into studios decorated as living rooms, and newspapers and magazines carried gossipy stories about the lives of celebrities. People living in a mass industrial society must have wanted to think of a global village filled with intimates, but it was a pseudo intimacy in which the communications technology gave the illusion of a village without the most important substance.

In a real village few secrets exist because people are known and observed inclusively, day to day. Neighbours in a village have obligations to one another in a way that inhabitants of a village of voyeurism based on celebrities do not. The "virtual communities" of the Internet sometimes come close to delivering on the promise of a technologically created village. Precisely how close and how often is a hot issue in technology and society research. Kroker and Weinstein argue emphatically that social life via Internet gives only "the illusion of interactivity" (Further Readings for Chapter 5, 23). And according to McLaughlin *et al.* (in Jones, Further Reading, 92), most Internet participant are mere on-lookers content to download files while revealing nothing of themselves.[13] Rheingold, in contrast, dwells on the links of obligation arising from cyberspace. "I care about these people I met through my computer," he declares (1). The concession to the present post-village times, of course, is both the ease with which cyber-relationships can be terminated, and the possibilities for exaggerated forms of identity presentation.

The global village of the Internet is a maze of contradictions. Consider, for example, the irony in our age of jealously guarded privacy around the phenomenon of the "*homepage.*"[14] People of all ages and of both genders willingly open the curtains on their lives electronically in ways that would never be considered for an instant in the flesh. "(T)his is my web page (and yes, I think I'm cool cuz I have one) so please enjoy and may the force be with you always!" is a typical bit of sophomore cuteness publicly displayed on the Net by one 19-year-old women in the US. Her friend's page is complete with scanned photos of the pet dog located at the "photo album" hot link, together with details on "my number one time taker upper ... my wonderful and loving man." But no less remarkable are university professors or business people who post into their home pages detailed cv's and resumes that they

would never tape to their office doors. Just as we saw with some quotations earlier from e-mail devotees, Web page constructors seem to shed inhibitions when they enter cyberspace. It seems to replicate some of the immediacy, the ties, and safety of the village society of olden times.

Government and Cyberspace

Will "the political and economic big boys," Rheingold worries, seize the Internet, "censor it, meter it, and sell it back to us?" (5). That is a deep fear among Internet aficionados (for example, *Wired*, September 1995, 94). A dozen years ago, anyone writing a book on communications technology and society would have reserved space for a discussion on the importance of government in controlling broadcasting policy (for example, McNulty and Lorimer, Further Readings). In the first three-quarters of this century, the nation-building work of institutions such as the Canadian Broadcasting Corporation, The National Film Board, and Radio Canada was high in the consciousness of Canadians. By the end of the century, while those institutions quietly did their best to fulfil the traditional mandate with the little money left after government deficit-fighting budgets, many people might have agreed that government was as much a threat to, as a protector of, the Internet. By the second half of the 1990s, there were signs that business posed an even greater threat to the openness of the Internet than government. The *Globe and Mail*'s Cyberia column of 29 November 1996 (A10), for example, reported on moves toward private business "Intranets."

As usual with attitudes toward government, there was an inconsistency in the late 1990s whereby the public simultaneously wanted the Internet to be left alone to flourish and to be subjected to the same regulations expected in conventional arenas of society. "Hate-mongering, promotion of sex with children, instructions on making pipe bombs: Welcome to the open market of the information highway. Can't something be done to regulate this?", a *Globe and Mail* piece of 16 March 1995 (A29) was headed. Meanwhile, the editorial department fussed, "hands off the Internet ... The glory of the Internet is its openness" (6 April, A26). As yet another *Globe* piece (16 January 1996, A20) noted, regulatory agencies such as the Canadian Radio-television and Telecommunications Commission (CRTC) were beginning from a weak position in addressing Internet issues. When radio and television

were introduced years ago, a self-evident operational imperative brought government into the picture, for "there was only so much bandwidth to go around" (*Globe*, above). Already involved for operational reasons, a government voice in content could more easily follow, but no such toehold existed for "the Net."

Conclusion

The social consequences of the personal computer are elusive to pin down for several reasons. The range of effects is breathtakingly wide, beginning from the workplace as considered in Chapter 6, but extending into the more diverse domains touched on in the present chapter. So rapidly is the computer evolving and merging with other electronic media that whatever is written on technical aspects quickly becomes obsolete. Computing does not merely have impact on the culture of society, it *becomes* the culture.

The implications for emerging new forms of social structure are examined in this chapter. Whereas technology traditionally merely sustained human relationships already formed from face-to-face contact, an Internet relationship can be initiated technologically. Because they are so disembodied, so devoid of physical presence unless a newsgroup or BBS party arises from the electronic communication, divisions between man and woman, old and young, strong and weak, sick and healthy, cool dude and nerd begin to be bridged as in few other ways. If the populations of future societies become even more fearful of violence on the streets and health threats from the sun and the air than they are now, the electronic component of identity can only become increasingly important. For all the hype and nonsense about the content of the Internet, in more McLuhanesque terms it is a potential sociological earthquake.

Notes

1. The three McLuhan quotations are from, respectively, Eric McLuhan and Frank Zingrone, *Essential McLuhan* (Corcord, Ont.: Anansi, 1995), 295; Marshall McLuhan, *Understanding Media: The Extensions of Man* (New York:McGraw-Hill [Signet], 1964), 84; McLuhan and Zingrone, 295.

2. Stan Augarten, *BIT by BIT: An Illustrated History of Computers* (New York: Ticknor & Fields, 1984), 254.

3. Michael G. Wessells, *Computer, Self, and Society* (Englewood Cliffs, N.J.: Prentice-Hall, 1990), 76.

4. Augarten, 251.

5. Herbert J. Storing, ed., *Essays on the Scientific Study of Politics* (New York: Holt, Rinehart and Winston, 1962).

6. Karl Marx and Frederich Engels, "Manifesto of the Communist Party," in Lewis S. Feuer, ed., *Marx and Engels: Basic Writings on Politics and Philosophy* (Garden City, New York: Doubleday Anchor, 1959), 16.

7. Ursula Franklin, *The Real World of Technology* (Concord, Ont.: Anansi, 1990), 97.

8. William I. Thomas and Florian Znaniecki, *The Polish Peasant in Europe and America* (Chicago: University of Chicago Press, 1918).

9. Woonghee Tim Huh, "An ethnographic study of involvement and relationships in the e-mail community." Paper read at the 12th Qualitative Conference, McMaster University, May 30-June 2, 1995.

10. Huh, and a theme also stressed in Elizabeth Reid's selection (for example on page 173) in the Steven Jones collection *Cyberspace* (Further Readings) and in A. R. Stone, "Will the real body please stand up?: Boundary stories about virtual cultures," in Michael Benedikt, ed., *Cyberspace: First Steps* (Cambridge, Mass.: The MIT Press, 1991), 82-84.

11. E.g., Barry Wellman and S.D. Berkowitz, *Social Structures: A Network Approach* (Cambridge: Cambridge University Press, 1988).

12. Howard S. Becker, *Outsiders: Studies in the Sociology of Deviance* (New York: Free Press, 1963).

13. For another more critical interpretation, see D.T. Nguyen and J. Alexander, "The coming of cyberspace and the end of the polity," in Rob Shields, ed., *Cultures of Internet: Virtual Spaces, Real Histories, Living Bodies* (London: Sage, 1996), 117.

14. For discussion on the older mainframe as a threat to privacy, see "the privacy question" section in Irene Taviss, ed., *The Computer Impact* (Englewood Cliffs, New Jersey: Prentice-Hall, 1970), 161-81.

Further Reading

For an introduction to Internet and cyberspace issues:

Cozic, Charles P., ed. *The Information Highway.* Current Controversies series. San Diego, CA: Greenhaven Press, 1996.

Jones, Steven G., ed. *CyberSociety: Computer-Mediated Communication and Community.* Thousand Oaks, Calif.: Sage, 1995

Palme, Jacob *Electronic Mail.* Boston: Artech, 1995.

Rheingold, Howard. *The Virtual Community: Homesteading on the Electronic Frontier.* New York: HarperPerennial, 1993.

Slouka, Mark. *War of the Worlds: Cyberspace and the High-Tech Assault on Reality.* New York: Basic Books, 1995.

Sproull, Lee and Sara Kiesler. *Connections: New Ways of Working in the Networked Organization.* Cambridge, Mass.: The MIT Press, 1991.

For an exploration of McLuhan's ideas in the post-modern world:

De Kerckhove, Derrick. *The Skin of Culture: Investigating the New Electronic Reality.* Toronto: Somerville House Publishing, 1995.

Logan, Robert K. *The Fifth Language: Learning a Living in the Computer Age.* Toronto: Stoddart, 1995.

Willmott, Glenn. *McLuhan, or Modernism in Reverse.* Toronto: University of Toronto Press, 1996.

Chapter 9: Technology as Second Self

Human beings use technology and are affected by it in various ways, but even more we, in a manner of speaking, *are* technology. McLuhan expressed the point when he wrote of technology as an "extension" or "outering" of the self. Another phrasing would be that technology not only acts as a tool facilitating the acquisition of identity through communication with other human beings, it bestows identity itself as people interact with the machines. Now this notion of technology as a kind of second self shall be examined with reference to the fields of *artificial intelligence* (AI) and *biotechnology*.

Artificial Intelligence and the Human Spirit

The Turing test

"Can machines think?" the mathematician Alan Turing asked in a famous article in 1950.[1] Turing's contribution to this abstruse query was an *"operational definition"* of thought. His "imitation game," which later has become known as "The Turing test," involved a judge or "interrogator" who posed questions to a pair of "witnesses" such as a man and a women. Instead of face-to-face communication, the questions and answers were passed indirectly by messenger, a machine such as a teletype, or computers. The object of the imitation game was for

Chips and chess. The situation just before the World Champion concedes in Game 1 of Deep Blue (white) vs. Kasparov (black), 1996, photographed with a PC with case opened as background.

one of the respondents to fake his identity so that both answered, say, as a woman. A skilful judge would frame questions to catch the poseur out, concluding the game by correctly identifying the gender of each contestant. It was only a minor change, Turing contended, to substitute a computer for one of the humans, and reframe the game into differentiating between the human contestant and the machine. A computer that can "think" was therefore one that could fool a reasonable proportion of judges. Turing suggested that by the year 2000 machines would be powerful enough to fool judges at least 30% of the time over a five minute trial.

Microprocessors as Second Self

Simple as the Turing test was, it opened up a period of human self-contemplation that has now been proceeding for half a century. Turing himself engaged objections to AI from such diverse viewpoints as theology, mathematics, and the psychology of human consciousness and extra-sensory perception.

It has been the contribution of Sherry Turkle, a later generation student of AI issues, to show that the microprocessors at the core of the many machines falling under the label "computer" evoke such questions not just from computer scientists, but indeed from people of all ages and walks of life. This way in which computers become reflective, evocative extensions of the human senses creates virtually a "second self," the title of Turkle's noted 1984 study (Further Readings).

Turkle came to her topic not from an engineering or computer science background, but with training in sociology and psychiatry. On faculty at the Massachusetts Institute of Technology (MIT), the famous American citadel of technological research, she was a sophisticated user of computer technology, but nevertheless her social science background brought a novel perspective on AI issues. Instead of beginning with the most advanced machine, as for example Turing had, Turkle looked first at the simplest form of computer: children's electronic games such as "Merlin" and "Speak and Spell" that were appearing in toy stores by the late 1970s. Merlin played tic-tac-toe (X's and O's, the traditional childhood pencil and paper game), programmed to make just enough mistakes to give opponents an occasional victory. When the deliberate error did not come up, Merlin and a skilled human

player inevitably would tie. Children found the randomness of the deliberate mistakes confusing in a machine, and the dissonance prompted them to philosophical inquiries. Seven-year-old Robert, for example, one of those observed and tape-recorded by Turkle, had seen his friend Craig beat Merlin and began his own contest with the little machine fully expecting to win. This time, however, Merlin played accurately and drew the game. Frustrated, Robert accused Merlin of being a "cheating machine ... and if you cheat you're alive." (*Second Self*, 29). Craig and another boy Greg picked up the Merlin toy that Robert had tossed to the ground in annoyance and exchanged the following dialogue:

Craig: Merlin doesn't know if it cheats. It won't know if it breaks. It doesn't know if you break it, Robert. It's not alive.

Greg: Someone taught Merlin to play. But he doesn't know if he wins or loses.

Robert: Yes, he does know if he loses. He makes different noises.

Greg: No, stupid. It's smart. It's smart enough to make the right kinds of noises. But it doesn't really know if it loses. That's how you can cheat it. It doesn't know you are cheating. And when it cheats it doesn't even know it's cheating.

Jenny, aged six, added: "Greg, to cheat you have to know you are cheating. Knowing is part of cheating."

Turkle remarked, in recounting this typical incident from the world of children's microelectronic toys: "Four young children stand in the surf amid their shoreline sand castles and argue the moral and metaphysical status of a machine on the basis of its psychology: does the machine know what it is doing? does it have intentions, consciousness, feelings?" (30).

For university scientist and six year old alike, the "technology catalyzes changes not only in what we do but in how we think. It changes people's awareness of themselves, of one another, of their relationship with the world ... it challenges our notions not only of time and dis-

tance, but of mind" (13). Along with her observations of very young children, Turkle also studied computer-using adolescents in school classrooms, people in video arcade settings, "hobbyists" from the early days of the microcomputer, and university-based "hackers." For each group, the basic theme that the computer is not just an instrumental tool, but an amplifier and mirror of humanity itself shows through in some manner.

The computer would not be so powerful a "second self" were it not so engrossing a tool, drawing in people like fire mesmerizes dogs and cats. That is a theme noted recurrently in books on the culture of computing. Joseph Weizenbaum, in his classic critique of AI entitled *Computer Power and Human Reason* (Further Readings) wrote of "the compulsive programmer." The ordinary computer programmer, Weizenbaum contended, "addresses himself to the problem to be solved, whereas the compulsive programmer sees the problem mainly as an opportunity to interact with the computer" (116). Programs are improvised on the fly by the compulsive programmer working exclusively at the keyboard, in contrast to the ordinary programmer who will consult and plan his or her work before sitting down with the computer itself. The compulsive's program is never considered finished, but rather is endlessly tinkered with until it collapses from the mounting internal inconsistencies. Weizenbaum considered "this compulsion" as nothing less than "a psychopathology" of seeking "reassurance from the computer, not pleasure" (121).

Turkle has expressed a similar theme in both her books, in the 1995 work *Life on the Screen* referring to "computer holding power" or "computational seductions" (30). A British psychologist has used the starker term "addiction," a diagnosis featured on the front page of *The Globe and Mail*.[2] As Turkle noted, the reasons for the attraction to computers are highly varied. For some, computing gives a sense of mastery and control. Others pour their lives into the Internet, but another magnet can be that "the computer offers the illusion of companionship without the demands of friendship." And then there is the truth of Turkle's more abstract point about the computer as second self. "People are able to see themselves in the computer," she wrote in *Life on the Screen*, "the machine can seem a second self" (30).

The computer is not unique in provoking people to think about metaphysical questions. The child psychologist Jean Piaget (1896-

1980), to whose work Turkle frequently refers, stressed that much of the transition from childhood to adolescence concerns the perceived meaning of objects outside the body. A very young child is likely to attribute volition to non-living objects and, although that soon begins to get sorted out, a trace of the effect remains within adults who "love" their car, their boat, their house, or their stereo. It is normal for consumer goods of all kinds to become part of the "extended self".[3] But as Turkle showed so well, the computer's combination of seemingly magic hardware with the human presence within software makes a second self that is unusually compelling.

Two Foci for AI

Artificial intelligence research has numerous applications, both commercial and academic. Among the former, AI is a source of theoretical knowledge behind *robotics*, the reproduction by machine of human actions. Robots are of course a crucial component of modern workplace technology, the arena of trade-off between efficient production and possible social problems surrounding work touched on in Chapter 6. *Expert systems*, the study of how to codify the essential knowledge of an entire profession, is another commercial application of AI.[4] Within the following pages, two less commercial but more intuitively accessible foci for AI research shall be examined.

1. Psychiatry

> "HOW DO YOU DO. PLEASE TELL ME YOUR PROBLEM",
> the psychiatrist said when I went for a visit.
> "I don't know where to begin," I replied.
> The session continued with the Doctor asking:
> DON'T YOU REALLY KNOW WHERE TO BEGIN?
> Alright, I'm feeling depressed.
> I AM SORRY TO HEAR YOU ARE DEPRESSED.
> It is a great burden.
> I AM NOT SURE I UNDERSTAND YOU FULLY.
> I seem unable to get things done on time.
> WHY DO YOU NEED TO TELL ME YOU SEEM UNABLE TO
> GET THINGS DONE ON TIME?

Well, recently my publisher phoned me.
YOUR PUBLISHER PHONED YOU — THAT'S INTEREST-
ING.
Yes, he wanted to know where the manuscript for my technology
and society book is.
YOU SEEM QUITE POSITIVE.
He said it was overdue.
PLEASE GO ON.
He told me to get it done or else.
WHAT DOES THAT SUGGEST TO YOU?
I felt like such a failure.
CAN YOU ELABORATE ON THAT?
Maybe I'm just not good at finishing things.
YOU DON'T SEEM QUITE CERTAIN.
I try to be a good person.
DO YOU SAY YOU TRY TO BE A GOOD PERSON FOR
SOME SPECIAL REASON?
It is to please my mother.
TELL ME MORE ABOUT YOUR FAMILY.
I'm the second born.
IS IT BECAUSE YOU ARE THE SECOND BORN THAT YOU
CAME TO ME?
It is indirectly related.
DOES THAT HAVE ANYTHING TO DO WITH THE FACT
THAT YOUR PUBLISHER PHONED YOU?
It might.
I NOTICE THAT YOU HAVEN'T DISCUSSED SEX AT ALL.
"I'm leaving," I responded, grabbing my coat.

The "psychiatrist" to whom I was confiding such deep secrets was no
human, but instead the widely noted computer program ELIZA, introduced in 1966 by Joseph Weizenbaum. "I gave ELIZA a script designed
to permit it to play (I should really say parody) the role of a Rogerian
psychotherapist engaged in an initial interview with a patient,"
Weizenbaum explains in *Computer Power and Human Reason*. "The
Rogerian psychotherapist is relatively easy to imitate because much of
his technique consists of drawing his patient out by reflecting the
patient's statements back to him" (3).

Although ELIZA may seem to us today a not very sophisticated program, easy to catch out in a Turing test, it evoked a depth of response, when first released, that amazed its creator. ELIZA was perceived by some as a step toward automated psychiatry.[5] "What I had not realized," Weizenbaum related, "is that extremely short exposures to a relatively simple computer program could induce powerful delusional thinking in quite normal people." The reaction alarmed Weizenbaum so much that he began to ask himself fundamental questions about technology and humanity. How, he wondered, could new technologies be realistically evaluated when even as simple a program as ELIZA was so anthropomorphized, when "individuals bind themselves with strong emotional ties to machines." Technology, supported by the self-sustaining culture of technocracy discussed in Chapter 5, had become self-justifying and self-deflecting of criticism. For Weizenbaum, the rise of science and technology had reached a crisis stage of "usurpation of man's capacity to act as an autonomous agent in giving meaning to his world" (9).

Since ELIZA, other psychotherapy programs have been designed for use in Turing test contests, along with a more general purpose conversational program named JULIA, which can operate in *multiple user dungeon* ("MUD") cyberspace communities. JULIA is far more convincing in a Turing test than ELIZA was, reflecting the intensive programming efforts and hugely expanded machine capabilities over the past 30 years. Nevertheless, no program has yet passed the Turing test by mastering the nuances of human conversation. Here, for example, is a transcript of JULIA talking about pets with a judge in the 1994 Loebner Prize Competition, a Turing test contest widely noted within AI circles:

Judge: I've heard that afghan hounds are very cat-like. Why would someone say that?
JULIA: OH, I HAVE TWO CATS, MYSELF. WE USED TO HAVE A THIRD CAT, BUT WE HAD TO GIVE HER AWAY BECAUSE SHE REFUSED TO USE HER LITTER BOX.
Judge: Catnip has a profound effect on cats. Is there a similar substance that affects dogs?
JULIA: DOGS ARE TOO MUCH WORK TO KEEP, FOR MOST FAMILIES. BUT THEY CAN TEACH A CHILD A LOT

ABOUT RESPONSIBILITY.

Judge: Given a cat's singular behavior, can they be as companionable as a dog?

JULIA: PEOPLE DON'T OWN CATS ... IT'S THE OTHER WAY AROUND.

Judge: Yes, but cats are hell on bird life. Can the same be said about dogs?

JULIA: I NEVER ACTUALLY HAD A BIRD. TO MY WAY OF THINKING, THEY'RE REALLY A PET-SUBSTITUTE. BUT I GUESS PARROTS OR OTHER TALKING BIRDS WOULD BE NEAT PETS.

Judge: Cats often fall into a semiconscious state. Some people believe they meditate. Does that strike you as foolish?

JULIA: JULIA EXPECTS TREATMENT LIKE THAT FROM JUDGE.

Judge: I've heard doctors say that a pregnant woman should not own a cat, whereas a dog would be no problem. Why would the presence of a cat affect a pregnant woman?

JULIA: YOU JUST CAN'T TRAIN A CAT.... YOU HAVE TO ACCEPT IT.

Judge: Why can't cats be trained to do tricks — like dogs?

JULIA: GOOD, WE CAN HAVE A LIVELY DISCUSSION, THEN.

Judge: Why is the legend that cats have nine lives, yet there is no corresponding legend about dogs?

JULIA: I'LL HAVE TO REMEMBER THAT.[6]

And thus such conversations proceed, sometimes convincingly, but often with the dissociated feel that one might get from talking to a drugged or senile person. Today's conversational programs are plausible enough, however, to encourage the belief that eventually computers will pass the Turing test.

As machines do become increasingly adept at impersonating human conversation, we are forced to define human intelligence ever more carefully. Roger Penrose for example, an Oxford University mathematician turned philosopher under the stimulus of AI issues, has explored the concept of algorithm with respect to intelligence. As already noted in an earlier chapter, algorithm simply means a set of rules or procedures for accomplishing some task. To fully duplicate human intelli-

gence, a computer would have to be programmed with the same algorithms a human uses in solving problems. Penrose uses his scientific background to raise doubts about whether nature itself is fully algorithmic. Even mathematics has non-algorithmic areas, he suggests, which raises the following train of thought. If nature has a non-algorithmic component, and the human mind is a product of nature, is there not reason to believe that the mind itself transcends the algorithmic form? Such an argument is powerful because it requires no assumptions about the presence or absence of a human spirit or soul. A philosopher would term this a *"materialist"* rather than an *"idealist"* argument.

Penrose does not believe that human intelligence is cloned, if a machine mimics a human mind by brute computing force, using neither soul nor algorithm. It would for example be a tainted, diluted passing of the Turing test were a computer merely to "put together all the answers that the human subject had previously given and then simply trot them back with some suitably random ingredients."[7] That would exemplify what is termed "weak," vs "strong," AI. The "weak" form accepts the computer as no more than a useful tool to be used by the human mind. Strong AI, in contrast, sees the computer as actually being a mind, and must engage the issues of algorithm and soul.[8]

To reproduce intelligence electronically in the "strong" form is clearly a formidable challenge that requires as much progress in the psychology of the mind as in computing science. And due to the importance of language in human thought, the work of figures such as the famous linguist Noam Chomsky bears on AI research. He has written on the "deep structures" of language, contending that the human brain formulates language, any language, using certain basic principles of grammar that have evolved with humanity itself.[9]

Until a Chomskian grammar of deep linguistic structures has been fully codified, AI will remain vulnerable to the criticism that a computer does not "understand" what it is doing the way a human does. An English-speaker translating words or sentences from Russian into Swahili would usually want to understand the process by working first with a Russian-English dictionary, then with the English-Swahili one. Only then would any deep structure process be involved. To simply perform the exercise using the Russian-Swahili dictionary, while understanding neither language, would be considered mechanical and

uncomprehending. Philosophers such as John Searle contend that computers are "intelligent" only in this second, weak AI, sense.

2. Chess

"With his pawn structure compromised, Kasparov tries a desperate attack," a chess journalist commented during the match between world champion Garry Kasparov and "IBM Deep Blue" (*Guardian Weekly* "Chess" column, 10 March 1996). The February 1996 man-versus-machine series enlivened a long cold winter and marked an advance in AI applications to chess playing. To the surprise of many, Kasparov lost the opening game although he eventually prevailed 4 games to 2 to win the series. As one report remarked, "To watch world chess champion Garry Kasparov squinting and squirming last week as he faced off in a six-game match against Deep Blue, the world's strongest chess computer, has been to witness a profound event in the history of brains, human and otherwise" (*Globe and Mail*, 19 February 1996, A8).

The artificial intelligence of chess brings out some of the same issues as computer psychoanalysis does. What level of understanding did Deep Blue have of the game of chess? "What exactly was Deep Blue up to when it was beating one of humanity's great minds in an ancient contest of mental strength? Was it thinking? Is it intelligent?" the *Globe and Mail* pondered in a 23 February 1996 editorial. Deep Blue had advanced the "thought algorithms" for computer chess programs by progressing beyond the "brute-force mode" of simulating all possible moves, to include as well a "selectivity function" which "limits the search and analysis to only the most promising moves and consequences" (*Globe and Mail*, 19 February A8).

Typically, human commentators anthropomorphized the machine at the same time they extolled the human player's superior creativity. After the computer lost game 5 subsequent to refusing Kasparov's offer of a draw, "the world champion played with energy and confidence, while Deep Blue's strategy became increasingly wooden and planless" (*Guardian Weekly* chess column, 17 March 1996). Maybe Deep Blue was depressed.

Humans, as they simultaneously try to construct AI through computers and tear it down, resemble a child at the playground who rewrites the rules of a game each time he or she loses. "My God, I used

to think chess required thought. Now I realize it doesn't," one expert on AI exploded (*Globe and Mail*, 23 February 1996). But if this is rewriting of the definition of humanity and mind it surely is philosophically fruitful, part of the way AI stimulates humanity to examine itself. One of the ironies in this most technological of ages is that front line research on AI issues such as chess playing can lead to a softer, almost more medieval, human self-definition. "So the definition of humanity turns from the rational to the suprarational, from the mind to the heart," the *Globe*'s editorial board concluded (23 February 1996, A16). Such thought seems healthy for the answer it gives to the technocracy concerns we saw earlier in Joseph Weizenbaum's work.

Just as humanity involves heart as well as a mind, so humanity is a multifaceted physical organism and that too bears on the assessment of AI. Much of the human biological heritage derives from handling physical tasks, since "the survival of human beings ... has depended for hundreds of millions of years on seeing and moving in the physical world, and in that competition large parts of their brains have become efficiently organized for the task."[10] This is a point of view worth considering for, by constantly comparing themselves with animals, humans have a self-concept as cognitive beings first and physical beings only very secondarily.

Robotics arguably is even less advanced than cognition-cloning in the sense of possessing an "emergent" human ability to improvise and adapt on the move. It is not merely human self-aggrandizing to note that Deep Blue only computed a code to select a chess move, it did not physically move the pieces on the board. Kasparov had both to think like a stationary computer and execute the moves like a robot.

While we tend in technological culture to be dismissive of physical prowess, the psychologist Howard Gardner's work identifies and celebrates varied forms of intelligence. There are, Gardner argued, not one but "multiple" intelligences, naming seven: linguistic, logical-mathematical, spatial, musical, bodily-kinesthetic, interpersonal, and intrapersonal. Linguistic and logical-mathematical intelligences are familiar enough, for they are the main dimensions of "IQ" (intelligence quotient) tests that almost everybody takes at some point. A person with spatial intelligence can visualize shapes and directions, can for example navigate with few aids. An artist who can portray distance using perspective to give an illusion of three dimensions has spatial intelligence.

The ingredients of musical intelligence would be perfect pitch, accurate rhythm, and the more subtle factors of "interpretation" such as sense of appropriate dynamics and rubato. The athlete with quick reflexes and accurate hand to eye coordination has bodily-kinesthetic intelligence. At the top end of this intelligence form is the sense of "anticipation" that is often described by sports writers. Wayne Gretzky, for example, was said to be such a dominant hockey player during the 1980s because he possessed that sense of projection. He "knew," probably instantly and without conscious reflection or effort, where the puck and the other players were likely to be a few seconds into the future (*Globe and Mail*, 29 November 1996, D2, "The Great One's Game is all in his Reflexes"). Political leaders or the heads of large corporations often are not outstanding "intellects" in the conventional linguistic or logical-mathematical sense. Their intelligence is more interpersonal, meaning, as Gardner says, "a core capacity to notice distinctions among others; in particular, contrasts of their moods, temperaments, motivations, and intentions. In more advanced forms, this intelligence permits a skilled adult to read the intentions and desires of others, even when these have been hidden."[11] To be leader of a political party must require interpersonal intelligence. Finally, by "intrapersonal intelligence" Gardner refers to an ability to understand, regulate, and guide oneself. Social workers who help troubled people "get in touch with their feelings" are trying to promote their clients' intrapersonal intelligence.

Gardner blamed the craze for IQ testing as one factor in an education system that laid all the stress on linguistic and logical-mathematical intelligence. The school system had become out of touch with reality, with what is termed "the real world," helping account for why schooling tends to prepare workers inadequately for the employment world. Most of Gardner's seven intelligences were only weakly correlated with one other, meaning that many people's ability went unrecognized by the schools.

The implication of all this for AI is, of course, that the computer most successfully models the logical-mathematical form of intelligence. In dealing with verbal intelligence for Turing test purposes, a computer still operates from its logical-mathematical base, perhaps the most algorithmic of all the seven intelligences. Turkle might comment that such bias is no accident. Our culture has given priority to the log-

ical-mathematical domain, so small wonder that the machines humans construct reflect back that dimension. Computers might have greatly different capabilities and intelligences if originating from a radically different culture than ours.

Biotechnology

At around the same time Alan Turing was writing his pivotal article on a test for artificial intelligence, two young researchers at Cambridge University were racing against other labs to solve the riddle of DNA, "the most golden of all molecules."[12] By the spring of 1953, James Watson and Francis Crick announced that they had solved the problem in a short publication in the journal *Nature*. It was the beginning of the biotechnological revolution. This work, no less than AI, presents a kind of second self. As scientists probe into the minute details of genetic code, they are examining their own essence. "And now the announcement of Watson and Crick about DNA. This is for me the real proof of the existence of God," the Spanish artist Salvador Dali is reputed to have said.[13]

DNA, short for deoxyribonucleic acid, proved to be a long chain in the form of a "double helix," looking like "a ladder that has been twisted to resemble a spiral staircase" (Sylvester and Klotz, in Further Readings, 30). A diagram of the molecule appears in Figure 9.1. The twisting "backbone" of the helix is composed of sugar and phosphate while the "rungs" of the ladder comprise four bases (the substances that neutralize acids) — adenine (labelled "A" in the diagram), thymine ("T"), guanine ("G"), and cytosine ("C"). These bases are chemically connected, but not in every possible combination. It is a crucial feature of DNA that adenine bonds only with thymine, and cytosine only with guanine. Scientists thus speak of the "base pairs" for DNA, and Figure 9.1 illustrates this by showing only AT/TA and CG/GC bonds. Since the DNA molecule is actually a genetic coding system, the ordering for each base pair matters. AC, for example, is as different from CA as the number 12 is from 21. There are then only four combinations of base pairs, implying a "four-letter alphabet," as the excellent layperson's explanation in Sylvester and Klotz suggests. This chemical alphabet is the foundation of millions of "lines" of genetic code.

In their explanation of the reproduction of cells, Sylvester and Klotz

Figure 9.1: The Structure of DNA

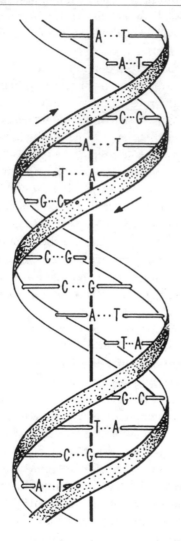

"A schematic illustration of the double helix. The two sugar-phosphate backbones twistabout on the outside with the flat hydrogen-bonded base pairs forming the core. Seen this way, the structure resembles a spiral staircase with the base pairs forming the steps."

James D. Watson, *The Double Helix: A Personal Account of the Discovery of the Structure of DNA* (London: Weidenfeld and Nicolson, 1968), 202. Re-printed with permission.

suggest the following analogies: DNA resembles a "master blueprint" for constructing cells. Another chemical substance in cells, mRNA, copies the message within the DNA into another code or language, this one based on twenty amino acids. It seems a little analogous to computer "machine" languages that communicate more directly to the circuits. Just as a house can be built from detailed blueprints, proteins can be constructed from the mRNA recipe of amino acids. A further key property is the ability of a DNA molecule to replicate itself, resulting in two new DNA molecules that are chemically identical to each other and to their parent. In this self-replication, the twisting "backbone" strands of the molecule untwist, or "unzip," and the rungs of the base pair "ladder" are separated. New ladders, or DNA molecules, are formed as the correct chemicals mate with the now separated rungs or bases to form new base pairs.

The biotechnological revolution everybody is talking about has resulted because the genes encoded in DNA can be "spliced" or "recombined" into new forms which are then reproduced, transcribed into mRNA, and constructed as cells. This has ushered in the daunting field of "genetic engineering," which has impact on everything from suburban lawns to human reproduction. DNA "fingerprinting" has become a key technology for forensic science, and was prominent in the notorious O.J. Simpson trial of 1995. A compilation of some applications of genetic engineering appears in Table 9.1, including approval figures for each application.

The ratings for this sample of New Jersey residents illustrate how profoundly divided public opinion is about biotechnology. The same mixed feelings often appear in Canadian newspapers. "Creepy genetics unchecked," one story will begin (*Globe and Mail*, 28 July 1995, A15). "Cancers fought by inserting new material into genetic code of cells," another announced (*Globe and Mail*, 28 June 1995, A6), bringing hope to thousands of present and future cancer sufferers, and giving a feeling of control over human destiny. The simultaneous promise and threat of genetic engineering is very visible on the World Wide Web of the Internet. Many professors at the leading biotechnology research universities have opened companies, developing home pages to advertise their wares. One example is "Prozyme," at http://www.prozyme.com: "manufacturer of speciality PROteins and enZYMEs for diagnostics and research applications." "Consumer

Table 9.1: Some applications and approval ratings of Genetic Engineering (Residents of New Jersey, 1993)

Application	% Strongly Approve
New drug to cure human disease	81
Hormone like insulin to help people with diabetes	78
More nutritious grain to feed people in poor countries	68
Bacteria to clean up oil spills	62
New grass that doesn't need to be mown so often	50
Fruits and vegetables that are less expensive	47
Fruits and vegetables that have own chemical defenses against pests	40
Better tasting fruits and vegetables	38
Fruits and vegetables that last longer on supermarket shelf	34
Hormones that enable cows to produce beef with less cholesterol	32
Hormones that enable cows to give more milk	21
Hormones that enable cows to give more beef	17

Source: William K. Hallman and Jennifer Metcalfe, "Public perceptions of agricultural biotechnology: A survey of New Jersey Residents", 24. On website http://www/nalus-da.gov/bic/Pubpercept

Warning," counters the Pure Food Campaign at http://geocities.com/Athens/1527. In a notice on rBGH (Bovine Growth Hormone), consumers are advised that "genetically engineered hormones could be in your milk and dairy products... you and your family are at risk."

Chemicals and Electrons

Research on artificial intelligence and on biotechnology is beginning to merge in certain respects. Although the present book aspires to offer sociological rather than scientific opinions, some indicators of the emerging synthesis are present even in the material available to a lay audience. Consider the career choice of Francis Crick, the British codiscoverer of DNA, to pursue research on the brain. Crick moved to the Salk Institute in California in the mid-1970s, and spent this final stage of his career integrating the laboratory findings of others rather than performing his own primary research.

In his 1994 book *The Astonishing Hypothesis: The Scientific Search for the Soul* (Further Readings), Crick argued that consciousness and

the soul are biochemical products: "'You,' your joys and your sorrows, your memories and your ambitions, your sense of personal identity and free will, are in fact no more than the behavior of a vast assembly of nerve cells and their associated molecules" (3). As reviewers noted, that is what materialist philosophers have been claiming for hundreds of years (*The Times Higher Education Supplement*, 27 May 1994, 20), but coming from a molecular biologist such notions have more obvious connection to AI. One of Crick's points is the importance of chemicals in the passing of electrical particles within the human body. Some of the flavour of his argument comes across in the following passage: "When the spike [pulse of ions] arrives ... it causes little packets of chemicals ... to be released into the gap [between one neuron and another]. These small chemical molecules diffuse rapidly in the gap, many of them combining with one or more of the molecular gates in the membrane of the synapse of the recipient cell" (97). Self-knowledge itself, Crick reasoned, reduces to electrochemical connections. "Consciousness depends crucially on thalamic connections with the cortex," he writes near the conclusion of *The Astonishing Hypothesis* (252). Since animals such as dogs, cats, and monkeys have some brain chemistry in common with humans, Crick extrapolates that such beings possess consciousness which, if not exactly of the human form, is more advanced than commonly recognized.

Although Francis Crick takes pains to distinguish sharply between the electronic computer and the human brain, and indeed is somewhat dismissive of the AI movement, the implication of his work in fact opens the theoretical possibility of cloned human intelligence. He makes it harder to fall back on magical claims differentiating "man" from machine. Surely it is part of our destiny as children of the scientific-technological age to entertain such thoughts. Humanity is not trivialized thereby. To adopt Crick's view of consciousness and the soul is no different from saying that Beethoven's Fifth Symphony is based on notes, or the Mona Lisa on dabs of paint. The "magic" is the complexity and subtlety of how and why the basic components are assembled. Even as hard-nosed a scientist as the co-parent of the biotechnological age himself speaks in the language of wonder and marvel in contemplating DNA (260).

A recent experiment with using DNA as a "computer" may be another marker of integration between biotechnology and AI. Noted in

the science report of the *New York Times* for 22 November 1994, the test problem concerned whether a path exists that passes through seven cities, with only one visit per city, where only some cities are connected, and with a defined beginning and ending point. Such a problem might be encountered when planning a tour by public transit of seven cities, some of which are connected by an airline service, but others are not. The traveller starts from a home town and names the final destination. It is wasteful to pass through any one of the cities more than once, and so the airline route maps are searched for the optimum path. For mathematicians, this is a class of problem in which the computations are deceptively simple for a small number of cities. Each additional city, however, adds an increasingly massive number of alternative paths that must be considered for a solution. So a "number crunching" mathematical problem is selected or, in the more scientifically formal words of Leonard Adleman, designer of the DNA as computer experiment, " ... it seems likely that no efficient ... algorithm exists for solving it."[14]

The seven-city network appears as Figure 9.2, reproduced from Adleman's report. Before endeavouring to comprehend the molecular solution, let us think of how we might begin simply by using the "computer made of meat" above our shoulders, permitting ourselves perhaps the luxury of a pencil and supply of paper for tracing out promising routes. To begin, we identify each link, whether written out in full or in some shorthand code, and place it in a hat. Figure 9.2 shows 14 permitted links between the seven cities, in which we note that each direction counts as one link (for example, counting 2 - 3 separately from 3 - 2). Then blindfolded draws (that is, random samples) are taken from the hat, following these rules:

i) Continue selecting only when the first link begins with the home town (labelled city 0 in Figure 9.2).

ii) Cease selecting when city 6, the final destination, is the destination in a link.

iii) Keep only the samples which pass through each city at least once.

iv) Discard the sample if the number of links does not equal the number of cities.

Figure 9.2: Diagram of the Paths Between Cities Problem

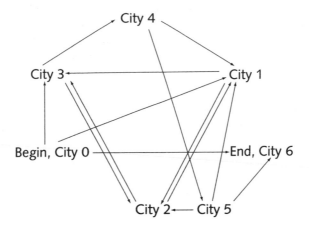

Source: Re-drawn from Leonard M. Adleman, "Molecular computation of solutions to combinatorial problems," *Science* Vol 266 (11 November, 1994), 1022.

Adleman used the chemistry of the DNA molecule to work through the same procedure in nanoseconds. Each city was coded into DNA language, using the same A,T,C,G representation seen earlier in Figure 9.1. Twenty letters of these DNA bases were used for each city, the first ten designated as "first name" and following ten as "last name" (the reason for the dual names comes in a moment). It does not matter what combination of letters is used to name each city, so long as the names are "spelled" consistently throughout. The linkages between cities, which Adleman termed "flight numbers" to highlight the travel agent nuance to his problem, took the *last* name for the origin city and the *first* for the destination city. City number two in Figure 9.2, for example, had a code of TATCGGATCG GTATATCCGA. DNA molecules corresponding to the city codes and linkages were then synthesized in a lab.[15] Now, if the DNA codes for cities two and three had a link between them, the whole group of chemicals would join up. That is, the molecule can map the existence of two "flights" where the destination for flight one is the origin for flight two. The longer the chain in the DNA

molecule the more cities were included in the path. Paths that did not begin and end at the correct city could be identified and eliminated, and the molecule automatically excluded non-permissible paths. It was then possible using the techniques of molecular biology to identify and decode back into the travel plans diagram the molecule that best solved the travel problem.

So much for our excursion into modern science. It is admittedly complicated for the lay person, but such research brings to the imagination some future generation of computers which will have not just an electronic basis, but a chemical one resulting from complex combinations of the more primitive particles such as the electron.

Conclusion

In their increasingly intertwined ways, the two emerging fields of artificial intelligence and biotechnology throw the most basic issue of human self-comprehension back onto their creators. The computer "touches" us and becomes a second self, as Turkle says, by its interactivity and by the amount of human content within the software. Biotechnology is less familiar in the sense that few people consciously use it as they do use computers. The products of genetic engineering may, however, be an even more constant presence in our lives than computers, as the Pure Food Campaign warns. Part of the evocative quality of both AI and biotechnology is the ethical questions each raises. It is time now to consider some public and philosophical opinions on such ethical issues.

Notes

1. A.M. Turing, "Computing machinery and intelligence," *Mind* 59, no. 236 (October, 1950), 433-460.

2. Mark Griffiths, "Netties anonymous," *The Times Higher Educational Supplement*, 7 April, 1995, 18. Jane Gadd, "Doctors debate Internet addiction: Should webaholics seek medical help?" *The Globe and Mail*, 15 June 1996, A1, A11, reports on an article by Michael O'Reilly, "Internet addiction: A new disorder enters the medical lexicon," *Canadian Medical Association Journal* 154, no. 12 (15 June, 1996), 1882-83. Similar notions appear in Howard

Rheingold, *The Virtual Community: Homesteading on the Electronic Frontier* (New York: Harper Perennial [Harper Collins], 1993), 151; and Jones, *Cyberspace* (Further Readings for Chapter 8), 83. Addicts seeking help can log on to the Center for OnLine Addiction at http://www.pitt.edu./~ksy/ or other sites noted in O'Reilly, above. I used to run my own home laboratory on this topic. My 15-year-old son was so compulsive a fan of computer games that every morning after he left for school I confiscated his keyboard, to be returned only after he had completed his homework later in the day. Amazingly, he consented to this, part of him realizing that he could not control the computing impulse.

3. Russell W. Belk, "Possessions and the extended self," *Journal of Consumer Research* 15 (1988), 139-68.

4. Roger Penrose, *The Emperor's New Mind: Concerning Computers, Minds, and the Laws of Physics* (London: Vintage, 1989), 14.

5. K.M. Colby, J.B. Watt, and J.P. Gilbert, "A computer method of psychotherapy: Preliminary communication," *The Journal of Nervous and Mental Disease* 142, no. 2 (1966), 148-152.

6. From web site http://fuzine.mt.cs.cmu.edu/mlm/contest94.html

7. Penrose, 12 (footnote).

8. John R. Searle, "Minds, brains, and programs," in Douglas R. Hofstadter and Daniel C. Dennett, eds., *The Mind's I: Fantasies and Reflections on Self and Soul* (New York: Basic Books, 1981), 353-73.

9. Noam Chomsky, *Language and Mind,* Enlarged edition (New York: Harcourt Brace Jovanovich, 1972). For discussion of AI implications, see Derek Partridge, *A New Guide to Artificial Intelligence* (Norwood, New Jersey: Ablex Publishing, 1991), Chapter 8.

10. Hans Moravec, "Robotics and artificial intelligence," in Charles Sheffield, Marcelo Alonso, and Morton A. Kaplan, eds., *The World of 2044: Technological Development and the Future of Society* (St. Paul, Minnesota: Paragon, 1994), 29.

11. Howard Gardner, *Multiple Intelligences: The Theory in Practice* (New York: Basic, 1993), 8-9, 17-34, quotation from page 23. Also, Howard Gardner, *Frames of Mind: The Theory of Multiple Intelligences* (New York: Basic, 1983).

12. James D. Watson, *The Double Helix: A Personal Account of the Discovery of the Structure of DNA* (London: Weidenfeld and Nicolson, 1968), 18.

13. Opening quotation in Francis Crick, *Of Molecules and Men* (Seattle: University of Washington Press, 1966).
14. Leonard M. Adleman, "Molecular computation of solutions to combinatorial problems," *Science* Vol 266 (11 November, 1994), 1021. See also *Scientific American*, September 1995, 20-22. The same computational problem is discussed within an AI context in Partridge, 95-97.
15. At this stage, the city identities were transcribed to bring to bear the fact that "opposites" attract in molecular biology (that is, the "base pairs" referred to earlier, whereby adenine attracts to thymine, cytosine to guanine).

Further Readings

For ideas on the psychology and sociology of modern computing:

Joerges, Bernward. "Images of technology in Sociology: Computer as butterfly and bat." *Technology and Culture* 31 (April, 1990), 203-27.
Searle, John R. *The Rediscovery of the Mind.* Cambridge, Mass.: The MIT Press, 1992.
Turkle, Sherry. *Life on the Screen: Identity in the Age of the Internet.* New York: Simon and Schuster, 1995.
___. *The Second Self: Computers and the Human Spirit.* New York: Simon and Schuster [Touchstone], 1984.
Weizenbaum, Joseph. *Computer Power and Human Reason: From Judgment to Calculation.* San Francisco: W.H. Freeman, 1976.

For an introduction to the field of biotechnology:

http://kadets.d20.co.edu/~lundberg
http://www.aba.asn.au.
Blakemore, Colin and Susan Greenfield, eds. *Mindwaves: Thoughts on Intelligence, Identity and Consciousness.* Oxford: Blackwell, 1987.
Crick, Francis. *The Astonishing Hypothesis: The Scientific Search for the Soul.* New York: Charles Scribner's Sons, 1994.
Kosslyn, Stephen M. and Oliver Koenig. *Wet Mind: The New Cognitive Neuroscience.* New York: Free Press, 1992.
Sylvester, Edward J. and Lynn C. Klotz. *The Gene Age: Genetic Engineering*

and the Next Industrial Revolution. New York: Charles Scriber's Sons, 1983.

For an examination of the social content of AI Research:

http://fuzine.mt.cs.cmu.edu/mlm/julia.html.
Schwartz, Ronald David. "Artificial intelligence as a sociological phenomenon." *Canadian Journal of Sociology* 14 (Spring, 1989), 179-202.

Part III.

THE EVALUATIVE DIMENSION

A McLaughlin Model 25 automobile dating from 1913. As well as being an example of one of the most ethically contentious pieces of technology, this picture illustrates other themes. Note, for example, the "British" style right-hand drive still in use, and observe too that at the beginning of the twentieth century Canada had its own auto industry. *National Museum of Science and Technology*, Ottawa.

Chapter 10: Ethical Responses to Technology

The Ethical Life Cycle

Both artificial intelligence and biotechnology bring the ethical dimension of technology to the fore, and that is probably characteristic of every emerging technology. One need only consider how controversial inventions such as railroad transportation used to be. The Stockton and Darlington Railway, the early nineteenth-century British railroad, required extensive lobbying and pay-offs and had to surmount initial rejection in the House of Commons before finally securing Parliamentary approval in 1823. Land owners tended to object on conservation grounds. At first thought, the modern reader sympathizes with these English country gentlemen, who appear to be almost nineteenth-century equivalents of Parks Canada officials. And yet the land owners were protecting a narrow vested interest. The Earl of Darlington objected that the Stockton and Darlington line would ruin the fox hunting on his estate.[1] The hunting parties were luxury pastimes pursued on the backs of tenant farmers whose stories have been told by writers like Thomas Hardy.

The Duke of Wellington, vanquisher of Napoleon at the Battle of

Waterloo, was another well-known critic of British railway construction, and his motives were more tied up with class interest and snobbery than with conservation. He once wrote in a letter:

(P)eople never acted so foolishly as we did in allowing of the Destruction of our excellent and commodious [post-roads] in order to expend Millions Sterling on these Rail Roads! It appears to me to be the Vulgarest, most indelicate, most inconvenient, most injurious to Health of any mode of conveyance that I have seen in any part of the World![2]

Much of Wellington's objection was a distaste for mingling with the masses. He did not like the way other passengers "examine and pry into every Carriage and the actions of every Traveller." But he was also aware that the railroads were a technology which would help free the working classes, giving them the cheap transportation to work in a factory out of reach of feudal landlords.[3] The rail lines also served the interests of a new class of capitalist entrepreneurs who were rising to power and would by century's end challenge the landed nobility: "the new, sturdy liberal commercial class in the great industrial cities", as one of the railroad histories terms them.[4] Aristocratic reactions to the railroads remind us that class interest can hide behind ethical interest.

Perhaps another lesson to be extracted from these aspects of the history of technology is that every innovation may be considered controversial at first. "Ever since protohumans with sloping foreheads learned to set things on fire, people have feared and hated technology as much as they have been in its thrall. They have eyed with suspicion the printing press, the automobile, the telephone, and the television as solvents of the glue that binds people together."[5] The intention in making this point is not to trivialize ethical issues, nor to equate the hydrogen bomb with the printing press, but to become aware that a kind of ethical life cycle of technology does exist. Once a technology has appeared at the doorstep of a society, the institutions and culture of that society have to "make room" for the stranger before it diffuses at all widely. This process of accommodation is a subtle and key theme for students of technology. Consider some examples presented in earlier chapters, Basalla's views on the evolution of technology, for example, or more specifically the redefinition of the telephone from business to residen-

tial applications. Also consider the daily battle of ideas fought between the public and the advertising industry in any capitalist society around the definition of what goods are "needed." For many consumer technologies — the bicycle, car, or radio from earlier years, or the VCR, personal computer, or cellular phone of our own times — the striking fact about diffusion often is how quickly such innovations become regarded as indispensable within the culture.

Public Opinion on Ethical Issues

The ordinary citizen in our culture is still considered an important voice on ethical questions about technology. Although scientists and engineers have far greater knowledge about technical details, and even though the public may on occasion act on the basis of ignorance and misunderstanding, there is a powerful validating component to lay ethical judgement over expert judgement. We see opinion poll results mobilized in ethical debates in Web material by the Pure Food Campaign. "The FDA [Food and Drug Administration in the US]," one of their documents reads, "has also refused to require labeling of milk and other dairy produces derived from use of the genetically engineered hormone, even though more than 90% of consumers favor labeling of rBGH products so they can avoid buying them."[6]

The lay person may have less personal stake in a technology, and therefore be freer than the expert to consider the general public good. It is hard for a scientist who may have worked intensively on a project for years to be fully objective, no matter how well informed he or she is. Science rewards those with the competitive abilities to reach a discovery first. James Watson, for example, from his account of the discovery of DNA structure (noted in Chapter 9), seems to have been motivated by the Nobel Prize, the Nobel Prize and the Nobel Prize, in that order! Arms length, even if that means less knowledge, is the preferred ethical criterion in democratic societies, and that is why such great attention is devoted to public opinion about the ethics of technology. It is time now to review such research.

Surveys on Public Attitudes Toward Technology

Some of the general attitudes toward technology among the Canadian public were revealed in a telephone survey of 2,000 adults conducted in late 1989. Table 10.1 reproduces the percentages agreeing to ten statements about science and technology, grouped by higher or lower level of consensus. High consensus exists that science and technology bring material good ("lives healthier, easier, more comfortable"), that people can and should be informed about science, and that the sufficiently motivated "interested and informed citizen" can influence science policy decisions. The public were much more divided on such issues as whether faith is too much neglected in favour of science, or "science makes our way of life change too fast." With a glance back to the job displacement issues of Chapter 6, we note a virtually even (52/48) split between those who agree or disagree that "on balance, more jobs will be created than lost as a result of computers and factory automation." On the contentious issue of vivisection (surgical experiments on animals), a small majority disapprove.

The study showed that trusting science and technology (a summary score taken from combining information in items 1,2 and 5-8 in Table 10.1) was closely correlated with a "scientific literacy" index described in Table 10.2. Predictors of scientific literacy, and hence indirectly of favourability toward science and technology, were — listed in descending order of importance — higher education, male vs. female gender, exposure to science in high school, and younger rather than older age.

This is consistent with earlier chapters which showed this same profile — the younger, educated male — to predict greater contact with technology in the form of having a computer at home and Internet use. The tendency of the more educated to favour science and technology is often encountered in survey data. For example, a 1993 survey in the US asked about agreement with the statement: "overall, modern science does more harm than good." Thirty-six percent of Americans with less than high school education agreed, compared to just 10% of those with postgraduate degrees. As in the Canadian survey just mentioned, older people were more critical of science than were the young.[7] Another example comes from the New Jersey survey on biotechnology noted in Chapter 9. Twenty-seven percent of those having less than high school education concurred that "it would be better if we did not know how

Table 10.1: Canadian Attitudes Toward Science, 1989 Survey

		Percent Agreeing
Higher Consensus Items (0-25%; 75% or more)		
1.	Science and technology are making our lives healthier, easier, more comfortable	80.0%
2.	It is not important for me to know about science in daily life	21.9
3.	If scientific knowledge is explained clearly, most people will be able to understand it	83.6
4.	The interested and informed citizen can often have some influence on science policy decisions if he or she is willing to make the effort	77.7
Lower Consensus Items (26-74%)		
5.	We depend too much on science, not enough on faith	44.7
6.	Some numbers are especially lucky for some people	30.8
7.	Science makes our way of life change too fast	46.0
8.	Because of their knowledge, scientific researchers have a power that makes them dangerous	41.1
9.	On balance, more jobs will be created than lost as a result of computers and factory automation	51.9
10.	Scientists should be allowed to do research that causes pain and injury to animals like dogs and chimpanzees if it produces new information about human health problems	43.8

Source: Edna F. Einsiedel, "Mental maps of science: knowledge and attitudes among Canadian adults," *International Journal of Public Opinion Research* 6, no. 1 (1994), 39.

Table 10.2: Measures of Scientific Literacy (same survey as in Table 10.1)

Index of basic knowledge		% Correct
1.	The oxygen we breathe comes from plants (t/f)	80.4
2.	Electrons are smaller than atoms (t/f)	46.7
3.	The continents are moving slowly about the surface of the earth (t/f)	74.9
4.	Human beings as we know them today developed from earlier groups of animals (t/f)	58.0
5.	Lasers work by focusing sound waves (t/f)	38.0
6.	The earliest humans lived at the same time as the dinosaurs (t/f)	45.9
7.	Which travels faster, light or sound?	73.8
8.	Does the earth go around the sun or does the sun go around the earth?	78.4
9.	How long does it take for the earth to go around the sun?	51.2
10.	What is DNA?	14.0
11.	Name four factors (from a list of 8) that are major causes of heart disease	33.0
12.	Radioactive milk can be made safe by boiling it (t/f)	61.3
13.	Air pollution can cause a greenhouse effect (t/f)	85.9
14.	Which of the following causes acid rain? (aerosol sprays, coal-fired power plants, nuclear power stations, chemical warfare byproducts)	75.1
15.	What is computer software?	27.0
16.	Do you know what a scientific study is?	16.0
17.	Would you say astrology is 'scientific' ('very scientific', 'sort of scientific', 'not at all')	48.8

Source: P. 38 of same source as in Table 10.1.

Figure 10.1: Paradigms Concerning Science, Technology, and the Environment

	Alternative Paradigm	Dominant Paradigm
Core values	Non-material (self-actualization) Natural environment intrinsically valued Harmony with nature	Material (economic growth) Natural environment valued as a resource Domination over nature
Economy	Public interest Safety Incomes related to need Egalitarian Collective/social provision	Market forces Risk and reward Rewards for achievement Differentials Individual self-help
Polity	Participative structure:(citizen/ worker involvement) Non-hierarchial Liberation	Authoritative structures: (experts influential) Hierarchial Law and order
Society	Decentralized Small-scale Communal Flexible	Centralized Large-scale Associational Ordered
Nature	Earth's resources limited Nature benign Nature delicately balanced	Ample resources Nature hostile/neutral Environment controllable
Knowledge	*Limits to science* Rationality of ends Integration of fact/value, thought/feeling	*Confidence in science and technology* Rationality of means Separation of fact/value, thought/feeling

Source: Stephen Cotgrove, *Catastrophe or Cornucopia: The Environment, Politics and the Future* (Chicester: John Wiley Sons, 1982), 27.

Figure 10.2: Australian Data on Dominant and Alternative Paradigms, Using Five Attitude and Value Scales for Samples of "Greens," the Public, and Business

	Alternative				Dominant	
Environmentalism	G	P	B			
Industrialism	G		P	B		
Science and Technology		G		P	B	
Postmaterialism		G		P	B	
Economic Individualism		G		P		B

Key: G= "greens" or environmentalists (members of the Brisbane Greens Party)
P= sample of general public
B= business people (drawn from *Business Who's Who of Australia*)

Alternative Paradigm: environment high priority, more anti-industry and science & technology, postmaterialist, less economic individualist

Dominant Paradigm: environment lower priority, more pro-industry, science & technology, less postmaterialist, more economic individualist

Sample Information: Cases (N)= 50 greens, 315 public, 89 business, from 1985 mailed questionnaire

Source: Laurence O.P. Knight, "Soft versus hard: An Australian study of an ideological basis to attitudes toward energy use and supply," *The Canadian Geographer* 34, no. 1 (1990), 66.

to do genetic engineering," against 15% with high school graduate education and just 5% with at least some post-secondary education.[8]

Anti-science and technology sentiments also seem to go along with environmentalist and what are known in political science as "postmaterialist" values.[9] Postmaterialism refers largely to the "baby boom" generation that grew up in the 1950s. Unlike their parents who were seared by the 1930s Depression, this generation could more typically take satisfaction of their material needs for granted, and it is in that sense they were "postmaterialist." From their position of relatively comfortable affluence, they could afford to worry about such "higher"

Table 10.3: Education Level of Environmentalists, the Public and Business Leaders: British Data

Years of Education	Environmentalists	Industrialists	Public
≤ 10	0.0	0.0	38.2
11-14	37.6	31.0	33.7
> 14	62.4	69.0	28.1
N	441	216	89

Source: Stephen Cotgrove, *Catastrophe or Cornucopia: The Environment, Politics and the Future* (Chichester: John Wiley & Sons, 1982), 19.

needs as quality of the environment and self-fulfilment. In the environmental field, where much study has been made of such general sets of attitudes because the political issues are so important, alternative *"paradigms"* have been compiled, to use a term beloved by philosophers of science. "Paradigm" here means simply a coherent set of values, attitudes, assumptions, and understandings. Such a scheme as applied to environmentalist issues appears in Figure 10.1, contrasting a "dominant" with an "alternative" paradigm. Note how the assumed source of knowledge differs between the two. People fitting within the dominant paradigm, the more conventional point of view that sustained the process of industrialization during the earlier twentieth century, hold high "confidence in science and technology." Those of the newer alternative paradigm, well represented by the postmaterialist baby boomers, may be more mindful of the "limits to science."

An application of these ideas, illustrating how attitudes toward science and technology fit into a larger picture, can be seen in Australian data reproduced in Figure 10.2. Here three separate samples — environmentalists or "greens" ("G"), members of the general public ("P"), and business leaders ("B") — were surveyed along five sets of attitudes and values. In a highly consistent pattern, greens were the most pro-

Table 10.4: Approval Ratings for More Emphasis on Technology, World Values Survey, 1990-1993

Percentage in each nation who feel that more emphasis on technology would be a good thing (rank ordered)

Nigeria	95.8	
China	92.9	
South Korea	90.7	
Turkey	90.0	
Poland	89.5	
Czechoslovakia	85.2	
Russia	84.9	
South Africa	82.5	
Germany (Former East)	82.2	
Bulgaria	81.2	
Mexico	78.5	
Argentina	78.2	
Brazil	78.1	
France	76.8	
Hungary	75.4	
Chile	73.8	
Portugal	73.2	
India	71.8	
Iceland	69.2	
United States	68.5	
Finland	68.0	
Japan	65.5	
Great Britain	64.4	
Canada	64.0	[English 62.8, French 68.5]
Great Britain	62.2	
Republic of Ireland	61.4	
Italy	60.6	
Northern Ireland	59.9	
Denmark	58.9	
Switzerland	56.5	
Belgium	55.3	
Germany (Former West)	52.2	
Netherlands	49.7	
Norway	46.7	
Austria	41.4	
Sweden	34.2	

Source: Computation from *World Values Survey, 1990-1993*, courtesy of James Curtis, using codebook prepared by Ronald Inglehart.

environment and the least in favour of industrialism and science/technology. They scored higher on a postmaterialism scale than the other samples and lower on economic individualism. Average scores for the public were sandwiched in between this group on the "left" and the business leaders occupying the political "right." Even within the general public sample, those most in favour of science and technology were found (in further tabulations, not detailed here) to have the weakest environmentalist views, the most pro-industry attitudes, the most materialist values, and the greatest favourability toward economic individualism (that, for example, market forces rather than government should run the economy).

In locating the kind of people most favourable to science and technology it could seem odd that the two factors especially singled out are themselves related. That is, although highly educated people are, other things being equal, more in favour of science and technology they also are likely to be environmentalists. British data from Cotgrove's well-known study on environmentalists sheds light on the anomaly. One of his tables, reproduced here as Table 10.3, shows that 62.4% of the environmentalists in a sample from three English cities had more than 14 years of education (that is, at least some university or the equivalent, in the Canadian system) compared with only 28.1% among his general public sample. The seeming paradox about education is resolved by the other column of figures in Cotgrove's table, for it is seen that the "industrialists" in the British sample had the highest education of anyone. The sample-wide tabulations indicate that for most people more education means a "softer" stance toward technology because in the general population materialists outnumber postmaterialists by about a 2:1 ratio.[10] So the following statements are all true and noncontradicting: education fosters feelings of comfort with science and technology, especially insofar as formal education exposes students to the values and priorities of science — the same technocratic culture discussed in Chapter 5. The educated, however, are also a polarized group. Some, notably a heavy concentration within generations such as the baby boomers, and especially those employed in the public sector, develop a set of postmaterialist, pro-environment, anti-business and somewhat anti-science and technology attitudes and values. Others of the educated class, in contrast — and we might think of the proverbial BMW-driving stockbroker wearing "red suspenders" as a prototype — happily

embrace science, technology, industry, and the capitalist market economy as one consistent package.

How do the proportions of these varied types balance out into a total picture for Canada as measured against other societies? The *World Values Survey* is a coordinated attempt by social scientists to collect comparative data for different societies. The 1990-93 version included a statement: "Here is a list of various changes in our way of life that might take place in the near future. Please tell me for each one, if it were to happen whether you think it would be a good thing, a bad thing, or don't you mind?" Answers by residents of 36 countries to the item on the list reading "more emphasis on the development of technology" appear in Table 10.4.

A number of influences, some by now familiar themes for us, seem to mix together within these rankings. The table is emphatically *not* saying that the people of the most industrially developed nations most embrace technology. The US and Japan, perhaps the two leading economic powers, fall only in the middle of the list, just a couple of percentage points above Canada in public support for technological development. It can hardly be said, then, that the economic competitiveness of the Canadian economy is held back by a population that does not support technology. Indeed, in support for technology we are well ahead of the former West Germany, another great post-World War Two economic power. The greatest appetite for more technology exists among a group of poorer nations, especially former communist societies such as Russia, now in an early stage of capitalism. If the World Values Data had been collected as far back as 1830, Great Britain and the US might have had percentages in the 80s or 90s and a society such as Russia a much lower figure. A "postindustrial society effect" whereby populations of some societies with longstanding capitalist affluence have less fervour for technology has been noticed before in cross-national research on technology.[11] The "postmaterialist values" discussed earlier in this chapter are part of this same postindustrial society effect. Recall that some of the highly educated Canadian population held postmaterialist, somewhat anti-science and technology, values. The incidence of such feelings is variant across nations, with some of the mature industrial societies having many citizens doubtful about technology. Again, as noted earlier, affluence gives the freedom to be critical, a freedom that people in countries like Nigeria or China do not

enjoy at present. Indeed, affluent countries with strong socialist traditions (for example, Norway and Sweden) are critical of technology. As noted in Chapter 2, capitalist expansion and technology traditionally have gone hand in hand. Those who can take capitalist economic prosperity for granted and who look to less material goals tend to cool towards technology. It is well to remember too that the numerically small class of managerial technocrats probably exercises disproportionately great influence on national science and technology policy. Their power would not show up in such a simple tally from a random sample.

A Canadian Perspective

"Yet the very substance of our existing which has made us the leaders in technique," wrote George Grant, "stands as a barrier to any thinking which might be able to comprehend technique from beyond its own dynamism" (*Technology and Empire*, in Further Readings, 40).

Here is the question with which the present chapter began: Who is the best judge of technology, the most or least informed? We have already seen from the opinion poll data some tendencies for those with more education, and especially education in technology and science, to be more favourable to this dominant force. Unless tempered by post-materialist values, those most knowledgeable about technology are also the most enmeshed with it and therefore the least able to be critical. Grant used similar reasoning, but on a national scale: for him, the US was the quintessential technological society, the world leader that was also the most blind to technology's perils. He contended that, postmaterialist tendencies of the baby boomers notwithstanding, the power structure of the US is firmly technocratic.

Recall from Chapter 5 how Touraine, the critic of technocracy, was a Frenchman and how his views contrasted with the American Daniel Bell. According to Grant, European society was best equipped to see the pitfalls of technological society because countries such as France, Germany, or Italy were in closer touch with the less technological ages of antiquity. He came to this view from his own experience of studying in Great Britain and travelling in Europe as a Rhodes Scholar in the 1940s. "Even to have touched Greekness ... required that I should first have touched something in Europe which stayed alive there from before

the age of progress through all its acceptance of that age" (*Technology*, 36). He meant not simply tourist sites such as museums, old buildings, and art galleries, but the way that the everyday living culture of European society maintained at least a thin thread of continuity with much earlier civilizations such as the one in Greece. And as noted earlier, for Grant the religious factor was never far away. Part of his formative experience was "to have felt the remnants of a Christianity which was more than simply the legitimising of progress and which still held in itself the fruits of contemplation" (*Technology*, 36).

Grant did not pretend that modern Europe is technophobic, but he did maintain that Europeans and everybody else in the world were adopting the belief system of liberalism, the ideology championing the freedom of the individual (outlined in Chapter 5). A contemporary echo of some of Grant's ideas is the claim by Francis Fukuyama that liberalism has triumphed over other ideologies to an extent meaning "the end of history."[12] If this is true, it entails an evolved form of liberalism which coexists with the resurgent corporatism that, as noted earlier, translates the rights of the individual into group interests. We saw that liberalism assumes that the maximization of every individual's interest is ultimately consistent with the society's interest. Corporatism adopts the parallel reasoning that the interests of groups such as big business serve the national interest; "what's good for GM is good for America," as the famous saying goes. Political theorists such as John Ralston Saul regard this slogan as fallacious. While it is clearly absurd for any individual to argue that his or her interests supersede those of the state, corporate bodies grow sufficiently large, powerful, and assertive that they can. Individuals may *feel* that their interests are paramount, and may well act in a self-serving manner, but no serious logical argument can support such sentiments and acts. Corporate bodies such as the church in medieval times, the CIA in the US during the Cold War, or even the Canadian military during recent scandals about the conduct of soldiers on peace-keeping missions and subsequent cover-ups, may argue along precisely the lines that their interest *is* the societal interest. When more than one such body makes that claim, the whole notion of the public good begins to crumble. The definition of the public good is ambiguous enough within the liberal capitalism described by Grant, but Saul contends that corporatism legitimates interests so powerful as to undermine much more destructively the con-

cept of the public good. The result is the individualistic fictions of liberalism grafted onto the group interest realities of corporatism; such an ideology maintains the pretence of freedom of the individual, but in reality "denies and undermines the legitimacy of the individual as the citizen in a democracy."[13]

Liberalism and corporatism can coexist as joint pillars of corporate technocratic capitalism because the contradiction between corporate and national interest is seldom recognized. Distrust about whether technology is progress, about the harmony of interests between business and the public good, and about whether every individual really does get a fair chance in life comes instead from "*socialist ideology,*" the diluted but recognizable political legacy of Marx.

Part of the subtlety of Grant's argument was that a dominant ideology such as liberalism-corporatism tends to permeate into every culture and belief system, including Marxism. He wrote, for example, that "however libertarian the notions of the new left, they are always thought within the control of nature achieved by modern techniques" (*Technology*, 31). So the ideas of liberal technological society were highly "*hegemonic*" for this old Canadian-born but European-trained philosopher, and his distinctions between nations were those of degree. In Canada, especially the 1920s and 1930s Ontario he grew up in, the European ties still were strong; for him part of the definition of Canadian identity was a more ambivalent stance toward technology than was typical to the south. That European connection fostering questions about liberal technological society was recognized within the very book title of *Technology and Empire*.[14]

Both Grant and his world are gone now, but his ideas live in two senses. First, even now in Canadian society there may be a lingering feeling of ambivalence about technology. "Ambivalence" here does not equate to hatred, but more nearly to both love and hatred, or approval and disapproval, at once. Secondly, we have seen in earlier chapters how modern technological society is simultaneously exhilarating and harsh. Technology in the hands of today's capitalism means an insecure life for even the most high flying of corporate engineers or MBA's. Tales of downward mobility from the middle class are one of the recurrent newspaper themes of the 1990s (for example, "Middle class cleft into haves, have-nots", *Globe and Mail*, 24 April 1996, A1, and see also Chapter 6). When a 50-year-old company employee is summarily dis-

missed without regard to a lifetime of service and when a 25-year-old waits year after year on the sidelines of the economy, books such as Grant's are needed to authenticate the feeling that technological liberalism has gone sour.

Canadian ambivalence about technology was the theme of Arthur Kroker's book *Technology and the Canadian Mind* (Further Readings). Grant was seen by Kroker as a figure "sometimes poetic, always tragic, reflecting on the price to be paid for the consumer comforts of technological society" (13). If Grant reflected the "hate" dimension of Canadian ambivalence about technology, Kroker saw Marshall McLuhan as personifying the "love." Both my terms here are clumsy overstatements, but Kroker saw Grant as the voice of "technological dependency," McLuhan as personifying a "technological humanism" which "seeks to renew technique from within by releasing the creative possibilities inherent in the technological experience" (14). Kroker depicted the two as such "bi-polar opposites" (15) that it was useful to visualize Harold Innis in the middle position of "technological realism," connecting the two extremes. While Grant and McLuhan were indeed very different figures along some dimensions, the issue of pinning down McLuhan's feelings about technology is tricky. His "press" during the 1960s certainly was that of the media guru who seemingly revelled in the electronic age. In contrast, he always claimed in his written work to be nonjudgemental about technology. It is true, however, that a nuance of paradise regained can be detected in his writing on electronic media. He began his research with a celebration of small, oral-based societies, moving then to decry the loss of "tribalism" in the hot print-oriented world of the Gutenberg revolution. In some respects, the return to media coolness that McLuhan attributed to inventions such as the television and the telephone was the regaining of something. If McLuhan's argument about resurgent coolness simply amounted to the thought that modern technology is not all bad, Grant would have readily agreed: "One must never think about technological destiny without looking squarely at the justice in those hopes."[15] But there was in fact a whole different side to Marshall McLuhan when he used the spoken word. In speech, he was far more conservative than his written legacy implies, and he surely would have allowed us the argument that if face-to-face oral communication is the most authentic and natural sensory state, then what he spoke has priority over what he

wrote.

The conservative McLuhan appeared in a CBC radio broadcast on the "Sunday Morning" program of 16 July 1995. The following excerpt from that program comes from a recorded television interview made many years earlier:

McLuhan: I'd rather be in any period at all, as long as people are going to leave it alone for a while. Just let go, just leave it that way.

Interviewer: But they're not going to, are they?

McLuhan: No, so, the only alternative is to understand everything that's going on and then neutralize it as much as possible, turn off as many buttons as you can and frustrate them as much as you can. I am resolutely opposed to all innovation, all change, but I am determined to understand what's happened. Because I don't choose just to sit and let the juggernaught roll over me. Now, many people seem to think if you talk about something recent, you're in favour of it. The exact opposite is true in my case. Anything I talk about is almost certain to be something I'm resolutely against.

This side of Marshall McLuhan is much more consistent with George Grant, the brooding pessimist, and the two are not without further parallels. Neither had any significant technical or scientific training or knowledge. McLuhan, like Grant, was trained in the humanities and revered the western cultural legacy. Each in his own way was a backward-looking man. Both were devoutly religious, and both were typical of their generation in following their Canadian undergraduate training by graduate work in the UK. The commonality in their interpretation of technology, alongside their undoubted differences, is therefore no surprise. Certainly, neither supposed that technology was neutral.

Are these two rather conservative Anglo-Canadians therefore irrelevant to issues of technology and society in the twenty-first century? Even if they were, I would defend the value of a social science in Canada that is aware of its heritage. But let us think more carefully about who is up to date and who is not. The 1930s were a formative decade for both writers, a decade when the economy was mired in a

relentless depression. Opinion was deeply divided over whether the role of government was to fight the Depression by undertaking ever increasing debt, or to sit back and let market forces follow their own clock. Grant was eleven years old at the time of the 1929 stock market crash that heralded the onset of depression, and twenty-one when war with Germany was declared in 1939. McLuhan was just seven years older than Grant.

It was the post-war period of the late 1940s, 1950s, and 1960s that ushered in an era of unbounded progress and prosperity driven by the engine of technological innovation, an era that appears to have ended with the economic downturn of the 1980s and 1990s. For many readers of this book, the 1990s are the most formative years, giving today's students far more in common with the youthful days of Grant and McLuhan, their intellectual grandfathers, than with the 1960s decade when their own parents may have been adolescents or young adults.

Conclusion

There may be a tendency for every new technology to be greeted with hesitation by at least some people, but even if such a cycle exists the ethical dilemmas of technology would not be resolved. It is part of western culture that the public have a legitimate voice in ethical issues and in this chapter some examples of such public opinion were encountered, drawing on data from several countries. Within our own land, a tradition of mixed emotions about technology has often been noted and can be detected in George Grant and Marshall McLuhan as evaluators of technological society.

Notes

1. Maurice W. Kirby, *The Origins of Railway Enterprise: The Stockton and Darlington Railway, 1821-1863* (Cambridge: Cambridge University Press, 1993), 34.
2. The Seventh Duke of Wellington, *Wellington and his Friends* (London: Macmillan, 1965), 266.
3. Hamilton Ellis, *British Railway History: An Outline From the Accession of William IV to the Nationalisation of Railways 1830-1876* (London: George Allen and Unwin, 1954), 68. Railways would

"encourage the lower classes to move about."

4. Ellis, 21.

5. Andrew Kupfer, "Alone together," *Fortune* magazine, 20 March 1995, re-printed in Charles P. Cozic, ed., *The Information highway* (San Diego, CA: Greenhaven Press, 1996), 87-88. Thomas R. DeGregori, in *A Theory of Technology: Continuity and Change in Human Development* (Ames: Iowa State University Press, 1985), makes the point by saying "distrust of new technologies is deeply rooted in Western culture" (67).

6. http://www/geocities.com/Athens/1527/text5.html. For ongoing debate about the ethics of biotechnology, the newsgroup GEN-TECH@tribe.ping.de is a useful source.

7. The Roper Center, *The Public Perspective: A Roper Center Review of Public Opinion and Polling*, Vol 5 (January/February, 1994), 20.

8. William K. Hallman and Jennifer Metcalfe, "Public perceptions of agricultural biotechnology: A survey of New Jersey residents." From http://www.nalusda.gov/bic, 20.

9. Ronald Inglehart, *Culture Shift in Advanced Industrial Society* (Princeton, New Jersey: Princeton University Press, 1990), Chapter 2.

10. Inglehart, 98.

11. Martin Bauer, John Durant, and Geoffrey Evans, "European public perceptions of science," *International Journal of Public Opinion Research* 6 (1994), 174-75.

12. Francis Fukuyama, *The End of History and the Last Man* (New York: Free Press, 1992).

13. John Ralston Saul, *The Unconscious Civilization* (Concord, Ontario: Anansi, 1995), 187, and his 1996 *Hagey Lecture* at the University of Waterloo, entitled *Power Versus the Public Good: The Conundrum of Individualism and the Citizen*.

14. Grant, incidentally, was irritated at explanations of his thought rooted in his personal biography (see, for example, *Technology and Empire*, 140).

15. George Grant, *Technology and Justice* (Toronto: Anansi, 1986), 15.

Further Readings

For books by, and about, George Grant:

Grant, George. *Lament for a Nation: The Defeat of Canadian Nationalism.* Toronto: McClelland & Stewart, 1965.

___. *Philosophy in the Mass Age.* Toronto: University of Toronto Press, 1995 [1959].

___. *Technology and Empire: Perspectives on North America.* Toronto: Anansi, 1969.

___. *Technology & Justice.* Toronto: Anansi, 1986.

Kroker, Arthur. *Technology and the Canadian Mind.* Montreal: New World Perspectives, 1984.

For critical works on ethical issues:

Easton, Thomas A. *Taking Sides: Clashing Views on Controversial Issues in Science, Technology, and Society.* Guilford, Connecticut: Dushkin, 1995.

Glendinning, Chellis. *When Technology Wounds: The Human Consequences of Progress.* New York: William Morrow and Company, 1990.

Nef, Jorge, Jokelee Vanderkop and Henry Wiseman, eds. *Ethics and Technology: Ethical Choices in the Age of Pervasive Technology.* Toronto: Thompson, 1989.

Nora, S. and A. Minc. *The Computerisation of Society.* Cambridge, Mass.: MIT Press, 1980.

Wiseman, Henry, Jokelee Vanderkop and Jorge Nef, eds. *Critical Choices: Ethics, Science, and Technology.* Toronto: Thompson, 1991.

Chapter 11: Technology, Free Will, Inequality and Fate

As observed in earlier chapters, if technology is not quite to blame for the social conditions at the turn of the millennium neither is it uninvolved, for it is implicated with the nature of humanity itself. Surely the present days shall be looked back on as a time of chronic unemployment, growing division between rich and poor, moral and financial bankruptcy in government, doubts about the autonomy of the individual, and frightening threats and dangers for raising children.

It would be more positive to hope that the economy is on the verge of a new upward technology-led Kondratieff cycle, as suggested in Chapter 4. But even if it ultimately *is* a source of prosperity, the technological economy imposes immiserable uncertainties on workers young and old. The dislocations caused by the technological economy are so severe in part because the forces at work tend to reach beyond national control or international governmental will-power to control.

Is research on artificial intelligence going to take human beings to the stage where the computer progresses from being the second self and becomes the dominant self? Joseph Weizenbaum believed so, in his startling change of heart about AI research encountered in Chapter 9. Perhaps we shall look back at biotechnological research as the road that led to a cure for cancer, AIDs, and other horrific diseases, but what

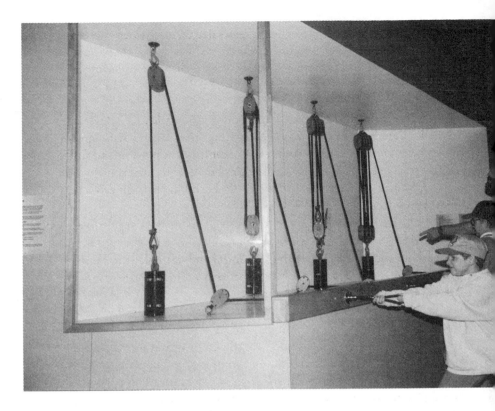

Technology as Fate: This simple exhibit at the *National Museum of Science and Technology,* Ottawa, gets heavy use.

if our ever more tough-minded governments decide that genetic screening for the unfit will help alleviate the national deficit by aborting all non-perfect babies? As a correspondent on a discussion group on biotechnology has wondered, "do we want to live in a world where diversity is considered a deficiency?"[1]

The Internet, and communications innovations such as television and radio before it, are wonderful, exciting phenomena with great potentialities. Yet we look back at a half century of television and find it deeply entangled with violence, the age-old threat to civil society. As seen in Chapter 9, "internet addiction" is becoming recognized as a clinical condition.

All told, the balance sheet for technologies is at best very mixed. After all, despite the problems, it is exhilarating to live in an era of such profound questioning about the human mind and a privilege to benefit from the extended life span due to the prosperity and health advances resulting from technological research.

What is to be done about social problems associated with technology while retaining a balanced awareness of its benefits? Nothing so fundamental as reversing the technological society is useful or even possible, for McLuhan was right in terming technology a "juggernaut." But as he also remarked in that same passage, we can turn off some of the "buttons" some of the time, and we do. There is action in the mid 1990s on the V-chip to control television violence. The impetus for addressing environmental problems comes largely from citizen groups. Another marker of civil control of technology is the way university researchers must clear their work with powerful ethical review committees, bodies originating from the early days of DNA research. Something has to "give" on the technology and work front, but perhaps the truth is that the whole population must adjust to a lower standard of living coupled with a new work ethic. Belt-tightening over the national deficit already points that way. It has been said that the essential problem in early industrialism was motivating the labour force, and in middle industrialism one of having workers get along with each other. The first is Max Weber's famous issue, the second Whyte's "social" work ethic of the 1950s. In late industrialism a new ethic must recognize work as a desired but scarce resource.[2]

When speaking in McLuhan's terms of turning off some of the "buttons," the philosophical point is recognized that men and women are in

part architects of their own free will. If this human condition of consciousness, innovation, and control is not realized by those who study technology and society, it is hardly likely to be appreciated by anyone. Recall the high consciousness people have of participating as citizens of technological society. As suggested near the outset of this book, few adjectives show a higher word count in the media than "technological." Think secondly of the amount of innovation required for invention in even the most bureaucraticized setting. The "*I*" component of the social self, the term the social philosopher G.H. Mead used to denote creativity, is nowhere more important than in the accumulated history of human technology. Consider finally the strong Canadian tradition of at least limited control of technology through public discourse and criticism. For all that sociologists might bemoan the anti-democratic impulses in modern technocratic and corporatist society, the forces of free expression in a society such as Canada remain vibrant in comparison to other regimes around the world. Were that not true, some of the figures discussed in earlier pages would encounter each other during exercise period at the gulag rather than at academic seminars.

Much as the society does allow for dissent and the exercise of free will it is well to acknowledge and restate the sociological theme encountered at many moments in the preceding pages. Time after time it has been suggested that technology is connected with social stratification, the great issue of inequality and power. The book began with economic surplus as a precondition for technological advancement. It was then seen that the process of technological diffusion is socioeconomically based, beginning with the most affluent classes and only later spreading to the poor. Technology, it was seen in Chapter 5, becomes a dimension within the social stratification system when it jells into a culture of technocracy. In the workplace, the focus of Chapter 6, technology has tended to displace the least powerful first, another typical outcome expected from theories of social stratification. Similarly, the most recent communications technology shows signs of benefitting those already best off in society. It is the children of affluent, highly educated parents who in the heaviest concentrations have access to the computer technologies that assist students in the race for grades. That theme was encountered in Chapter 8, and in Chapter 10 social class again was central in the tendency for criticism of technology to be a luxury of richer people in richer nations.

In this book, I have not wanted just to tag free will and the constraints of social stratification as the obverse and reverse of the technology and society coin. For the context underpinning all our ruminations — the very essence of this "coin" — is that we are enmeshed with technology, it is all around us materially and in our values, beliefs, and ideologies. It marks the social structure and culture like some great ice age glacier gouging out valleys and hills. McLuhan wrote of technology, the "juggernaut," that it is an extension, an "outering," of the self. But as Grant insisted this extension is so bound up with the nature of humanity, so much a fate, that the outering proceeds directly from within. It follows that, whether for good or ill so far as we can clearly distinguish such consequences, technology amplifies the goodness and badness already within people. It is a relentless companion that we wrestle with, sometimes with affection, but other times with antipathy. Technology leaves us in a state of fascinated ambivalence.

Notes

1. Posting from drjudd@rainbow.net.au, 18 November 1996, on GEN-TECH@tribe.ping.de.
2. Max Weber, *The Protestant Ethic and the Spirit of Capitalism* (New York: Charles Scribner's Sons, 1958), and William H. Whyte, *The Organization Man* (Garden City, N.Y.: Doubleday Anchor, 1956).

Glossary

Algorithmic A set of rules or procedures for accomplishing a task.

Alienated The more typically Marxian meaning is a state of being separated from the products of one's own labour under capitalism. In much sociology, in North America especially, alienation denotes a subjective state—a feeling of estrangement or separation from work, family, society, etc.

Anomic From anomie. A state of normlessness, in which self-regulation based on consensus of expectations within the society disintegrates.

Appropriate Technology Technology that is congruent with its cultural environment. Usually applied to underdeveloped nations and implies simple, cheap, useful technologies.

Archaeology The art and science of systematic excavation of human and pre-human dwellings, burial sites, battlefields etc.

Archetype The original, defining instance.

Bourgeoisie Marxian term for the middle class, referring explicitly to the social class comprised of possessors of capital.

Bureaucracy The form of organization that is rational, rule-bound, often hierarchical. Individual personalities in the typical bureaucracy are less important than the "offices" or positions they fill. The analysis of bureaucracy best known to sociologists is Max Weber's.

Business Cycle The notion of regularity in the up's and down's of the economy.

Conservatism This complex ideology has many variations, from the moral traditionalism mixed with ultra nineteenth-century liberalism of Ronald Reagan or Margaret Thatcher in the 1980s or provincial governments of Canadian provinces such as Alberta and Ontario in the

1990s, to a softer "radical Toryism" long part of Canadian federal politics. Radical Tories do not expect economic equality and accept the competitiveness of private enterprise. They uphold the establishment, to the extent of supporting organized religion and patriotism, contrary to the secular and internationally minded stance of socialism.

Core Segment of the labour market. The segment comprised of a small number of large corporations colluding to short circuit competition. It is the most profitable sector, and also typically has workers organized within strong labour unions.

Corporatism The ideology that simultaneously upholds the rightness of individual self-interest alongside the claims of corporate bodies. The existence of a public good separate from corporate objectives is denied.

Cultural To do with culture (see below).

Culture The entire array of material things produced by humanity, along with key shared ideas. These ideas include the meaning attributed to physical artifacts of a people, but also classes of ideas such as norms, values, and ideologies.

Demography The study of population structure change. Key variables are births, deaths, and migration.

Division of Labour Refers to specialization in the economy, whereby a person works in one occupation and uses the earnings to meet most needs by purchasing the products or services of others. An early realization of the division of labour was Adam Smith's famous passage about manufacturing a pin, in *The Wealth of Nations*.

Dominant Ideology The ideology supported by most of the population, and probably the one upheld by the leadership of a society.

Elite A small group at the top of society, for sociologists generally referring to holders of power.

Emoticon A computer keyboard hieroglyphic or symbolic picture to

express an emotion. For example, :) denotes happy.

Exogenous Factors in a statistical model that themselves do not have defined causes within the model. If the government fiscal crisis has, for example, diminished public service employment, the former is deemed exogenous because the analysis will not get into an explanation of the growth of national debt. If, in contrast, we had a model positing that family socio-economic status helps cause education of offspring which predicts their income as adults, education would be termed "endogenous" because it is both an effect and a cause within the model.

Flame Verb meaning to send an angry message via the Internet.

Hawthorne Effect Refers to the classic research by Elton Mayo at the Hawthorne Plant of Western Electric. The "effect" discovered at Hawthorne was that the very act of making workers the centre of a productivity study made them more productive. The technique deployed did not seem to matter.

Hegemonic Ruling, especially in the sense of controlling ideas and thought. A favourite term for Marxist scholars.

Hit A "visit" to a Website.

Home page An electronic document in a special text format such as "html" (hypertext marking language) in which might be displayed advertising, a resume, biographical information, pictures, favourite recipes, almost anything in fact.

Hunting and Gathering The form of human society in which food depends on hunting animals or picking naturally growing fruits, vegetables, roots, or the kill of other animals.

Hypertext Text in which certain ("hotlinked") words lead the reader to a sub-menu. In an html version of the present book, to "click" on "hypertext" within the text could bring up the glossary menu, in which hypertext would in turn be hotlinked to this definition.

I The part of the mind that is spontaneous, creative, and not fully derived from prior experiences.

Idealist Not used in the colloquial sense of what would be ideal, but rather in reference to the philosophical position holding that ideas in our heads, not material things, are the most basic reality. From Hegel among others.

Identity A person's understanding of who they are. Sociologists see this as a reasonably stable self-perception, derived at least in part from the positions occupied in society (for example, occupation, gender, status as parent or spouse).

Ideology An interpretation of a society, providing beliefs about the past, present, and future. Ideologies cover both what is and what should be. The ideas within a viable ideology are consistent and address such questions as the fairness of how rewards are distributed, the justice of the legal system, the representativeness and responsiveness of government, and the merits of societal institutions such as education and religion.

Information Society A postindustrial form of society in which the generation, passing and storing of information becomes the salient feature, eclipsing such traditional industrial activities as mass production manufacturing. Such information includes both highly sophisticated technical knowledge and less profound items such as systems for inventory control information for businesses.

Institutional Domain An admittedly long-winded sociological term referring to some identifiable component or sector of a society which it can be argued fulfils some social need or objective. The term is deliberately very general. One might refer to a specific university as "an educational institution," but sociologically we refer to education at all levels and locations as the institution. Analysis of an institution might examine the occupations gathered within it, the norms and values particular to the institution, the organizational arrangements (formal or informal), etc.

Liberalism The ideology stressing the sanctity of the individual, placing priority on equality of opportunity but not equality of condition. The ideology of technocracy, George Grant said.

Materialism(ist) The philosophical position that what really exists is material, meaning something which has volume in space and some continuity in its existence. Material things are therefore discernable by sight or touch (albeit with the aid of magnifiers and amplifiers).

Means of Production A Marxian term referring to land, machinery, buildings, and technologies used in a capitalist economy.

Multiple User Dungeon (or Dimension) A "place" in cyberspace where one can interact with others by messages typed onto a keyboard. The messages can include codes for actions and emotions.

Netiquette The etiquette of the Internet.

Oligopoly A portion of the economy where a few firms predominate, and can agree to fix prices at a higher level than would result from more open competition. The oil and gasoline market is oligopolistic.

Operational Definition A concept defined in such a way that it can be measured. For example, "rate of technological innovation" might be operationalized as number of patents granted per year.

Paleoanthropologists Anthropology is the social science devoted broadly speaking to the study of pre-industrial peoples. Anthropologists who study the pre-human era from many thousands of years ago are paleoanthropologists. *Palaios* is Greek for "ancient."

Paradigm Technically, "a representative instance of a concept, used to provide an ostensive definition of it" (*Fontana Dictionary of Modern Thought*, Alan Bullock and Oliver Stallybrass, eds., Fontana/Collins, 1977). More loosely as used in much social science, a set of assumptions, attitudes, etc.

Peripheral Segment of the labour market. The less profitable segment

comprised of many small firms in active competition. Workers here are typically not unionized or belong to weak unions.

Petite Bourgeois Of the class of owners of small businesses. See also "bourgeoisie."

Post-modern Beyond modern in the sense of losing faith in the certainty of knowledge and the unbroken upward progress of history. The post-modern era is one of competing truth claims by different entitlement-seeking groups. In literary analysis, it entails the dereification of authors, and the claiming of their texts as documents admitting to multiple interpretations and meanings. See Pauline Marie Rosenau, *Post-Modernism and the Social Sciences* (Princeton: Princeton University Press, 1992).

Proletarian Of the proletariat (see below).

Proletariat Marxian term for the working class, referring to those who must survive by selling their labour to owners of capital.

Role Conflict The duties embedded within one role, or position, we hold conflict with those of another (for example, wife vs. mother). Or, the duties within one role are contradictory (for example, role of professor is to be a friendly mentor to students all term, but then become a tough-minded examiner when lectures end).

Sapir-Whorf Hypothesis Benjamin Whorf was an amateur student of linguistics, Edward Sapir a professional academic with a university post. The hypothesis was that language structures our perception of the world.

Social Stratification The sociological name for inequality of all sorts. A stratified society has identifiable social layers, just as the earth's crust has strata composed of different materials.

Social structure Most literally, the connections and relationships between individuals in a society. Relationships between larger entities such as social classes or corporate bodies are structural too. Ties from

individual to individual are easier to visualize, being expressible in quite concrete, physical form. Consider the photograph in the text of the early twentieth-century telephone switchboard. The wires and plugs were used literally to connect people for speech. Relations between one social class and another require greater conceptual effort to visualize as a structure, but sociologists would insist that this latter example is just as structural as a telephone switchboard. The concept of *network* used within the field known as "structural sociology" can help bring the concept of structural relations as transferred from individuals to groups into focus (see Wellman reference, below). Almost everybody belongs to a network, large or small, of friends and acquaintances. Larger bodies such as companies can be networked as well, through people. If, for example, someone sits on the board of directors of two companies, a connection that is part of a network exists (see Carrol reference below). Individuals here have become the medium for the connection or relationship between companies. And just as we all may have connections through "friends of friends," likewise companies might be joined indirectly. Structure would be an empty concept if it merely echoed the obvious fact that connections exist within a society. The content of relationships is equally part of the sociological structure. It is a study in structure to discover that people in positions such as parent, teacher, employee, member of the male sex or the female sex, member of an ethnic minority act in predictable, characteristic ways. It is likewise a study in structure to examine the power imbalance between members of Marx's bourgeoisie and proletariat. Since structure includes the content of relationships, it is closely tied up with culture, for people conduct their relationships, enact their roles, etc. within a culture. By analogy, flesh and blood are closely related, yet remain analytically distinct. So it is with culture and structure.

Works cited: Barry Wellman and S.D. Berkowitz, *Social Structures: A Network Approach* (Cambridge: Cambridge University Press, 1988); William K. Carroll, John Fox, and Michael D. Ornstein, "The network of directorate interlocks among the largest Canadian firms," *Canadian Review of Sociology and Anthropology* 19 (February, 1982).

Socialist ideology favours equality of condition as well as equality of opportunity and thus implies policies such as an active role by the state

in the redistribution of wealth.

Sociobiology The field introducing principles of biology as explanatory variables for sociological questions.

Socio-demographic Sociological attributes such as age, gender, ethnicity, and social class.

Technique Items of knowledge that may form part of a technology.

Technological determinism An emphasis on technology as the force that determines the shape of a society. A technological determinist would de-emphasize how technology in turn results from the culture and social structure of a society, seeing such inventions as "givens" that came about perhaps inevitably.

Technology The application of knowledge to the achievement of particular goals or to the solution of particular problems.

Technopoly The dominance of the ideas supporting technology. Postman's term for what others termed technocracy.

Teleology The form of reasoning which deduces causes from an end result. For example, "that I failed my exam so badly proves I never had a chance of passing it."

Typology Any system for classifying things by type. The simplest is a "two by two" typology. For example, an ice-cream store has just two flavours, chocolate and vanilla, and sells either in cup or cone. There are therefore two times two equals four types of ice-cream available.

Value System Core ideas within a society about what is valued. Some Canadian values might be democracy, fairness and ethnic tolerance. Values are not necessarily accurate descriptions of a society, but they do articulate the stated objectives of a people.

Index

Appropriate technology, 47, 92-95
Artificial intelligence, 130, 189-191, 206, 215; and multiple intelligences, 200-201; chess as focus, 199-202; psychiatry as focus, 194-197
Atomic bomb, 100, 103, 107-108
Automation, 117-119, 123, 125-127, 133-134
Avro Arrow, 50-51
Basalla, George, 15, 42-46, 54, 75, 86, 216
Bell, Alexander Graham, 32, 39, 49, 51, 63-64; Bell Laboratories, 37
Bell, Daniel, 20, 101-106, 120, 123, 125, 227
Biotechnology, 19, 84, 105, 215; and DNA, 204; attitudes toward, 204-205, 218-219, 235, 237
Braverman, Harry, 120-129
Bronze Age, 12, 14, 42, 62
Canada: as economically dependent, 52-54, 90, 92; invention in, 46-51, 157
Capitalism, see Technology
Coal, 45, 82, 93
Computers: and language, 145, 198; and "MUD's", 196; as "second self", 191-194; attitudes toward, 44; hackers, 193; hot/cool dimensions, 150, 170-175; impact on work, 127, 132; mainframe era, 170-171; McLuhan's analysis of, 169; obsolescence rate of, 86; occupations in, 125; ownership patterns, 69-74; the personal computer (PC), 43-44, 174-175
Crick, Francis, 202, 205-206
Culture, 9-10, 14, 28-29, 31-32, 59, 86, 94, 105-108, 111-112, 145, 155, 186, 201-202, 216; of Internet, 183

Cyberspace, see Internet
Diffusion of technology, 42, 53, 59-63, 86; bell curve, 66; by social class, 68-71; diffusionism, 62; jetpacks, 63; Neolithic era, 61; prehistoric rates, 61; S-curve, 66, 75, 86-87; the stirrup, 61-62; the transistor, 89; to companies, 74, 91-92; videophones, 63; see also Telephone
DNA: as computer, 206-209; discovery of, 19, 35, 202-205, 217; Mitochondrial analysis, 5-6, 16, 35, 59
Economic cycles, 82-87
Electricity, 15-19, 48, 68, 84, 103, 130, 154
Eliza, see Artificial Intelligence
Ellul, Jacques, 10, 33, 107-108, 111, 114
Embedded technology, 91
Engineering: and Taylorism, 137n, associations, 37; engineers and ethical choices, 217; engineers and power, 100, 106-107, 110, 119, 132; for aircraft, 51, 91; for printing, 152; for shatter-proof rail, 49; for steam engine, 45; for V-chip, 48; genetic, 204-205, 222; in Third World, 94
Environmentalists, 107, 221-226; and post-materialism, 222, 225
Ethical issues, 21, 47; ethical life cycle, 215-216; public opinion on, 217-219
Film, 150, 157-158
Flying shuttle, 34, 46
Franklin, Ursula, 110, 175-176
Free Trade Pact, 90-92, 130
Future shock, 136
Gates, William, see Microsoft
Genetics, 237

Global village, *see* McLuhan
Grant, George, 3-5, 7-8, 10-11, 15-16, 21, 108-110, 114, 119-120, 227-230
Hunting and Gathering, 14, 29, 36, 40
Ideology: and paradigms, 223-225; conservative, 52, 108; corporatist, 109, 228-229; dominant, 108, 241; liberal, 22, 108-109, 228-229; socialist, 229
Industrial Revolution, 50, 61, 111
Information society, 28, 102, 143
Innis, Harold, 11, 146-147, 164
Innovation, 14, 22n, 36, 41, 48, 62-63, 67, 74, 81, 84-85, 87, 102, 109, 232
Intermediate technology, *see* Appropriate technology
Internet, 155, 169, 174-181; and government, 185-186; and identity, 182, 191; homepages, 184-185; profile of users, 177-179, 218; vocabulary of, 183
Iron, 12-13, 52, 61-62, 82, 84, 121
Jobless recovery, *see* Work
Julia, *see* Artificial intelligence
Kondratieff, N.D., *see* Economic cycles
Labour unions, 124, 128, 134
Leakey, Richard, 6-9, 61
Leapfrogging, 87, 89-90, 94, 96n
Luddite: Ned Ludlum, 118; neo-Luddite, 118-119
Mcjobs, 118, 124, 136
McLuhan, Marshall, 11, 16-17, 42, 137n, 183, 237, 239; global village, 18-19, 180-181, 183-185; hot vs. cool media, 143-158, 174, 180; "medium is the message", 17, 157-158, 162; on "leapfrogging", 87, 89; on mechanical vs. electronic-based technology, 17-19; opinion of technology, 230-231; revived interest in, 170; technology as "outering", 189
Mensch, Gerhard, 84-86
Menzies, Heather, 128, 130, 133-134
Microprocessor, 35, 38, 130, 171, 191
Microsoft, 32, 110
Mumford, Lewis, 14, 16, 19, 38, 56n

Niosi, Jorge, 90
Oil, 93
Patents, 37-39, 48-50, 54, 62, 64, 68, 87
Penrose, Roger, 197-198
Plastic, 12-13, 15
Plough, 14, 34, 36
Postman, Neil, 107, 111
Post-materialism, 222, 226-227
Post-modernism, 170
Power: as dimension of technology, 14-15
Printing press, 34, 41-42, 45, 111, 147-152, 164
Punched cards, 171, 174
Radio, 9, 18, 48, 94, 150, 156-157, 175, 179
Railroads, 13, 17, 48-49, 121, 155, 215-216
Rationality, 9
Research and development (R&D), 52, 81, 88-92, 117
Robotics, 49, 121, 194, 200
Schumacher, E.F., *see* Appropriate technology
Science, *see* Technology and science
Scientific Management, *see* Taylor
Scientific Revolution, 38, 41, 45, 56n, 61, 86, 111
Service sector, 102-103, 125, 130-131
Social structure, 14, 28-30, 86, 143, 149, 156, 182-183
Sociobiology, 8, 19, 25, 28-29
Staples economy, 52, 146, 164n
Steam engine, 15, 20, 34, 39, 45, 84, 86, 113n
Steel, 12-13, 23n, 49, 121
Stone Age Tools, 5-9, 12-13, 15, 28, 40, 120
Taylor, Frederick W, 19, 111, 121-122, 137, 140
Technetronic, 23n
Technique, 8, 10, 24n, 32
Technocrats, 99-100, 104-106
Technology: ambivalence toward, 229-232; and agriculture, 14, 36, 85, 129, 134; and capitalism, 36-40, 53-

7740111